Advance Praise for
State of the World 2014: Governing for Sustainability

"This volume offers a variety of informed and often passionate voices on the interface of environmental degradation and risk with conceptions and models of governance that, if we can summon the will, would promote sustainable management of the global commons. A clear, lively, thought-provoking book, which serves well as a reasoned call to action."

—**David M. Malone,** Rector of the United Nations University

"Achieving sustainable ways of living is inextricably linked to how we organize work in the future. *State of the World 2014* makes an important contribution by illustrating how trade unions, far from being outdated, will be at the forefront of a just transition. It is a challenging compilation—coming at exactly the right time."

—**Sharan Burrow,** General Secretary of the International Trade Union Confederation

"For thirty years, the *State of the World* report has helped to map the gathering and then accelerating storm of environmental, climate, and resource crises. Identifying itself firmly with the collective interest of humanity as a whole living in harmony with nature, the annual report has sought to balance authoritative reporting of the increasingly bleak health of the environment with sustainable pathways out of the accumulating crises. In a world of competing sources of authority and power, the pursuit of atomized individual and national self-interests will court planetary disaster. This year's *State of the World* report has its focus on governance: how, in a world without world government, we can and must make enforceable rules for using finite resources democratically, equitably and, above all, sustainably, with fallible governments and imperfect markets working together for the common good."

—**Ramesh Thakur,** The Australian National University, Editor-in-Chief, *Global Governance*

"*State of the World 2014* can be read as a 'State of the Wealth' report. Never before has wealth commanded so much power or been so concentrated—even to the point of threatening civilized life. Wealth becomes unable to offer, not just a better future, but any future. Therein lies its weakness and the hope that the major governance shift that sustainability requires can be brought about."

—**Roberto Bissio,** Coordinator of Social Watch

About Island Press

Since 1984, the nonprofit organization Island Press has been stimulating, shaping, and communicating ideas that are essential for solving environmental problems worldwide. With more than 800 titles in print and some 40 new releases each year, we are the nation's leading publisher on environmental issues. We identify innovative thinkers and emerging trends in the environmental field. We work with world-renowned experts and authors to develop cross-disciplinary solutions to environmental challenges.

Island Press designs and executes educational campaigns in conjunction with our authors to communicate their critical messages in print, in person, and online using the latest technologies, innovative programs, and the media. Our goal is to reach targeted audiences—scientists, policymakers, environmental advocates, urban planners, the media, and concerned citizens— with information that can be used to create the framework for long-term ecological health and human well-being.

Island Press gratefully acknowledges major support of our work by The Agua Fund, The Andrew W. Mellon Foundation, Betsy & Jesse Fink Foundation, The Bobolink Foundation, The Curtis and Edith Munson Foundation, Forrest C. and Frances H. Lattner Foundation, G.O. Forward Fund of the Saint Paul Foundation, Gordon and Betty Moore Foundation, The Kresge Foundation, The Margaret A. Cargill Foundation, New Mexico Water Initiative, a project of Hanuman Foundation, The Overbrook Foundation, The S.D. Bechtel, Jr. Foundation, The Summit Charitable Foundation, Inc., V. Kann Rasmussen Foundation, The Wallace Alexander Gerbode Foundation, and other generous supporters.

The opinions expressed in this book are those of the author(s) and do not necessarily reflect the views of our supporters.

State of the World 2014

Governing for Sustainability

Other Worldwatch Books

State of the World 1984 through *2013*
(an annual report on progress toward a sustainable society)

Vital Signs 1992 through *2003* and *2005* through *2013*
(a report on the trends that are shaping our future)

Saving the Planet
Lester R. Brown
Christopher Flavin
Sandra Postel

How Much Is Enough?
Alan Thein Durning

Last Oasis
Sandra Postel

Full House
Lester R. Brown
Hal Kane

Power Surge
Christopher Flavin
Nicholas Lenssen

Who Will Feed China?
Lester R. Brown

Tough Choices
Lester R. Brown

Fighting for Survival
Michael Renner

The Natural Wealth of Nations
David Malin Roodman

Life Out of Bounds
Chris Bright

Beyond Malthus
Lester R. Brown
Gary Gardner
Brian Halweil

Pillar of Sand
Sandra Postel

Vanishing Borders
Hilary French

Eat Here
Brian Halweil

Inspiring Progress
Gary Gardner

State of the World 2014

Governing for Sustainability

Tom Prugh and Michael Renner, *Project Directors*

Gar Alperovitz
Katie Auth
Petra Bartosiewicz
David Bollier
Peter G. Brown
Colleen Cordes
Cormac Cullinan
Antoine Ebel
Sam Geall

Judith Gouverneur
John M. Gowdy
Monty Hempel
Isabel Hilton
Maria Ivanova
Matthew Wilburn King
Marissa Miley
Evan Musolino
Nina Netzer
Thomas Palley

Lou Pingeot
Tatiana Rinke
Aaron Sachs
Jeremy J. Schmidt
D. Conor Seyle
Sean Sweeney
Burns Weston
Richard Worthington
Monika Zimmerman

Lisa Mastny, *Editor*

ISLANDPRESS

Washington | Covelo | London

Dedication

We are in a race between tipping points in nature
and our political systems.
     —Lester R. Brown, *Plan B* (2008)

The year 2014 marks the fortieth anniversary of the Worldwatch Institute and the thirtieth anniversary of the *State of the World* series, as well as the eightieth birthday of Lester Brown, the man who founded them both. Dedicating this book to Lester is especially apt because it focuses on governance, a topic that he has long recognized as the most powerful obstacle to creating a sustainable future.

When Lester created Worldwatch in 1974, solar panels cost 30 times as much as they do today, and wind power was used mostly to pump water. The first Macintosh computer would not be launched for another decade, and the World Wide Web for nearly two decades. But Lester was convinced that strong winds of change were blowing in fields as diverse as energy, communications, health care, security, and urbanization, and that they would combine to transform the human prospect in profound ways.

Transformational change, as Lester had learned with the Green Revolution, always brings side effects. Often, these side effects are unfortunate, and occasionally they are tragic. Lester wanted to build an agile institution that could anticipate those changes and help shape them in the public interest. He recruited a small band of synthesizers—people who could write clearly about complicated subjects for a general audience—to survey the primary literature for problems and opportunities while they were still small dots on the horizon. He took enormous delight when, in the second year of Worldwatch's existence, its five senior staff racked up more coverage in the *New York Times* than the entire Brookings Institution.

Lester's early work assessing India's agricultural situation resulted in broad policy shifts that saved millions of lives. His book *Who Will Feed China?* (1995) made him a household name in that vast country. His works on redefining national security helped bring about a shift in the way that military leaders and diplomats around the world view environmental issues. Among his myriad honors, Lester has won a MacArthur Fellowship, the United Nations Environment Prize, the Blue Planet Prize, and 25 honorary degrees. He has no plans to retire.

At the same time, Lester leads a life that is true to his values. He lives modestly and eats a healthy diet. In 2009, he placed third in the 75–79 age group in the Cherry Blossom National Championship 10-mile race in his hometown of Washington, D.C.

Over the last 40 years, Lester has written more nonfiction books than most Americans have read. His books are filled with original ideas that range across an incredibly broad canvas. It is altogether fitting that this book, addressing the most important institutional challenges to a sustainable future, be dedicated to Lester R. Brown.

—Denis Hayes
    President, Bullitt Foundation
    Founder, Earth Day Network
    Former Senior Fellow, Worldwatch Institute

Acknowledgments

Acknowledgments are meant to thank the people who have had important roles in making a book possible. With *State of the World* there are always many such people, and this presents the obvious problem of where to begin. But this year, there is absolutely no doubt about where to begin: with Linda Starke.

As inaugural editor, Linda was present at the creation of *State of the World* when it was launched in 1984. She also edited every subsequent edition through 2013, when she decided it was time to cut back on her workload. The 2014 *State of the World* is thus the first edition in the report's history to have been produced without the benefit of her sharp eye, her legendary skills as a production manager, and her strong, reasoned opinions. That is a remarkable record of accomplishment, and to the extent that *State of the World* has maintained a reputation for clear writing, thought-provoking content, and responsible scholarship, Linda deserves a huge share of the credit.

Stepping into Linda's role is Lisa Mastny, who has already built a reputation for being nimble and meticulous in her editing of multiple other Worldwatch research reports. Continuity and the report's crisp, accessible look are provided by long-time graphic designer Lyle Rosbotham, whose involvement with *State of the World* stretches back more than a decade.

Worldwatch Institute and its projects, including *State of the World*, have benefited over the years from the invaluable financial support of a variety of institutions and foundations. This year, we would like to extend our deepest appreciation to the following: Ray C. Anderson Foundation; The Asian Development Bank; Carbon War Room; Caribbean Community Secretariat (CARICOM); Climate and Development Knowledge Network (CDKN); Del Mar Global Trust; Embassy of the Federal Republic of Germany in the United States; Energy and Environment Partnership with Central America (EEP); Estate of Aldean G. Rhymer; Garfield Foundation (discretionary grant fund of Brian and Bina Garfield); The Goldman Environmental Prize; The William and Flora Hewlett Foundation in partnership with Population Reference Bureau; Hitz Foundation; INCAE Business School; Inter-American Development Bank; International Climate Initiative (ICI) of the German

Federal Ministry for the Environment, Nature Conservation and Nuclear Safety (BMU); Steven C. Leuthold Family Foundation; The Low-Emissions Development Strategy – Global Partnership (LEDS-GP); MAP Royalty Inc. Sustainable Energy Fellowship Program; the National Renewable Energy Laboratory (NREL) and the U.S. Department of Energy; Organization of American States; The Population Institute; Randles Family Living Trust; V. Kann Rasmussen Foundation; Renewable Energy Policy Network for the 21st Century (REN21); Serendipity Foundation; The Shenandoah Foundation; Town Creek Foundation; Turner Foundation; United Nations Foundation; United Nations Population Fund (UNFPA); Johanette Wallerstein Institute, Inc.; and Weeden Foundation.

Many individual and business donors make our work—and especially this year, this book—possible. We are grateful to them all and wish there were room here for all their names. Among the many whose financial contributions and in-kind donations were especially valuable, we would like to thank Ed Begley Jr., Edith Borie, Stanley and Anita Eisenberg, Robert Gillespie, Charles Keil, Adam Lewis, John McBride, Leigh Merinoff, MOM's Organic Market, Nutiva, George Powlick and Julie Foreman, Peter and Sara Ribbens, Peter Seidel, Laney Thornton, and three anonymous donors. Among the Worldwatch Board of Directors, we especially thank L. Russell Bennett, Mike Biddle, Edith Eddy, Robert Friese, Ed Groark, Nancy and Jerre Hitz, Isaac van Melle, David Orr, John Robbins, and Richard Swanson.

State of the World has found a good home at the highly regarded sustainability publisher Island Press, which is publishing and distributing the report in English for the third year in 2014; thanks once again to Emily Turner Davis, Maureen Gately, Jaime Jennings, Julie Marshall, David Miller, Sharis Simonian, and the rest of their fine team. We also owe a profound debt of gratitude to our international publishing partners for their commitment and hard work in translating, distributing, and communicating the results of the report year after year. Specifically, many thanks to Universidade Livre da Mata Atlântica/Worldwatch Brasil; Paper Tiger Publishing House (Bulgaria), China Social Science Press; Worldwatch Institute Europe; Gaudeamus Helsinki University Press (Finland); Organization Earth (Greece); Earth Day Foundation (Hungary); Centre for Environment Education (India); WWF-Italia and Edizioni Ambiente; Worldwatch Japan; Korea Green Foundation Doyosae (South Korea); FUHEM Ecosocial and Icaria Editorial (Spain); Taiwan Watch Institute; and Turkiye Erozyonla Mucadele, Agaclandima ve Dogal Varliklari Koruma Vakfi (TEMA), and Kultur Yayinlari Is-Turk Limited Sirketi (Turkey).

A number of individuals deserve special note for their indispensable roles in helping to inform *State of the World*, give it a strong international sensibility, and make it available to broad audiences around the globe: Burcu

Arik, Eduardo Athayde, Ana Belén Martín, José Bellver, Gianfranco Bologna, Melanie Gabriel Camacho, George Cheng, ZsuZsa Foltanyi, Tetyana Illiash, Cyril Ivanov, Haibing Ma, Kwangho Min, Anna Monjo, Marco Moro, Bo Normander, Soki Oda, Mamata Pandya, Ioannis Sakiotis, Kartikeya Sarabhai, Tuomas Seppa, Martín Vázquez, and Yun-Chia.

As always, the people for whom no thanks can be adequate are this year's chapter and text box authors. This group of outstanding scholars, activists, and journalists gave generously of their time and expertise, coped graciously with our editing requests, and delivered strong content in a timely fashion. They each found a place in their busy lives for contributing a piece of this, the 40th Worldwatch Institute anniversary edition of *State of the World*. We are deeply grateful.

Tom Prugh and Michael Renner
Project Directors
www.worldwatch.org

Contents

BOXES

TABLES

FIGURES

Units of measure throughout this book are metric unless common usage dictates otherwise.

Foreword

David W. Orr

> *If men were angels, no government would be necessary. If angels were*
> *to govern men, neither external nor internal controls on government*
> *would be necessary.*
>   —James Madison[1]

Long before the climate crisis was "the greatest market failure the world has ever seen," it was a massive political and governmental failure. The knowledge that carbon emissions would sooner or later threaten the survival of civilization was known decades ago, but governments have done very little about it relative to the scale, scope, and longevity of the problem. The reasons for their lethargy are many, but one in particular stands out.[2]

For half a century, a concerted war has been waged against government in Western democracies, particularly in the United Kingdom and the United States. Its origins can be traced back to the more virulent strands of classic liberalism once arrayed against the entrenched power of royalty. Its present form was given voice by Ronald Reagan, who reoriented the Republican Party and much of U.S. politics around the idea that "government is the problem," and by Margaret Thatcher in Britain, who ruled in the conviction that there was "no such thing as society," only atomized self interests. Other forces and factions joined in an odd alliance of ideologists, media tycoons, corporations, and conservative economists such as Friedrich Hayek and Milton Friedman.

Many other factors contributed to the hollowing out of Western-style governments. Particularly in the United States, wars and excessive military spending contributed greatly to deficits, impoverishment of the public sector, and declining credibility of public institutions. The rise of multinational corporations and the global economy created rival sources of authority and power. Electoral corruption, gerrymandering, and right-wing media contributed to public hostility toward governments, politics, and even the idea

David W. Orr is the Paul Sears Distinguished Professor of Environmental Studies and Politics at Oberlin College in Ohio.

of the public good. The Internet helped as well to partition the public into ideological tribes at the expense of a broad and civil public dialogue.

But the war against government is not what it is purported to be. Indeed, it is not a war against excessive government at all, but a concerted campaign to reduce only those parts of government dedicated to public welfare, health, education, environment, and infrastructure. But conservatives virtually everywhere support higher military expenditures, domestic surveillance, larger police forces, and exorbitant subsidies for fossil fuel industries and nuclear power along with lower taxes on corporations and the wealthy.

The upshot is that the public capacity to solve public problems has diminished sharply, and the power of the private sector, banks, financial institutions, and corporations has risen. As a countervailing and regulatory force, the power of democratic governments has eroded, and with it much of the effectiveness of public institutions to foresee, plan, and act—which is to say, govern.

A different pattern has emerged in China, which joins capitalism and authoritarian government. For a time, at least, it has been rather more effective at solving problems associated with rapid growth, building infrastructure, and deploying renewable energy. As the climate and environmental crisis grows, however, so too the traffic jams, air pollution, water shortages, and public dissatisfaction. It remains to be seen whether the marriage of authoritarianism and public engagement can work over the long term.[3]

Elsewhere, the number of failed states with tissue-thin governments is growing under the weight of population growth, corruption, crime, changing climate, and food shortages. Poverty and the lack of basic services, including education, contribute to a sense of hopelessness that feeds the anger that drives young men, in particular, into radical groups, further threatening stability. The foreseeable future offers little respite. We face what John Platt once called "a crisis of crises," each amplified by the others. A rapidly warming Earth occupied by 10 billion people and 193 nation-states, some armed with nuclear weapons, some clinging to ancient religious and ethnic hatreds, and still others holding fast to their economic and political advantages, threatens the survival of civilization.[4]

Warmer and more acidic oceans will be less capable of supporting humankind. Massive storms, rising seas, higher temperatures, and disassembling ecologies will disrupt food production, public health, water systems, urban settlements, transportation, electricity supplies, and the capacity to meet a growing number of emergencies. Climate destabilization will grow worse for many decades to come. Presuming that we stabilize carbon dioxide (CO_2) levels in the atmosphere by, say, 2050, the effects will last for centuries, perhaps millennia, and no society, economy, and political system will escape the consequences. That is where we are headed.[5]

What's to be done? Of many possibilities, three stand out. First, avoiding the worst that could happen will require sharp reductions of CO_2 emissions trending toward zero by mid-century. We are possibly close to a threshold beyond which climate change will be uncontrollable no matter what we do. To avoid that possibility, we will have to quickly sequester the remaining reserves of fossil fuels that cannot be safely burned. To do so, the choices are roughly to:

a) confiscate fossil fuels from their present owners; or
b) compensate their owners, rather like the British ended slavery in the Caribbean in the nineteenth century;
c) rapidly deploy alternative technologies and thereby render fossil fuels uncompetitive;
d) geoengineer the atmosphere in order to lower temperatures and buy us time to think of something better to do; or
e) some combination of the above.

The particularities and perplexities of various policies aside, if civilization is to last, we must permanently remove reserves of coal, oil, tar sands, and natural gas from the asset side of the economic ledger, but without collapsing the global economy.[6]

A second and related priority will be to reform the global economy to internalize its full costs and fairly distribute benefits, costs, and risks within and between generations. By one reckoning, a majority of the costs of economic growth has been offloaded on the poor and disadvantaged. Most of the accumulation of CO_2 presently in the atmosphere is from the industrialized nations.[7]

There is little prospect of a peaceful transition to a better future without achieving a much more equitable distribution of wealth in an economic framework calibrated to the laws of entropy and ecology. But that economy will be a great deal more like the "stationary state" predicted by John Stuart Mill in 1848 than the "casino capitalism" or "turbo capitalism" of the post–World War II era. A sustainable and fair economy will be one that pays its full costs, creates no waste, and deals far more in public goods and necessities such as housing, education, public infrastructure, and collective goods than in financial speculation and consumerism.[8]

A third and related priority will require a significant change in how we relate to future generations. Economist Kenneth Boulding once facetiously asked, "What has posterity done for me…lately?" The answer, of course, is "nothing." But a decent regard for posterity is inseparable from our own self-interest, as Boulding argued. Yet posterity presently has little or no legal standing, and so its right to life, liberty, and property exists—if at all—under a darkening shadow of the effects of the behavior of previous generations, mostly our own.[9]

We have long assumed that benefits flowing from one generation to the next were overwhelmingly positive. But that is no longer as true as it once was. The burdens imposed by a worsening climate and associated environmental havoc place the lives and fortunes of our descendants in great jeopardy. They will have no defense unless and until foundational environmental rights are codified in law, solidified as a core value in politics, and embedded in our culture.

Other challenges loom ahead. Soon, millions of people will have to be relocated from sea coasts and from increasingly arid and hazardous regions of Earth. Agriculture everywhere must be made more resilient and freed of its dependence on fossil fuels. Emergency response capacities everywhere must be expanded. The list of necessary actions and precautionary measures is very long. We are like a ship sailing into a storm and needing to trim sails, batten hatches, and jettison excess cargo. But how will we decide to do comparable things in the conduct of the public business?[10]

We have four broad pathways, each with many variations. The first is to let the market manage by the mysterious workings of the proverbial "invisible hand." There are many purported advantages of doing so. In theory, markets require no political consensus, government programs, or public planning. In the right circumstances, they are agile, creative, and adaptable. But markets always perform far better in neoclassical textbooks than they do in reality. The truth is that they have a consistently poor record of foresight, or concern for the disadvantaged, or fairness, or whales, or grandchildren, or democratic institutions…unless it turns a profit.

Unsupervised markets work against the interests of the larger society. As Karl Polanyi once warned, "To allow the market mechanism to be sole director of the fate of human beings and their natural environment, indeed, even of the amount and use of purchasing power, would result in the demolition of society." In sum, markets do many things well, but for things that cannot be priced, they are inept and autistic to human needs and ecological imperatives.[11]

The second alternative is to bolster public institutions and governments at all levels. Indeed, in the face of climate change, subnational governments are becoming more agile with alliances between states, provinces, and regions. Cities are coming together in creative ways to implement climate actions that presently cannot be taken at national levels. The results are often more effective, cheaper, and better fitted to particular situations than national policies. Networks of agencies and nongovernmental organizations stitched together by electronic media are capable of rapid interdisciplinary responses to the challenges. But inevitably, these efforts are limited because they are contingent on the powers and policies associated with sovereign national governments.[12]

A third pathway, then, is to create and maintain effective, agile, accountable, and democratic central governments. Centralized governments alone have the capacity to respond at the scale necessary to effect changes appropriate to the "long emergency." They alone can wage war, grant or withhold rights, control currencies, manage fiscal policies, respond to large-scale crises, regulate commerce, and enter into binding international agreements. With respect to climate change, only central governments can effectively price or control carbon for an entire country. Only effective central governments can command the resources required to mobilize entire societies.[13]

But a yawning chasm exists between current performance and the quality of governance necessary to meet the exigencies of the long emergency ahead. As James Madison put it, "The great difficulty is this: You must first enable the government to control the governed; and in the next place, oblige it to control itself." Governments today cannot consistently control themselves because they are decimated by a plague of corruption that devours the public interest in virtually every political system. It infects the media, economy, banking system, and corporations. This is the fountainhead of our political misfortunes, and of most others.[14]

The solution is not so much new government agencies as it is, in political philosopher Alan Ryan's words, "the slow implementation of better governance by weeding out corruption and ignorance." And that will require a rigorously enforced separation between money and the conduct of the public business. The struggle to separate money from policy making and law will, in time, come to be seen rather like historic battles against feudalism, monarchy, and slavery.[15]

There is, however, a caveat leading to a final pathway. Little or no improvement of politics or governance is possible where ignorance, ideological superstitions, and indolence reign. Effective government, in its various forms, will require an alert, informed, ecologically literate, thoughtful, and empathic citizenry. Whether and to what extent this will be democratic remains to be seen. The limitations of democracy as practiced in consumer-oriented, corporate-dominated societies are well known. Unreformed, they will be more debilitating under the conditions we will experience in the twenty-first century.

But our past successes, notably those of World War II and the Cold War, have bred overconfidence that democracies will succeed in dealing with an entirely different kind of threat, one with time-lags between causes and effects and with deadlines beyond which loom irrevocable, irreversible, and wholly adverse changes. Relative to climate change, David Runciman writes that the "long-term strengths [of democracies], if anything, make it harder. That is why climate change is so dangerous for democracies. It represents the potentially fatal version of the [over] confidence trap."[16]

Even so, is a new birth of democracy possible? Is it possible to create new and more effective forms of citizenship in the twenty-first century? Is it possible to use television and the Internet to organize an active and strongly democratic society, from neighborhoods to planetary politics? Is it possible for nongovernmental organizations and diverse, cross-cultural citizen networks to accomplish what present forms of politics and governance cannot do? Time will tell.

What we do know is that citizens, networks, corporations, regional affiliations, nongovernmental organizations, and central governments will all have to play their parts. The twenty-first century and beyond is all-hands-on-deck time for humankind. We have no time for further procrastination, evasion, and policy mistakes. We must now mobilize society for a rapid transition to a low-carbon future. The longer we wait to deal with the climate crisis and all that it portends, the larger the eventual government intrusion in the economy and society will necessarily be, and the more problematic its eventual outcome.

We have entered the rapids of the human journey. Whether we can avoid capsizing the frail craft of civilization or not will depend greatly on our ability and that of our descendants to create and sustain effective, agile, and adaptive forms of governance that persist for very long time spans. One hopes that these will be strongly democratic, but there is no guarantee that they will be, especially over times far longer than that of the Chinese empire or the Catholic Church. It's never been done before. But that could be said prior to every major human achievement as well.

Introduction

CHAPTER 1

Failing Governance, Unsustainable Planet

Michael Renner and Tom Prugh

In early November 2013, Typhoon Haiyan struck the Philippines, the strongest cyclone to make landfall in recorded history. It killed thousands of people, displaced more than 4 million, and left 2.5 million in need of food aid. Hitting just before the round of climate negotiations known as the 19th Conference of the Parties (COP) to the United Nations Framework Convention on Climate Change (UNFCCC), it was yet another reminder of the climate-charged superstorms and other disasters that lie in store if countries do not act with due haste to reduce greenhouse gas emissions. It prompted the Philippines' chief negotiator at COP 19, Yeb Sano, to announce that he would fast until conference participants made "meaningful" progress.[1]

Cold, hard data reinforce the sense that humanity is at an unprecedented crossroads that requires a sharp departure from politics and business as usual. In 2012, global emissions of carbon dioxide (CO_2) from fossil fuel burning and cement production climbed to a new peak of 9.7 billion tons, and they were projected to reach 9.9 billion tons in 2013. The 2.7 percent average annual increase in emissions during 2003–12 was almost triple the rate of the previous decade. In early 2013, the concentration of CO_2 in the earth's atmosphere for the first time crossed the threshold of 400 parts per million.[2]

The chances of limiting global temperature increases to 2 degrees Celsius (3.6 degrees Fahrenheit) within this century are "swiftly diminishing," in the judgment of Achim Steiner, executive director of the United Nations Environment Programme. This goal was endorsed by governments in 2010 as a "safe" maximum to avoid the worst consequences, although some regard it as still too high. Yet under current government policies, global greenhouse gas emissions still will be 8 to 12 billion tons higher than the maximum allowable in 2020, likely leading to a warming of 3.7 degrees Celsius or worse. The International Energy Agency (IEA) projects that current policies could raise temperatures by as much as 6 degrees Celsius.[3]

Although governments pay lip service to the goal of keeping climate

Michael Renner and **Tom Prugh** are codirectors of the *State of the World 2014: Governing for Sustainability* project.

change within tolerable limits, they have fallen far short of needed action in many ways. International climate governance has been marked by increased wheel spinning in recent years, and policies in several countries now represent a weakening of earlier commitments. An analysis by Climate Action Tracker warns of a "major risk of downward spiral in ambition, a retreat from action and recarbonization of the energy system."[4]

Recent actions by Australia's new government, for example, could cause that country's greenhouse gas emissions to increase 12 percent by 2020 (instead of being reduced 5 percent from 2000 levels, as pledged earlier). Japan abandoned its 2020 target for cutting national emissions to 25 percent below 1990 levels in favor of a much less ambitious cut of 3.8 percent. Canada barrels ahead in developing its carbon-intensive tar sands deposits. And the Polish government opted to welcome an "international coal and climate summit" staged by the World Coal Association at the very same time that it hosted the most recent round of international climate talks. For the climate conference itself, Poland accepted corporate sponsorship from leading car manufacturers, oil companies, builders of coal power plants, and steel manufacturers.[5]

Climate change is certainly not the only factor undermining sustainability, but no other phenomenon carries such risks to the survival of planetary civilization. Climate change interacts with and exacerbates many other issues of concern for environmental integrity and human well-being—such as water availability and food production, biodiversity, health, disaster protection, and employment. It has far-reaching socioeconomic and political implications. The international governance processes for climate protection and for sustainable development (the Rio+20 conference and its aftermath) proceed largely on separate tracks, but the year 2015 will be a key milestone for both of them.

Climate Policy's Tower of Babel

Environmentalists have long clung to the belief that science would drive government action on climate change and other global environmental challenges. This flows from an assumption that the picture that emerges is so self-evident and compelling that no one could seriously dispute the need for action. Yet, as Monty Hempel points out in Chapter 4 of this book, knowledge alone is not enough, and indeed things have turned out differently.

For one, climate science is so complex that it is far from easily communicated to the general public. Scientific consensus-building naturally tends to err on the side of caution and understatement. In a 2012 commentary, Kevin Anderson and Alice Bows argue that climate change scenarios all too often are subjugated to orthodox economic views that regard unimpeded growth as the inviolable goal: "When it comes to avoiding a 2°C rise [in average global temperatures], 'impossible' is translated into 'difficult but doable', whereas 'urgent and radical' emerge as 'challenging'—all to appease the god of economics (or, more precisely, finance)." With the exception of outspoken individuals like James Hansen—who served as head of NASA's Goddard Institute for Space Studies until 2013—most scientists have been reluctant to engage in the fierce, polarized political debates of how society should respond to distressing scientific findings.[6]

Nomad Tales

An empty coal train heads back to the mines, Maitland, NSW, Australia.

Meanwhile, a well-oiled machinery of climate denialists has managed to sow doubt (or worse) about the ever-strengthening climate science consensus, helping to reassure those whose inclination is to disbelieve the science. At a time of global economic crisis, denialists have been able to stoke fears among the general public that sustainability policies are at odds with concerns about jobs and incomes. Such efforts have been amplified by a media that often perpetuates a false equivalency between climate scientists and "skeptics."[7]

If the science of climate change is hard to comprehend, so is the human process that has emerged over the last two decades around efforts to address it. The structures and processes under the UN's climate regime are largely indecipherable to the majority of the people on this planet. A veritable climate-speak Tower of Babel has arisen, replete with a proliferating number of acronyms that range from AAUs, AWG-LCA and AWG-KP to CDM, CERs, and GCF; from LULUCF, NAMAs and NAPAs to QELROs, REDD and REDD+; and on to RMUs, SBSTA, and SD-PAMs—to name only a few. The UNFCCC's own glossary of acronyms comprises more than 180 entries.[8]

Clearly, negotiations among the world's 189 member states that are party to the UN climate convention, as well as the various regional or interest groups with which they align, are by their nature a complex undertaking. Although not as large as the environmental mega-summits such as the 1992 Rio Earth Summit and 2012's Rio+20, the annual high-level climate conferences have become massive gatherings. The first COP, held in Berlin in 1995, drew 1,925 participants (not counting media representatives). By 2013, the number of participants registered to attend COP 19 in Warsaw had expanded almost ninefold, to 9,135. Media interest, however, shrank dramatically, falling from 2,044 journalists attending in 1995 to 971 in 2013.[9]

A more fundamental problem than the sheer numbers is the politics that is driving—or more often, blocking—the climate talks. Relative to the massive carbon reductions needed, two decades of international climate negotiations have yielded precious little in the way of tangible progress, but plenty of frustration. In 2009, the high expectations for COP 15 in Copenhagen, Denmark, led climate activists to speak of "Hopenhagen." But following the sobering failure that ensued, "Nopenhagen" became the more apt moniker, leading to searching questions about whether the following year's meeting in Cancún, Mexico, would be a "Can-cún" or "Can't-cún." Word play aside, the deadlock on key issues has persisted. In effect, the negotiators keep kicking the problem farther down the road, always in the hope that the success that eludes them one year might come within reach the next year.[10]

Various forces have prevented greater success. A recent analysis indicates that the top fossil fuel-producing countries hold 25–30 percent of the high-level (officer) posts in the bodies of the UN climate convention, a disproportionate share given that these countries account for only 16 percent of UNFCCC members. Since 2009, coal exporters have been particularly well represented.[11]

Although individual country positions vary, industrialized countries on the whole have been unwilling to abandon their materials-intensive and wasteful lifestyles, whereas emerging economies are intent on avoiding any mandatory commitments that could block their chance of emulating the West's consumerist model. There is much inertia, and outright resistance, from various sides to a meaningful and binding carbon reduction agreement, and it comes foremost at the expense of the most vulnerable and poorest countries.

The United States, historically the largest carbon polluter, insists on the kind of "flexibility" that is poison for a binding global climate treaty. Speaking at London's Chatham House in October 2013, U.S. Special Envoy for Climate Change Todd Stern said that "rather than negotiated targets and timetables, we support a structure of nationally determined mitigation commitments, which allow countries to 'self-differentiate' by determining the right kind and level of commitment, consistent with their own circumstances and capabilities." (See Chapter 11 by Petra Bartosiewicz and Marissa Miley for an account of the failure to establish a more aggressive U.S. policy.)[12]

China's leaders stake their legitimacy on providing a steady and growing flow of goods and services for a population that has no real say in political decision making. They are opposed to any international agreement that would impede the country's economic growth. Yet China's unprecedented pace of economic expansion has translated not only into skyrocketing CO_2 emissions, but also into an environmental devastation and threat to public health that increasingly is becoming the main rallying cry of domestic popular activism. (See Chapter 12 by Sam Geall and Isabel Hilton.)

Confronting Petro-Power

If runaway climate change is to be avoided, a global pact to leave the bulk of the world's proven fossil fuel reserves in the ground is indispensable. The currently proven reserves of oil, natural gas, and coal contain about 3 trillion tons of CO_2. Two-thirds or more of this can never be touched if there is to be any hope of avoiding a destabilized climate. Yet this climate reality runs headlong into a global capitalist economy whose *raison d'être* is endless growth and that therefore demands an ever-expanding flow of energy.[13]

The additional fossil fuel extraction capacity represented in such forms of "extreme energy" as tar sands, Arctic and deepwater deposits, shale oil and gas (unlocked through hydraulic fracturing or "fracking" technology), and mountaintop-removal coal will lock society into an unsustainable energy system for decades to come. The 2012 exploration and development expenditures of 200 fossil fuel companies listed on stock exchanges worldwide are estimated at $674 billion. (This compares with renewable energy investments of $244 billion the same year.) Global exploration and production spending for oil and gas has increased 2.4-fold since 2000, and the IEA projects that by 2035, a cumulative $14.7 trillion may be spent for such purposes, with another $3.1 trillion for refining and distribution—triple the projected spending on renewables.[14]

Fossil fuel companies have every incentive to extract as much as possible of the extremely valuable reserves they have on their books. Leaving the bulk of the world's fossil fuel deposits untouched will require quasi-revolutionary change. Nothing like this has ever been attempted in human history, and it likely will require a combination of regulation, litigation, shareholder activism, and dogged divestment and civil disobedience campaigns. Any such effort runs fundamentally counter to the interests of powerful and politically well-connected companies—not just the fossil fuel producers themselves, but also carbon-intensive sectors such as power utilities, motor vehicle manufacturers, and the petrochemical industry. (To overcome such opposition, there will need to be some sort of compensation or other transition arrangement, although this is too complex an issue to be addressed here.)

A recent analysis by Richard Heede found that just 81 private and state-owned corporations are responsible for about 40 percent of cumulative carbon emissions since the start of the Industrial Revolution, while 9 centrally planned states contributed another 21 percent. (See Table 1–1.) In 2012, just 25 companies were behind 58 percent of worldwide "upstream" oil and gas investments. These include privately owned companies such as Exxon-

Table 1–1. Carbon Emissions by Type of Entity, 1751–2010		
Entity	Cumulative Emissions	Share of Global Total
	Billion tons of CO_2-equivalent	Percent
50 Investor-owned corporations*	314.8	21.7
31 State-owned corporations	287.7	19.8
9 Nation-state carbon producers[†]	311.8	21.5
Subtotal	914.3	63.0
Total, World	**1,450.3**	**100.0**

*Fossil fuel and cement producers.
[†] Current or former centrally planned states (includes the Soviet Union and post-Soviet Russia as two separate entities).
Source: See endnote 15.

Mobil, Chevron, Royal Dutch Shell, and BP, as well as wholly or partially state-owned firms such as Petrochina, Brazil's Petrobras, Russia's Gazprom, Mexico's Pemex, and Norway's Statoil.[15]

It is no secret that these private firms act solely at the behest of a narrow class of shareholders. The state-owned firms at least nominally serve a broader public interest; in many countries, nationalization was an outcome of historic power struggles over who benefits from the extraction of fossil fuels. Still, state ownership does not necessarily translate into policies in the public interest. State companies may be run in ways that are functionally no different than private companies. Or they may be controlled by unrepresentative regimes that channel revenues into repression or corrupt practices, as Evan Musolino and Katie Auth write in Chapter 17. Fossil fuel revenues can be used responsibly, as Norway has shown. But the full costs of climate change will eventually surpass any benefits that may be derived from continued exploitation of fossil fuels.

It is worth noting that underlying and propping up this web of powerful corporate actors, whose interests so often clash with the public interest, are the wishes, desires, and buying power of hundreds of millions of people. The lure of consumerism (aided by massive advertisement spending) has proven to be almost irresistible around the planet, and many people define themselves more in terms of their material possessions than in terms of being active citizens.

Automobiles are a case in point. They remain one of humanity's key status symbols and are often seen as an embodiment of freedom and individualism. Yet all but a tiny share of the world's motor vehicles run on oil-derived fuels, and total vehicle registrations exceeded 1 billion (for the first time) in 2010. Not only does this vast and growing fleet represent enormous ongoing demand for carbon-based fuels, but it is also a critical factor in locking society onto a dangerous energy path. The vehicle fleet turns over very slowly (every 12–15 years in the United States, and even more slowly during recessions), so that consumer choices and buying behavior embed a great deal of capital in vehicles and the infrastructure that accompanies them, committing society to their long-term use.[16]

Markets to the Rescue?

Fighting the interests of fossil fuel companies is a colossal undertaking, not least because we live in an era in which corporations and markets are seen as near-sacrosanct forces. A *laissez-faire* attitude often described as neoliberalism has prevailed. Deregulation and privatization have heralded an increasingly globalized economy and the emergence of globe-spanning corporations whose influence and power often trumps that of governments, communities, and labor unions.

The view that government is the problem and private markets are the solution has carried over into the design of environmental policy. Governments, academics, and many mainstream environmental groups have put considerable hope into the assumption that, with the proper signals, markets would ride to the rescue and drive a clean economy transition. Specifically, this has found expression in proposals for carbon markets and so-called cap-and-trade systems. In principle, the idea of imposing a cap on emissions and of putting a price on carbon is sensible. The actual manner in which this has been implemented—specifically the European Union's Emissions Trading System (EU ETS), which had 88 percent of the world's carbon trading volume in 2012—raises fundamental questions about whether salvation will be found solely in market-based mechanisms.[17]

ETS carbon prices have nosedived repeatedly. In the scheme's first phase (2005–07), prices plummeted from a peak of around €30 ($38.70) per ton in April 2005 to a mere €0.10 ($0.14) per ton in September 2007. This was due largely to an overly generous allocation of emission allowances and exemptions—itself the product of industry lobbying clout. Although the EU insisted that it was learning by doing, the experience was replicated in the second phase, when prices once again collapsed, from about €25 ($36.75) per ton in 2008 to between €5 and €10 ($6.40–12.80) per ton in 2012. Prices stayed below €5 ($7) per ton during 2013, and absent regulatory intervention, analysts expect that they will remain low for the entire third phase (2013–20). Johannes Teyssen, the CEO of Germany's largest utility E.ON, commented in 2012, "I don't know a single person in the world that would invest a dime based on ETS signals." Fixing the system—if it can be fixed—

would require dramatically reducing the supply of carbon certificates and lowering the overall cap on emissions.[18]

As national or regional emissions trading schemes are being adopted elsewhere around the globe—most recently in China and Mexico—fundamental governance lessons from the EU ETS experience need to be taken to heart. As a recent Climate Action Tracker report argues, "The new systems yet have to prove that their implementation will actually reduce emissions." Fresh thinking also needs to be applied to related approaches such as the Clean Development Mechanism. A recent examination by Germany's *Der Spiegel* likened this approach, under which rich polluters have bought often questionable or even fraudulent carbon "offsets" in poorer countries, to the selling of indulgences. It is a practice that keeps carbon markets flooded with certificates and prices low.[19]

Market-based mechanisms such as carbon trading seem to relieve governments of the difficult political decisions needed to alter unsustainable production and consumption structures. Trading of emission permits, for example, allows governments to avoid imposing a politically unpopular carbon tax. Yet carbon markets cannot possibly function without the kinds of extensive rules and regulations that have come to be excoriated as "command-and-control" policies. And there are other governance-related reasons for skepticism. Emissions trading favors—and often enriches—a "carbon priesthood" of corporations, traders, and financiers. The arcane nature of such systems prevents meaningful public engagement.

Moreover, the dogma of market worship has marginalized a large body of knowledge about the management of common-pool resources that points to the fruitful possibilities of controlling global carbon pollution by managing the atmosphere as a commons. This work—which for general audiences emerged into the light of day only when a major scholar of commons management, political economist Elinor Ostrom, won the 2009 Nobel memorial prize in economics—soundly refutes the argument that privatization of common resources such as the atmosphere's waste-absorption capacity is the preferred, or only, way to address the problem. (See Chapter 2 by D. Conor Seyle and Matthew Wilburn King and Chapter 9 by David Bollier and Burns Weston for more on Ostrom's work.)

Since Adam Smith, economists have argued that markets, even though driven by selfish, short-term motivations to maximize private gain, ultimately serve the public interest. This view springs from an idealized set of exchanges that assumes that all players have the same information and that markets will eventually self correct. But it conveniently overlooks the fact that some market players grow to become far more powerful than others. Markets do not have a social conscience, environmental ethic, or long-term vision, and therefore market dynamics and the public interest do not necessarily align. Although market tools could incentivize companies to go beyond the minimum of meeting a particular law or regulatory standard, markets as such are a poor arbiter of processes that decide whether civilization thrives or perishes.

Making Democracy Safe for Markets

For all the governance faults one can find in the political sphere—whether it be the sclerosis of bureaucracies or the lack of vision among those holding elected office—governance of the economic sphere is not even nominally democratic, and thus suffers from a basic defect. Market signals and impulses increasingly make business leaders slaves to the quarterly bottom line, irrespective of whether that bottom line is congruent with a company's, let alone society's, longer-term well-being. Businesses increasingly regard labor as a cost item to be minimized, driving a relentless process of automation and putting pressure on employment and wages. Yet gains in labor productivity are less and less shared with the workforce. Sidelining social and environmental factors by relegating them to the status of "externalities," as economists are trained to do, is equivalent to shutting one's eyes to realities that one prefers not to see.

This narrow, short-term view is reinforced by the demands of a bloated finance sector. Thomas Palley (Chapter 16) writes that the rising influence of finance has been an engine of an economy that gobbles up growing amounts of scarce resources even as it distributes the product in ever more unequal ways. The result is vast wealth gaps, which have given rise to the notion of the 1 percent versus the 99 percent. A key task will be to govern the finance sector in ways that facilitate the transition to a more equitable and sustainable economy and to inject a greater degree of accountability into the private sector.

Collective bargaining and related structures (including the so-called works councils that represent worker interests at factories in several European countries) historically have been among the tools to introduce at least a modicum of democracy in the workplace, and have been essential for raising wages. But these processes have weakened as union representation has declined in countries where it once was strong (while it never gained much of a foothold in other countries). Aided by globalization, multinational corporations are able to force concessions from labor and from governments alike; workers often accept wage or benefit cuts for fear of jobs being moved offshore; local, regional, and national governments vie for industries by offering big tax giveaways or other "sweeteners."

Governance in the economic sphere—determining what gets produced, how, and who benefits—has a powerful influence on society's ability to achieve social and environmental sustainability. But economic governance

Abhist Veljajiva

Opening ceremony of the EU's eighth Asia-Europe Meeting at the Royal Palace in Brussels.

also carries over directly into the political sphere. The concentration of wealth and power essentially narrows the ranks of those with an effective voice in decision making and in public discourse. The drafting of legislation by lobbyists is not uncommon, for example, and there has long been a revolving door for people moving between positions in government and business. In Brussels, an estimated 15,000 lobbyists seek to influence European Union rule making, according to the Alliance for Lobbying Transparency and Ethics Regulation.[20]

The electoral and political decision-making processes of some countries (including the United States) have been captured by powerful interests opposed to decisive action for sustainability. This became clear during the fight over U.S. climate legislation in 2009. According to the Center for Responsive Politics, during 2009 the $22.4 million spent by pro-environmental groups on federal lobbying efforts was dwarfed by oil and gas industry expenditures of $175 million. The floodgates of private money influence were opened wide by the U.S. Supreme Court's "Citizens United" ruling in 2010, which allowed unlimited political spending by corporations, associations, and trade unions. Political advocacy groups spent more than $300 million on the 2012 presidential campaign, up from $79 million during the previous election.[21]

A growing threat to democratic governance is also found in investor/state dispute-settlement clauses that are included in many bilateral investment treaties. These allow companies investing abroad to challenge a broad array of health, environmental, social protection, and other laws. Instead of applicable domestic courts, such claims are adjudicated via private dispute-settlement tribunals, where secretive panels of trade lawyers can overrule the will of parliaments. According to Corporate Europe Observatory, a Brussels-based watchdog group, more than 1,200 such treaties have been signed by member states of the EU alone.[22]

The number of claims for compensation brought by multinational corporations under such clauses keeps rising and involves billions of dollars. According to the UN Conference on Trade and Development, at least 62 new cases were initiated against host countries in 2012—the highest number ever filed in a single year. Cumulatively, the number of claims reached 518 as of May 2013, filed against 95 different countries. Of the 244 cases that have been concluded so far, 42 percent were decided in favor of the state and 31 percent in favor of the investor, while another 27 percent were settled. Thus, corporations do not always win cases they initiate, but sometimes the mere threat of a claim or its submission has been enough for legislation to be abandoned or watered down. The Transnational Institute sees "a permanent tension between investor rights and public welfare interests."[23]

Beyond bilateral treaties, the North American Free Trade Agreement has been used extensively for bringing investor claims. But two multilateral treaties with broader reach are currently being negotiated. If passed, they will essentially be models for the rest of the world. These treaties—the Transatlantic Trade and Investment Partnership between the United States and the European Union, and the Trans-Pacific Partnership between the United States and countries in the Asia-Pacific region—are being negotiated in secret, shielded from public discussion and parliamentary scrutiny, even as corporate lobbyists are playing a key role. Like the existing bilateral treaties that enshrine investor-state dispute-settlement mechanisms, these new treaties would further limit the ability of governments to make rules in the public interest.[24]

What Are Governments For?

Given the emphasis on market mechanisms and investor rights, it is easy to lose sight of the fact that the job of governments is…to govern. Governing means drawing up the rules by which society functions. These may entail both mandates and incentives, and the best policy is ultimately one that combines a rich repertoire of appropriate tools. There is a proper role for markets, too. But the voluntary measures that have been embraced so eagerly in recent years are increasingly at odds with the planetary climate and sustainability emergency.

A broad range of governmental actions can steer economies toward climate stability and environmental sustainability. This includes rising energy efficiency standards for industrial equipment, buildings, motor vehicles, and consumer goods. Binding emission limits are another option, such as proposed carbon pollution standards for U.S. power plants that would effectively rule out conventional coal units. In many instances, such policies already exist but need to be made more ambitious and stringent.[25]

Government regulation and market mechanisms can be combined in imaginative ways, as Japan's Top Runner program has shown since 1998. Efficiency standards for a wide range of products are set by committees composed of representatives from the manufacturing industry, trade unions, universities, and consumer organizations. They identify the most efficient model in a given product category. It becomes a baseline that all manufacturers must meet within 4–8 years, when the process is then repeated. This approach drives continuous innovation but also provides time for lagging manufacturers to catch up or to invent an even more efficient product.[26]

Governments can contribute to greater sustainability by reorienting their own procurement budgets and infrastructure projects—for example, away from additional road building and toward public transport; away from sprawl and toward denser cities (which thereby tend to gain in livability).

Another field where government action is needed is in redirecting public financial streams from unsustainable to sustainable economic activities. This includes phasing out fossil fuel subsidies and ending financing by international development banks and national export credit agencies for fossil fuel projects. According to a preliminary analysis by the Natural Resources Defense Council, in just the 2007–13 period, the top four funders alone—Japan, the United States, Germany, and South Korea—have provided $37.7 billion for coal projects in developing countries. But Denmark, Finland, Iceland, Norway, Sweden, the United Kingdom, and the United States have now announced that they will no longer finance coal projects abroad.[27]

An often-overlooked and potentially promising trend with respect to functional government is the apparent shift of impetus for action on critical sustainability issues from national governments, which have often dithered, to local and regional governments. As Monika Zimmermann discusses in Chapter 14, in the last 20 years or so local governments have radically stepped up their organizing, cooperation, and degree of commitment to addressing issues such as climate change. It is perhaps no coincidence that local and regional governing bodies are both closer (in distance as well as in degree of bureaucratic separation) to the people and communities governed, and less likely to be captured by special interests.

Governance from the Bottom Up

Governmental structures and decision-making processes diverge widely by country, but the common challenge is how to imbue them with a greater degree of foresight, accountability, transparency, and responsiveness. Can humanity devise governance institutions and processes—both political and economic—that are able to overcome the barriers to greater sustainability? It is an empirical question that we will likely see answered in the coming years, as either we rise to the challenge or nature imposes something like sustainability upon us. John Gowdy, in Chapter 3, argues that there is in fact an evolutionary basis for the dilemma we seem to have backed ourselves into—which suggests that failing to devise institutions that can mitigate our worst genetic tendencies will take us down nature's pathway to sustainability, with whatever costs and disruption to human civilization it sees fit to inflict.

Although global society has largely ignored them, for years there have been alternatives to the dominant worldview that the natural world is a platform for living situated in a warehouse of resources that are ours for the taking. Ecological economists and others repeatedly have made the case for operating within Earth's system limits. Other eloquent voices have urged consideration of perspectives on the human place in the world that would enable and support this mode of operation.

In Chapter 4, Monty Hempel asserts that teaching ecoliteracy, while necessary, will not be enough by itself to achieve an Earth-centered worldview; it will need to be combined with ethics training and appeals to action. Richard Worthington cautions in Chapter 5 that we cannot rely on the digitization of everything to solve the problems we face, absent concerted action in other, especially political, spheres. And a trio of chapters, by Peter Brown and Jeremy Schmidt (Chapter 6), Cormac Cullinan (Chapter 7), and Antoine Ebel and Tatiana Rinke (Chapter 8), urge us to rein in our worst tendencies in order to free up ecological space for the rest of creation, and to expand the circle of stakeholders to include the voiceless: other creatures, indigenous cultures, and youth and the generations to come.

David Bollier and Burns Weston (Chapter 9) urge that humanity infuse ecological governance with a commons- and rights-based approach that is anchored in laws and policies drawn up at the local and national scale. The plodding pace of international talks on climate protection and sustainable development has made many civil-society activists weary of a mismatch between promising rhetoric and paltry outcomes. Maria Ivanova (Chapter 13) points to outcomes of the Rio+20 conference that are nonetheless significant for shaping global governance in coming decades. It would be a mistake for civil society to retreat from these processes, but Lou Pingeot (Chapter 15) cautions against the rising corporate influence on them.

In the face of governmental inertia and corporate capture of many decision-making processes, strong and persistent bottom-up political pressure is needed more than ever. It was grassroots mobilization under the banner of nationwide Earth Day celebrations that helped bring about landmark U.S. laws such as the Clean Air Act and Clean Water Act in the early 1970s, when the United States was an environmental policy pacesetter. But over time, parts of the environmental movement have grown comfortable with a more establishmentarian orientation that cherishes mainstream respectability, ample funding, and access to the corridors of power. Chapter 11, by Petra Bartosiewicz and Marissa Miley, explores how a small group of well-funded mainstream environmental groups preferred an elite approach to passing cap-and-trade legislation over grassroots mobilization—a strategy that ultimately failed.

Elite environmentalism runs the danger of being disconnected from environmental justice perspectives driven by the devastating real-world impacts on communities of mining projects, petrochemical plants, or other toxic facilities sited near poor neighborhoods, or, for that matter, dubious green solutions such as large-scale biofuels plantations associated with land grabbing and displacement of small farmers. Aaron Sachs (Chapter 10) insists that we not lose sight of the injustices of today's world when we worry about the coming storms and floods and heat waves in a future warmer world. Successful social movements throughout history, Sachs reminds us, have incorporated a strong sense of ethics.

Undoubtedly, new grassroots movements are emerging, and new energy is being unleashed—keeping true to the view that the whole point of civil society organizations is to be a thorn in the side of the powerful. This is part of a broader phenomenon of spreading popular protests driven by a range of grievances and demands—irrespective of the political governance system in question. A recent study analyzing 843 protests between January 2006 and July 2013 in 87 countries found a steady increase in protests from 59 in 2006 to 112 during just the first half of 2013. Many of the protests—ranging from marches and rallies to acts of civil disobedience—involve issues that are of relevance to a more sustainable and equitable society. The lack of "real democracy" is a major motivating factor and is seen as an underlying reason for the lack of economic and environmental justice. (See Table 1–2.)[28]

Referring to what he calls an "emerging fossil fuel resistance," Bill McKib-

Table 1–2. Worldwide Protests by Selected Grievance or Demand, 2006–2013		
Category (total number of protests)	**Selected Grievance or Demand**	**Number of Protests**
Economic Justice and Austerity (488)	Jobs, wages, labor conditions	133
	Inequality	113
	Agrarian/Land reform	49
	Fuel and energy prices	32
	Food prices	29
Failure of Political Presentation (376)	Real democracy	218
	Corporate influence, deregulation, privatization	149
	Transparency and accountability	42
Global Justice (311)	Environmental justice	144
	Global commons	25
Rights (302)	Commons rights	67
	Labor rights	62

Note: The report distinguishes among a total of 34 specific types of grievances/demands.
Source: See endnote 28.

ben, founder of 350.org, observes that in the last few years a new grassroots movement "has blocked the construction of dozens of coal-fired power plants, fought the oil industry to a draw on the Keystone pipeline, convinced a wide swath of American institutions to divest themselves of their fossil fuel stocks, and challenged practices like mountaintop-removal coal mining and fracking for natural gas."[29]

The proposed Keystone XL pipeline, slated to carry Canadian tar sands to the Gulf of Mexico, has emerged as a lightning rod of resistance in the United States. Similarly, opposition by native peoples and others in British Columbia has put on hold the proposed Northern Gateway pipeline (intended to carry tar sands over a distance of 1,177 kilometers to an export terminal and eventually to Asian markets). Several planned coal export terminals in the Pacific Northwest also have drawn strong local opposition due to environmental and health concerns. In Europe, France and Bulgaria have banned fracking, and opposition to the controversial practice is rising in the United Kingdom. In the autumn of 2013, EU lawmakers provided initial approval of a measure requiring extensive environmental audits before fracking can go forward. In China, pollution may be the single largest cause of social unrest, as Sam Geall and Isabel Hilton explain in Chapter 12. Since 2007, waves of social unrest there have halted numerous large industrial and infrastructure projects.[30]

Distributed Leadership

McKibben thinks that the new fossil fuel resistance movement is beginning to win some victories, "not despite its lack of clearly identifiable leaders" but rather "because of it." Like the "distributed generation" system that renewable energy technologies enable, human society needs to develop forms of distributed leadership. Along these lines, McKibben sees greater value in a more dispersed opposition network than a highly centralized one that relies critically on the vision and actions of a small handful of leaders. He observes, "often the best insights are going to come from below: from people… whose life experience means they understand how power works not because they exercise it but because they are subjected to it."[31]

Climate and other sustainability questions cannot be seen solely through the prism of environmentalism. The fight for sustainability needs to incorporate dimensions of social justice, equity, and human rights.

The far-reaching impacts that a transition to a more sustainable society holds for the lives of billions of people implies that governance needs to be as democratic, transparent, and accountable as possible, and this imperative extends to the workplace. Unions find themselves on the defensive in many countries, but the labor movement needs to be an active participant in the transition toward sustainability. Beyond the demand for a socially just transition that has become a rallying cry among union activists, Judith Gouverneur and Nina Netzer argue in Chapter 21 that a fundamental reorganization of work needs to be undertaken so that available work is better shared in a sustainable economy.

Sean Sweeney (Chapter 20) discusses the difficulty of transforming the energy system at a time when fossil fuel corporations push ahead with additional carbon-intensive projects. He argues in favor of greater "energy democracy" that gives workers, communities, and the public at large a more meaningful voice in decision making. Fossil fuel corporations are among the largest companies in the world. Like their counterparts in other sectors of the economy, they have acquired a "too big to fail" aura, yet they elude meaningful democratic accountability at a time when their decisions affect virtually everyone on the planet.

Beyond the energy sector, economic governance reforms could include accelerating the creation of so-called "benefit" corporations. Colleen Cordes (Chapter 19) examines this still-new phenomenon of companies that orient themselves toward a broader array of stakeholders, including their employees and the local communities within which they operate. Gar Alperovitz (Chapter 18) discusses the detrimental effects of large wealth and income gaps and notes that, because of the socialization of technological gains, those gaps are mostly undeserved by those at the top *and* the bottom—a point that even some mainstream economists concede. Community wealth-building strategies—including cooperatives, worker-owned firms, community development corporations, community development financial institutions, social enterprises, community land trusts, and employee-owned enterprises—can pool capital in ways that build wealth, create living wage jobs, and anchor those jobs in communities.

Finally, it seems clear that the antidote to the ills of concentrating wealth and power that are so instrumental in thwarting efforts to achieve sustainability is deconcentrating—devolving—wealth and power. Chapter 22, the concluding chapter, is a meditation on the material in this book and on the variety of political and economic means available to achieve that end. In particular, we argue that a more engaged citizenry is key, not only to the success of specific movements such as the resistance to the fossil fuel domination that drives climate change, but to all dimensions of sustainability. It is no longer enough for people everywhere to struggle for nominally democratic polities, and then to hand off power and responsibility for their ongoing operation and integrity to others. That seems inevitably to invite corruption and the appropriation of the machinery of governance for private ends.

People everywhere must strive to don the mantle of citizenship and commit to persistent engagement in the governing of their workplaces, communities, and nations. Concentrated power and wealth will forever seek to fulfill only its own narrow interests—even as the biosphere and civilization are corrupted and perhaps destroyed. Only a steady popular commitment to engaged governance can prevent this outcome. The quest for environmental sustainability, social equity, and a deep, deliberative culture of citizen engagement are thus closely intertwined goals.

CHAPTER 2

Understanding Governance

D. Conor Seyle and Matthew Wilburn King

Over the past 30 years, the idea of governance (versus government) as a critical framework for understanding human society has taken root in the scholarly and policy communities. In the 1990s, political economist Elinor Ostrom's Nobel Prize-winning work introduced the idea that systems created by local communities could lead to sustainable governance of natural resources. At the same time, scholars of international relations began to appreciate the way that many global systems were fairly well governed even in the absence of formal international institutions. International organizations such as the World Bank and UNESCO began to realize that the quality of governance in the places they operated was a major factor in the success, or failure, of their programs. The result is an increasing shift in the research community toward talking about governance as a critical piece of understanding human collective behavior. (See Figure 2–1.)[1]

But how could this concept apparently be all things to all people? What exactly is "governance"? Why is it a valuable lens for looking at human behavior, and how does it add to the global policy discussion about how to create a more sustainable, peaceful world? What is the result of all of this academic and research interest, and do the theories developed bear any relationship to the nitty-gritty details of how the world is governed today?

What Is Governance?

As a basic term, "governance" refers to the processes by which any complex activity or system is coordinated. Its roots are found in the Latin *gubernare*, an adaptation of the Greek word for the steering of a ship, *kybernan*. Any system in which many separate pieces must work together toward some end has some form of governance: early steam engines, for example, were made safer by the installation of a "governor" that maintained a constant speed and kept the engine from damaging itself. The specifics of governance mechanisms are diverse and can range from consciously designed devices

D. Conor Seyle is associate director of research and development at One Earth Future Foundation. **Matthew Wilburn King** is president of the Living GREEN Foundation.

like those used by steam engines, to hodgepodge and decentralized systems brought about through evolution. Ants, for instance, have evolved instincts that drive them to cooperate in ways that appear to be highly organized and well governed, despite having no central decision-making structure. (See Chapter 3.)[2]

The same basic definition holds true for human society: human social groups are complex systems and so require governance systems to accomplish collective goals. The scholarly literature offers a variety of definitions of governance of human groups. A simple one is that governance encompasses any mechanism that people use to create "the conditions for ordered rule and collective action." A more elaborate definition defines governance as "the constellation of authoritative rules, institutions and practices by means of which any collectivity manages its affairs." An attempt to define governance at the level of the state describes "the exercise of economic, political, and administrative authority to manage a country's affairs at all levels."[3]

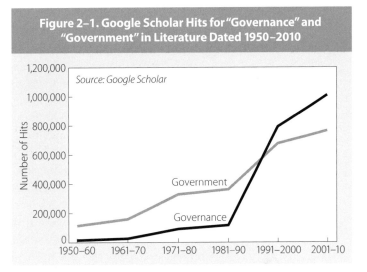

Figure 2–1. Google Scholar Hits for "Governance" and "Government" in Literature Dated 1950–2010

In this chapter, we take governance to mean the formal and informal mechanisms and processes that humans use to manage our social, political, and economic relationships with each other and the ecosphere. These mechanisms and processes are embodied in social institutions and reflect social norms, values, and power relationships.

Governance therefore includes any system that humans use to make and enforce collective decisions. Consider the diversity implied in this range: families have governance systems that help establish things like bedtimes and table manners; communities have governance systems for natural resources, such as rivers, that regulate water use or establish catch limits; businesses have governing boards that set company policies; and cities, states, and countries have governance systems that set up the political means by which behavior within the system can be regulated. In short, the diversity of human social groups and their responses to collective problems leads to an array of systems to govern. And obviously, no one structure is able to effectively govern all of the different domains of human behavior.

As a result, the discussion of governance is necessarily complicated. The diversity of research on governance says more about the ways that it can vary than its universal traits. Governance systems can be structured as hierarchies with centralized and structured chains of communication, as networks with no chains of command but distributed collective decision making, or as hybrids of these two systems. They can be participatory or have few decision makers whose decisions are rigidly enforced. The jurisdiction of governance can be defined by physical terrain, or by issue: the Fédération Internationale de Football Association (FIFA), for example, controls no territory but exists to govern international competitive soccer, regardless of where it is played.

The scope of the system governed can vary from the ultra-local to the global. Some governance systems control access to the water in a single lake, while others govern activity on all the high seas. Governance systems can be carefully designed, or they can be accidents of history. The "rational design" movement in the study of international relations, for example, has called for international organizations to consider carefully how their respective institutions should be structured to accomplish their specific goals. An alternate perspective posits that systems develop through evolutionary pressures— with systems that work persisting and multiplying, and those that do not work facing both internal and external pressures for reform.[4]

Underlying this complexity, all governance systems have some basic elements: they must have some way of making decisions on behalf of the collective or allowing collective decisions to be made, and some way of ensuring that the decisions that are made are carried out. Governance systems are all ultimately variations on those two themes, and the dizzying array of specific structures reflects the diversity of problems that human society faces.

"Good" Governance

The one truism that can be drawn from the research on governance is that governance is different from place to place and from system to system. There is no "one-size-fits-all" solution to governance problems. As a result, it may be less helpful to think about how to define governance and instead to think about how good governance should be assessed. Because "good" is in the eye of the beholder, it requires some kind of yardstick. To some, governance is seen as good when it protects human rights, or when it leads to sustainable governance of natural resources, or when it is seen as legitimate by the people who are governed, or when it is efficient and effective. Each of these is a slightly different conception of good governance, but they share common elements that can help us find a useful definition of "good."

Good governance protects human rights. One common perspective is that governance systems must promote the well-being of the governed, or at minimum not violate their foundational human rights. (In some perspectives, those rights are expanded to include the rights of other creatures and ecosystems in general; see Chapter 7.) Although the list of what constitutes human rights has been debated for hundreds of years and is not yet fully agreed upon, political scientists Allan Buchanan and Robert Keohane claim that "there is agreement that the list at least includes the rights to physical security, to liberty (understood as at least encompassing freedom from slavery, servitude, and forced occupations), and the right to subsistence."[5]

In this conception, governance is good when it protects, if not actively promotes, basic conceptions of human rights. How exactly a governing body protects human rights depends on the system itself, but commonly it relies upon legal protections—an approach that requires a system with processes in place to hold decision makers accountable to some agreed-upon rule of law or process. Decisions that are made inclusively and with the full consent of those affected is also a way that human rights can be protected (even in the absence of formal legal structures), when stakeholders are able to explain how decisions can affect their rights and prevent serious damages to them.

Good governance governs by consent. Another criterion that many scholars use to judge governance, and one with a rich tradition in political thought, is the degree to which it reflects the consent of the governed. In this conception, because governance requires that individuals hand off decision-making authority to a superordinate institution, good governance must operate based on the decision of the governed to relinquish that authority; otherwise, it is a system imposed on the governed by force. This way of thinking extends back to John Locke's claim in 1690: "That which begins and actually constitutes any political society, is nothing but the consent of any number of freemen capable of a majority to unite and incorporate into such a society. And this is that, and that only, which did, or could give beginning to any lawful government in the world."[6]

Locke's claim is that it is only right for one person to make decisions on behalf of another when the other person agrees to allow the governor to do so. The questions of whether and how this consent is given have important implications for governance. Governance systems are effective only when their decisions are put into practice, and if a citizen does not give consent to be governed by a system, then he or she may not go along with the decisions that the system makes. On a large scale, this kind of lack of consent can manifest itself as strikes, resistance movements, and the kinds of civil disobedience that can bring down governments.

Tom Tyler, a psychologist and lawyer at New York University, has studied for more than 30 years the question of what leads people to see systems as appropriate (or "legitimate") governors. He has found consistently that it is fairness, not personal benefit, that leads people to consent to governing processes and procedures. If people believe that the governance system has made a decision in a manner that they see as fair and just, they are more willing to accept that decision—regardless of whether they personally benefit from the decision or not.[7]

This suggests that the core question that humans ask when assessing legitimacy is not, "Will I benefit from this?" but, "Is this system fair?" According to Tyler's work, the answer is found in the extent of people's feeling that the system operates without preference toward any one group, that it treats constituents with respect and dignity, and that decision makers are benevolent or at least not actively malicious. As with the protection of human rights, creating a system with these characteristics often relies upon the establishment of rules and procedures to which decision makers are held accountable.[8]

The research has also found that one particularly important element is the constituents' perceptions of "voice," or the degree to which they feel that their perspectives have been taken seriously by the governors. This finding is not absolute: in larger systems, such as national governments, the question of personal benefit does appear to influence perceptions of legitimacy, possibly because in such systems it is harder to have a deep understanding of the processes and of how an individual's voice is represented. Survey data on governmental legitimacy have found it to be correlated with procedural elements, including good governance, protection of civil liberties, as well as more benefit-related correlates like poverty-reduction efforts and personal financial satisfaction.[9]

Good governance governs sustainably. Another conception of good governance is that governance systems should lead to long-term sustainability of resources. Jared Diamond's *Collapse: How Societies Choose to Fail or Succeed* illustrates both what good governance looks like in this conception and the pitfalls of bad governance. Throughout history, quite a few societies and social systems have realized (sometimes too late) that their mechanisms for collective decision making have led them down a garden path to a point where key resources necessary for survival are being depleted.

In some cases, the result is complete ecological collapse—and a crisis for the humans who caused it. According to current historical understanding, the *moai* (giant stone heads) of Easter Island reflect a religious system of governance that encouraged clan competition and felling trees to build these representations of their ancestors. The outcome was the overconsumption of trees, an ecological catastrophe, and a descent into war and starvation. One definition of good governance is simply any system that avoids catastrophic collapse of the resources the system depends on.

What sustainability looks like in practice is a contentious question. As noted earlier, however, Elinor Ostrom won a Nobel Prize for her serious and systemic attempt to explore what sustainable systems look like in the case of common-pool resources. In *Governing the Commons: The Evolution of Insti-*

tutions for Collective Action, Ostrom reviewed the manifold examples of small, locally organized governance systems around the world that have managed resources sustainably, often for hundreds of years. She consistently found that sustainability is possible and that sustainable management of local resources is often built with bottom-up processes that emphasize social connections and local control rather than large, centralized institutions.[10]

McKay Savage

Rice terraces north of Ubud on Bali, Indonesia.

Ostrom found that systems that relied on social connections and close links between those who exploited the resources and those who made decisions were often effective—a finding that surprised the many people who had believed that centralized decision making was the only way to achieve sustainable management. Her work identified a set of design characteristics that defined good, sustainable systems for the management of common-pool resources, including clear boundary rules; resource access by members that is commensurate with their work contribution; support for rights to organize, modify the system, and resolve disputes; and the presence of monitors with the capacity to punish violators.[11]

Other studies in sustainable governance have found similar conclusions: that local systems, rather than large centralized governance, appear to be particularly well suited to the sustainable management of resources; and many of Ostrom's "design elements" have been supported in other research. The research is not yet clear, however, on which of these elements are more important than others, or how different types of resources may change the needed structures. What is clear is that the claim that common-pool resources must ultimately face a "challenge of the commons" as individuals loot the shared resources is empirically false: many societies successfully avoid this fate.[12]

Good governance allows specialization. Good governance can unlock dramatic benefits for the people within that system. Most notably, by allowing different elements of the governed system to focus on specialized tasks that collectively support the goals of the system, there can be an increase in efficiency and the productivity of the system as a whole. This argument is at the root of Adam Smith's analysis in *The Wealth of Nations*: he argued that international trade allowed different countries to specialize in their output. In other words, if cotton could be grown more easily in India than in the

United Kingdom, and wheat more easily in the U.K. than in India, then a system that allowed each country to produce its own specialty and trade with the other was more likely to result in everyone having enough food *and* clothing. This was admittedly an oversimplification of the complexities associated with trade and production, but a logical claim.[13]

A similar process is thought to have occurred as humans transitioned from bands of hunter-gatherers to settled producers of agriculture. The resulting increase in food production meant that specialized farmers could produce enough food to support specialized soldiers, tradespeople, priests, and governors. This system, however, required more complex governance systems to distribute the resources, which may have played a role in the development of more complex civilization.[14]

From this conception, good governance is that which increases the efficiency of human groups and collective productivity. Although this argument is somewhat coldly analytical when considered in the light of the questions of human rights and legitimacy discussed earlier, over a long period of time, it may be the ultimate criteria by which governance systems are judged. One argument advanced by some scholars is that governance can be seen through an evolutionary lens: governance systems that meet the needs of their constituents and that facilitate effective performance of collective tasks persist and allow their members to thrive.[15]

In contrast, systems that are unable to accomplish their goals are unstable and are more prone to conflict. They face internal pressure to transform and external competition from other groups. As a result, these systems fade away or are transformed, as largely happened with absolute monarchies in Europe. If this model of governance change is correct, then the effectiveness of collective groups is the ultimate metric by which good governance is judged. Even in this case, however, there is evidence that some of the same elements seen in prior conceptions of good governance matter: mature democracies characterized by representative decision making and strong rule of law appear to be simply more effective than other governments in many ways.[16]

Putting it all together: what is good governance? As with the structure of governance, the assessment of what good governance is reveals more questions than answers. "Good" is a relative term and depends on the yardstick being used. When translating these general questions into practical assessments of governance, however, the yardsticks begin to converge on some basic recurring principles. Whether concerned about human rights, legitimacy, or even sustainability, it appears that good governance systems need to be inclusive and participatory: they need to allow the members of the system to change the rules when needed and to have a voice in the collective decisions that are made.

Whether concerned about legitimacy, human rights, or effectiveness, systems need to be accountable to processes that guarantee fair treatment and to establish predictable rules that are applied equally to all members of the collective. And ultimately, as found by Ostrom and reinforced in the concerns about human rights, systems need to be in place to resolve disputes and to sanction those who would violate the rules and collective values of the group. Although the specific way that "good" governance is defined may vary by observer, the characteristics of good governance do not differ as much as may be expected.

The Future of Governance

So what does all of this imply for the state of the world and the future of governance? It is all too obvious that regardless of how good governance is defined, many systems fail to live up to the standards in place. Bad governance that ignores the consent of the governed or that harms people or the planet persists in many parts of the world, at all levels. At the global level, the systems of governance are often patchwork, inefficient, and in some cases missing entirely.

The increasing focus on governance as a topic of study is encouraging, however. By developing a better understanding of what governance is, how it works, and how it can be improved, the possibility that we may create better governance in the future is greatly increased. There are positive signs in this regard, with international institutions such as the World Bank and UNESCO now treating the development of good governance and state capacity as part of their overall work.

A potentially promising development is the proliferation of nonstate actors on the international stage. Increasingly, international governance is reaching out to the private sector and to civil society to forge international systems. In 2000, the creation of the United Nations Global Compact involved both the private sector and nongovernmental organizations (NGOs) in the promotion of business activities to support sustainable development and human rights. This represented one of the first times that the United Nations systematically reached out to the private sector to support its organizational goals. Since the boom of international NGOs began in the 1950s, a host of new and increasingly well-organized groups has emerged to represent stakeholders and different perspectives on the international stage. The result of the increased role of nonstate actors, and the willingness of state actors to engage them, is the proliferation of institutions that incorporate multiple sectors of society into their decision-making processes.[17]

These multi-stakeholder, networked institutions lack some of the legal authority of traditionally treaty-based international law, but they also reflect a fact of the global world: an increase in globalization means an increase in transnational problems. New systems that bring together states, NGOs, and the private sector may represent a new approach to governance that will help these different sectors work together to solve problems. And because they are stakeholder driven, these systems may have built-in support for legitimacy, as well as specific issue expertise—factors that will help them achieve their governance goals.[18]

The future of governance is hard to predict, but one thing is abundantly clear: addressing the challenges of an increasingly integrated and populated planet requires good governance. In the absence of good systems for resource distribution and conflict management, the future holds dark clouds. Yet the many examples of good, sustainable, legitimate governance that exist at many levels internationally do give reason for hope. They underscore that if the world can fix the existing problems with current governance systems, then the problems of the future may be easier to resolve than we now may think.

Political Governance

CHAPTER 3

Governance, Sustainability, and Evolution

John M. Gowdy

How has it come to pass that humans so completely dominate Earth's biophysical processes that we are now on the brink of a major shift in the state of the biosphere? Why, in the face of impending ecological disaster, does it seem so difficult to make the basic societal changes needed to ensure our long-term survival? The answers to both of these questions lie deep in our evolutionary history. Framing governance in terms of this history can help us solidify the successes we have achieved at the individual and community levels and, more importantly, inform us about the changes in governance that are needed if we are to gain control of our destiny as a species.[1]

Governance systems comprise the formal and informal ways that humans manage relationships with each other and with the natural world. In an evolutionary framework, governance can be viewed on three different levels: the individual, the community, and the global socioeconomic system. At the individual level, behavioral science has made great progress in identifying regularities in human decision making. These regularities have been used successfully to design policies to promote sustainable behavior such as recycling and the use of energy-saving appliances, as well as other efforts to "nudge" people to make better personal choices. At the community level as well, strategies for the successful management of human and natural resources have been identified and incorporated into public policy.

At the highest level in the hierarchy, however, the world socioeconomic system thus far has proved highly resistant to the fundamental changes needed to avoid system-wide collapse. Each of these governance levels and the interaction between them can be explored from the perspective of evolution. A missing piece in governance has been the failure to recognize that these levels are sometimes in conflict. The same behavior may be rational at one level but irrational at others.

Contemporary behavioral science, neuroscience, and evolutionary theory have shown that human behavior is a combination of genetic, devel-

John M. Gowdy is a professor of economics and a professor of science and technology studies at Rensselaer Polytechnic Institute in Troy, New York.

opmental, and cultural factors. None of these can be understood in isolation, but recent advances in understanding behavior reveal how they are intertwined, giving insights into the behavioral adjustments and policy formulations needed to manage social transitions, including the transition to sustainability. Today, research efforts in a number of disciplines are beginning to coalesce into what evolutionary biologist David Sloan Wilson calls a "science of intentional change."[2]

Evolutionary biology reveals a middle ground between the position that human behavior is rigidly determined and the "blank slate" tradition of the standard social science model. Our evolutionary history has instilled in the human species the ability to make rapid and complex adaptions, via culture, to special circumstances. Culture is what makes us human, and it offers the greatest hope for our species to successfully make the transition to a sustainable presence on planet Earth.

Governance and Individual Behavior

More than 25 years of research have shown that human beings are not the paragon of super-rationality that we like to think we are (and that much public policy is based upon). Policies that acknowledge this insight can be used to nudge behavior in directions that benefit individuals as well as society as a whole.[3]

A now-classic example is organ donation. The share of people agreeing to be organ donors varies greatly within Europe, from 4 percent in Denmark, 12 percent in Germany, and 17 percent in the United Kingdom to nearly 100 percent in Austria, France, and Poland. The difference is accounted for by the fact that, in the first three countries, people are asked on their driver's license application to "Check the box below if you *want* to participate in the organ donor program." People in the other countries are given the choice: "Check the box below if you *do not want* to participate in the organ donor program." Because organ donation is a rather complicated moral decision that most people would prefer not to think about, the fall-back, do-nothing choice is appealing.[4]

Public policy in this case is merely changing the wording of the question. In effect, the person who designed the questions is really the one who made the choice about organ donations. Other public policy experiments have shown that knowing that one is being observed can positively affect decision making. In an experiment with a public utility program designed to prevent blackouts, participation in the program tripled when participants knew that their behavior was being observed. Observation was four times more effective than offering a monetary incentive.[5]

In another experiment, in California, door hangers were left on customers' houses indicating how much electricity they used compared to their neighbors. These customers reduced their energy consumption by 10 percent compared to customers given door hangers that offered only energy-saving tips. Residents who used less energy than the average actually increased their consumption, however—a "boomerang effect" that disappeared when smiley faces were added to the door tags.[6]

Other, more long-term interventions have used insights from behavioral psychology to design educational curricula and course content. An example is the Good Behavior Game, which begins by having students themselves establish norms of good behavior. After these norms are set, groups of students compete to be good. Controlled experiments have shown that the positive effects of the game last until adulthood even when the game is played only in the first two years of school.[7]

A promising area of research is the role of evolved behavior in people's consumption decisions. Consumption behavior is motivated by two desires: to meet basic wants and to gain status. Basic wants can be satisfied, but wants that are driven by status considerations are essentially insatiable. As a result, growth in industrialized economies, with their large populations of middle-class consumers, has become a zero-sum game, contributing little to individual well-being and perhaps even undermining it. Reducing material consumption is thus a necessary component of governance for sustainability, and understanding the evolutionary dynamics behind human behavior may help us design polices to channel behaviors such as status-seeking onto more socially and environmentally benign paths.[8]

Some progress is being made in this area. Neuroscience and behavioral economics, for instance, have all but demolished the rational-actor model of standard economic theory. Economists no longer rely solely on the price system as a policy tool. And although the focus of most research is still on individual behavior, neuroscience research has confirmed the existence of the social brain. The human brain evolved to allow us to function together in social groups: a growing body of evidence indicates that humans are unique among mammals in their degree of sociality. Our ability to solve resource management problems collectively is a manifestation of our uniqueness, and it offers another ray of hope that our species may achieve a sustainable way of living.[9]

Governance at the Community Level

For most of our existence, humans have lived in small groups within the confines of local ecosystems. Cooperation evolved because those groups that worked together survived, while those that did not perished. Institutional rules for cooperation and living within biophysical limits emerged early in pre-agricultural human societies and ensured the viability of these small groups. The rules of the game changed with the widespread adoption of

Members of several families of San people (Bushmen) at their settlement in Namibia.

agriculture some 10,000 years ago as we broke out of the confines of small communities and local ecosystems, but our evolutionary history still leads us to cooperative behavior.

The obsessive focus of economic theory on "self-regarding behavior" led economists to deny the possibility of cooperation except in cases where it was directly and immediately beneficial to both parties. Because of the free-rider problem (when someone receives a benefit without paying his or her share of its cost), economists assumed that the only options for successfully managing common-property resources were rigid, top-down control or the complete assignment of private property rights to individuals. But recent work has shown that cooperation is widespread in the natural world, including among humans, and all successful human groups have a variety of rules to punish free riders and encourage altruism. This is part of community governance, and it is accomplished by social pressure as well as formal sanctions.[10]

Elinor Ostrom, co-recipient of the 2009 Nobel Prize in economics, drew on her experience in small-scale societies around the world to identify eight principles for the successful management of common-property resources: (1) a strong group identity, (2) fairness in distributing costs and benefits, (3) consensus decision making, (4) effective monitoring of effort and rewards, (5) graduated sanctions, (6) rapid and fair conflict resolution, (7) sufficient autonomy when the group is part of a larger system, and (8) appropriate coordination between groups. Ostrom and her colleagues identified these principles from meticulous studies of the effectiveness of different systems of common-property management. When the principles are in place, local communities do a remarkable job of protecting their resource bases even under intense outside pressure.[11]

Sustainable human communities existed for some 2 million years (counting *Homo erectus* as human), so it is not as if we cannot do it. Sustainable indigenous human cultures still exist, although they are being eliminated at an alarming rate. Sustainable communities are also appearing within the world capitalist system, as subcultures are being "re-engineered" as eco-communities built around the needs of humans and ecosystems, not markets. The success of movements for sustainable local agriculture and local currencies, and the continuing resistance of native peoples to assimilation into the market economy, attest to our ability to confront the global system.[12]

Can the human propensity for cooperation and community building be harnessed sufficiently to scale up and challenge a global system built on competition and accumulation? Perhaps, but we need to be realistic about what we are up against. (See Box 3–1.) In small communities, the "good of the group" corresponds to the "good of individuals within the group." This is not necessarily true for very large groups. The positive benefits of cooperation are undeniable, and numerous recent books have touted our cooperative nature. But the optimism of those on the cooperation bandwagon is often pushed too far. For example, evolutionary biologist Mark Pagel writes:

> Modern societies differ vastly from the small tribes that once competed to occupy Earth. But the old psychology plays out well in our globalized, multicultural world. Our species' history is the progressive triumph of cooperation over conflict as people recognized that cooperation could return greater rewards than endless cycles of betrayal and revenge.[13]

There are reasons to temper this optimism. First, hunter-gatherer societies did not practice the kind of intergroup warfare that characterizes agricultural and industrial societies: within-group conflicts were significant, but warfare with other groups was largely absent. Human history is not a simple story of progress from savagery to civilization. Secondly, Pagel and others equate the impersonal interconnections involved in producing and consuming world economic output with cooperation. But coordination in producing economic surplus is not the same as cooperating for the common good.[14]

Stressing the virtues of cooperation can be a more nuanced approach to human nature than the "selfish gene"/"economic man" worldview, but the dark side to human cooperation must be understood if we are to realistically assess our present circumstances. The leap to agriculture and state societies some 8,000 years ago represented a rare but highly successful evolutionary transition to "ultrasociality," a type of social organization seen in only a handful of species, including ants and termites. Ultrasociality is characterized by a full-time division of labor, specialists who do not aid in food production, sharing of information, collective defense, and complex city-states. The profound social and environmental consequences of such complexity have resulted in what biologist E. O. Wilson calls "the social conquest of Earth." The ultrasocial human economy is the uppermost level in the governance hierarchy, and it is the most problematic.[15]

The Emergence of Human Ultrasociality

With the appearance of agriculture came a basic change in the economic organization of human societies: the switch from producing for livelihood to

Box 3–1. Can Networked Governance Help?

If evolution seems to have backed us into a corner when it comes to existential threats such as climate change, does it also offer a way out?

The failure of traditional human governance institutions to come to grips with climate change—to perceive the threat, formulate a coherent and flexible response, and then enact it with vigor and discipline—is all too plain. Nearly all climate scientists now agree that climate-warming trends over the past century can be attributed mainly to human activity, and it is no longer a matter of scientific dispute that climate change poses real challenges for current and future generations.

Humanity has been aware of climate change for decades, yet for the most part neither individuals nor institutions have been able to respond at the appropriate scale or speed. We have failed to significantly reduce carbon emissions or our reliance on fossil fuels, a triumph of short-term interest in sustaining or raising current levels of energy consumption over our long-term welfare.

The paradox is that our evolutionary history has equipped us for long-term planning and action. Humans possess a highly advanced capacity for mental "time travel" and are arguably unique in the degree to which we can recall past events and anticipate future scenarios. To an extent, at least, we can imagine and predict multiple, complex outcomes and act accordingly in the present to achieve desired outcomes in the future. This general capacity is very old; the first direct evidence for it is found in the 2-million-year-old stone tools shaped by our distant ancestors.

Moreover, humans regularly do make long-term plans: we invest in retirement accounts, establish trust funds and endowments, and buy insurance, for example. While these plans sometimes have long-term impacts on society, however, they frequently yield results that will directly affect only the individuals themselves or the next

one or two generations. Evolutionary theory suggests a reason for that, too: we care most about our genetic relatives—our great-grandparents, grandparents, parents, children, grandchildren, and great-grandchildren, or an approximate span of 140 years that includes both past and future family members. Beyond that, most people do not care much about the past or the future.

To embody and act upon those concerns that extend beyond family to others and to times beyond our own lifespans, humans have created institutions. Governments are preeminent among the institutions that are supposed to perform this role, but, as noted earlier, they have not been effective at addressing climate change. Humans are creatures of culture—the product of learned human behaviors and actions that cannot be attributed directly to genetic inheritance. Governance is a cultural phenomenon and evolves similarly to physical traits: behaviors can be transmitted and can change over time.

We are now seeing the emergence of a kind of governance that departs from the centralized, top-down structures that we have relied upon so far to solve problems. *Networked* systems of governance are a shift toward a more self-organizing approach that brings together dispersed individuals from the state, civil society, and private sectors that have a shared interest. Each acts independently yet remains connected through exchanging information, planning for future events, and cooperating as is useful.

Systems of networked governance arose soon after World War II and have been growing ever since as an adaptation to meet the global challenges and complex problems that existing systems, which frequently are slow and hampered by the politics of entrenched interests, have failed to address adequately. Networked systems of governance make it possible for small groups to act quickly and in locally appropriate ways, testing solutions that can then be passed

Box 3–1. continued

on to other groups with similar aims. Harvard political scientist Joseph Nye has described these networks as a cultural adaptation that is evolving slowly to supplant the formal mechanisms of international cooperation.

Some current examples of networked governance addressing the challenge of sustainability include the Extractive Industries Transparency Initiative, the Roundtable on Sustainable Palm Oil, the Marine Stewardship Council, the Equator Principles, and the Forestry Stewardship Council. Each has been successful to varying degrees because they facilitate collaboration among a wider range of actors including the private sector, governments, international organizations, and nongovernmental organizations to achieve a common vision in the absence of regulation.

Networked governance may be the very type of social evolutionary development or adaptation that will make it possible for us to counter our inherent biases so that we can begin to reorder our lives in a way that moves us toward a more sustainable future. As systems of networked governance become more prevalent and stand (or fail) the test of time, we can help drive their evolution by exploring ways that they might be replicated at varying scales to share lessons learned and encourage adoption of good governance practices. The survival and evolution of cultures rely on the inheritance of learned behaviors, including cultures of good governance. (See Chapter 1.) Networked systems of governance are currently the most versatile, agile, and adaptive systems available to meet the challenges ahead of us. The task now is to identify and strengthen these new systems as they are emerging.

—Matthew Wilburn King
President, LivingGREENNetwork.org
Source: See endnote 13.

producing for surplus. Natural selection among competing groups of early agriculturalists favored those societies that were the most efficient in producing economic surplus, and those that could take advantage of increasing returns to larger size. This led to (1) human domination of ecosystems, (2) explosive population growth, and (3) highly hierarchical societies. The population increase after agriculture was unprecedented in the prior 200,000-year history of *Homo sapiens*, exploding from about 4 million to over 200 million in a few thousand years. A second population explosion, from under 1 billion in 1800 to over 7 billion today, came when fossil fuels and the Industrial Revolution ushered in the Anthropocene—the Age of Humans. (See Chapter 6.)[16]

With the transition to agriculture, and the later transition to industrial society, our place in the natural world changed drastically. Today, the total dry weight of human biomass is about 125 million tons. The dry weight of our domesticated animals is about 300 million tons. The weight of all other vertebrates is only 10 million tons. In only a few thousand years, humans made the transition from being just another large mammal living in the confines of local ecosystems, to a species dominating the planet's biophysical systems.[17]

We were not the first species to make the transition to ultrasociality. Ultrasocial insects also dominate their ecosystems. Worldwide, ants and ter-

Arian Zwegers

The treeless Easter Island landscape of Rano Raraku.

mites account for about 2 percent of Earth's insect species but 50 percent of the insect biomass. The specifics of the human transition to ultrasociality may differ from ants, but it was driven by the same impersonal forces of natural selection at the group level, and the results in terms of ecosystem dominance and the effect on individual autonomy are strikingly similar.[18]

For both humans and social insects, with the adoption of agriculture, the nature of the group changed from a collection of individuals cooperating to achieve mutually beneficial outcomes to something akin to a single organism centered on a narrow economic purpose, namely, the production of agricultural surplus. In ultrasocial species, the flourishing of the group is often at odds with the well-being of particular individuals in the group.[19]

In terms of governance, a key insight is that ultrasocial societies are reinforced by what American social scientist Donald Campbell called "downward causation." In the current global economy, the goal of economic growth is reinforced by layers of human institutions, including religions, political philosophies, hierarchical control of basic resources, and the influence of power and money. This is not to say that counter-currents do not exist; in fact, much of human history after large-scale agriculture can be seen as a struggle between those elites who resist interference with the system's "natural" drive for accumulation at all costs and those who want to make the system a servant of humanity, not its master.

Author Jared Diamond asks, what was the Easter Islander who cut down the last tree thinking of? He suggests, "Jobs, not trees!" or, "Technology will solve our problems, never fear, we'll find a substitute for wood." In any case, the broader answer is that he or she was thinking in the mode of the dominant ideology of the ultrasocial system that was the Easter Island economy and culture. The tree-cutter's culture, like other groups selected after agriculture, flourished (for a while) because it outcompeted other groups in the race to produce surplus. Groups that were the most cohesive and most focused grew faster and were selected over others. Cultural group selection favored groups that had customs and beliefs that were conducive to growth.[20]

Ultrasociality is an evolutionary outcome, and evolution cannot see ahead. The failure of the world socioeconomic system to address climate change is a good example of downward causation at work. Nothing substantial has been done to stop greenhouse gas emissions because growth and accumulation fueled by cheap fossil energy are driving the system, and the growth imperative is supported vigorously by the cultural beliefs and political institutions that evolved to reinforce it. Money generated by fossil fuels flows through the political system to thwart any attempt to limit their use. No serious threats to the global economic system from climate change have yet appeared, so the system has not adjusted even though we may have locked ourselves into catastrophic change in the not-too-distant future.

Cooperation Versus Accumulation

Grassroots movements have been instrumental in pushing governance toward sustainability. But many well-meaning efforts have gotten off track by trying to reconcile sustainability with the dominant ideology of growth and accumulation. The governance hierarchies are sometimes in conflict: the mandate of the global economic system to grow and accumulate trumps lower-level efforts. Progress has been made in designing policies to shape individual behavior and to guide institutional change at the community level. But at the top of the hierarchy—the global socioeconomic system—little has been accomplished toward redesigning institutions to promote sustainability and individual well-being. It is the imperative of growth and accumulation that ultimately drives individual decisions.

An example of this is shale gas extraction in the United States. This extraction contributes to climate change, disrupts local communities, and may cause numerous environmental problems. But it is the needs of the world socioeconomic system that dictate shale gas extraction and use. The decision to use the resource almost seems out of human hands. Journalist Richard Manning observed about drilling efforts in North Dakota's Bakken shale formation:

> Once we had the Bakken's [production technology] recipe right, there were no decisions left to be made, save the hundreds or thousands of piecemeal decisions made over kitchen tables when people sign leases. You might hate the idea of oil rigs on the family ranch, but if you don't sell, someone else will, and it's all going to hell anyway, so might as well sign. We do not decide whether to drill oil. Price decides. Price and how much is in the ground.[21]

An evolutionary perspective can help us focus on trajectories and dynamic paths to sustainability, not just static milestones like a steady-state economy, zero population growth, or limiting atmospheric carbon dioxide to some fixed level. These are laudable objectives, but unless we understand the forces driving human expansion, policies to achieve those objectives will continue to fail.

Consider a simple thought experiment: suppose the human population could miraculously and painlessly be reduced to a few hundred million, and Earth's forest and ocean ecosystems restored. If we kept the current dominant socioeconomic system of growth, accumulation, and expansion, within a few decades we would be right back where we are now: more than 7 billion people and many of Earth's life-support systems teetering on collapse.[22]

So far, the global capitalism juggernaut has had the evolutionary advantage in terms of natural selection. But just because the system is the outcome of "natural" forces does not mean it is desirable. If we value the future of our species, and the rights of the other species that share this planet with us, we should assert human intentionality and eliminate the worst aspects of the global economy. There has always been resistance to the power of the system, but it must be informed with a recognition of the power of the system as a highly evolved, interrelated whole. The question for governance is whether we can gain control over a global system that has made us, in E. O. Wilson's words, "a danger to ourselves and the rest of life." So far, it is an open question as to whether the power of human agency will be sufficient to confront the magnitude of our predicament.[23]

The global economy acts "as if" it were a superorganism driven by the forces of natural selection to survive and expand. Like an ant colony, it works by rules that have evolved to facilitate the production of economic surplus. And like an ultrasocial insect society, the needs of the superorganism tend to override the well-being of individuals within the colony. Humans are not ants, however, and examples abound of human agency actively overriding the worst abuses of the economic system. For example, by a variety of measures, the most successful societies in providing for the well-being of their citizens are the Scandinavian countries. Those countries have long histories of difficult but successful struggle against the powerful economic interests that are always fighting against attempts to limit the power of the market.

To achieve sustainability, we must, one way or another, design institutions to assert control over the global economy. Can this be achieved successfully by a world government acting in the best interests of individual humans and the rest of the natural world, or is such a system bound to degenerate into a self-serving dictatorship of the few? Can a bottom-up revolution successfully challenge the political and military power of the ruling elite and control the abuses of global capitalism? Can Ostrom's rules for successful community governance be scaled up to the level of the global economy?

These questions are difficult to answer. But the way to begin to address them is through an understanding of the evolutionary dynamics that created the current human enterprise. One thing is certain: if we continue to let the blind, mechanical forces of ultrasocial evolution determine our future, our prospects look bleak.

CHAPTER 4

Ecoliteracy:
Knowledge Is Not Enough

Monty Hempel

In the early 1990s, Oberlin College professor David Orr coined the term "ecological literacy" (or ecoliteracy) to describe people's ability to understand the complex natural systems that enable and support life on Earth. It embodied the implicit assumption that if humans were more ecoliterate, then we would be more likely to respect the limits of those systems and to create communities that operate harmoniously within the natural world— the key requirement of sustainability. Colleges and universities around the world have since launched hundreds of programs that aim to raise the level of ecoliteracy among students and, to some extent, within society at large.[1]

Yet the results have been mixed, and serious questions remain. For example, is ecoliteracy just a green version of scientific literacy? Is improving ecoliteracy the key to stronger environmental governance? Will it enable us to address the host of pressing sustainability problems that we face—especially runaway climate disruption?

These questions were on the minds of 10 American college students during a recent research expedition in the Republic of Palau, a coral Eden of nearly 300 islands located about 800 kilometers east of the Philippines. When asked if the climate was changing in Palau, a local fisherman replied: "Of course it is. We all know it. But it doesn't matter what we think unless you guys in the U.S. know it, too." He pointed to coral bleaching events, seasonal shifts in rainfall and wind direction, and rising spring tides. Then he observed, slowly shaking his head, "It's not science; only what every eagle ray already knows."[2]

The spotted eagle ray is a totem spirit for some traditional Palauans. It symbolizes a kind of indigenous, place-based knowledge about ecology that may yet serve to refine and strengthen Western understanding of ecoliteracy and its role in environmental governance. Roaming the waters of a coral reef, the spotted eagle ray may or may not be a good indicator of climate conditions, but the fact that many islanders know the ray's habitat

Monty Hempel is Hedco Professor of Environmental Studies and director of the Center for Sustainability at the University of Redlands. He is also a documentary filmmaker and president of Blue Planet United (www .blueplanetunited.org).

needs, including water temperature, suggests a form of cultural ecoliteracy that could prove useful in monitoring climate impacts on coral reef ecosystems.

The islanders' knowledge of the eagle ray represents the pre-scientific era of ecology—a time when practical understanding of one's bioregion was highly valued, even necessary for survival. It was a time when being able to "read" one's environment was essential for securing food, water, personal safety, and other requisites of a more self-sufficient way of life. (Granted, this practical knowledge was often used destructively to exploit natural resources.) Today, this kind of ecoliteracy has disappeared in most places, and with it the fundamental sense of connection that people had with the natural world.

Restoring ecoliteracy to this connective role and fortifying it with the power of science and widespread recognition of global interdependence is perhaps the greatest challenge of this century. Meeting that challenge will require both scientific learning and visceral learning about humanity's place in the great web of life. It also will require forms of governance that can effectively apply this blend of ecological and emotional intelligence in the creation of more sustainable communities—place-based communities that are green, prosperous, fair, and "glocally" embedded in international and transnational networks of ecoliterate citizens.

Developing emotional connections to the natural world—to wild places, natural beauty, native plants, wildlife, and healthy ecosystems—is at least as important for protecting the environment as breakthroughs in environmental science, policy, and management. Weaving together attachment to place with scientific knowledge about that place (and its relationships with other places) is vital for effectively managing the environmental challenges we face. Such braided concepts of ecoliteracy hold enormous promise for improving environmental governance, particularly in response to a global set of interlocking, slow-motion crises, beginning with climate disruption.

Restoring and Expanding Ecoliteracy

Scientists are able to assess the health of our planet to an unprecedented extent, due largely to advanced ground sensors and satellite monitoring. We have never before had so many pertinent facts at our disposal, not to mention mountains of data. At the same time, science and technology have introduced so much complexity into the systems that shape our lives that truths about what really matters in the long term may be harder than ever to discover.[3]

Ecoliteracy is the principal way that we make sense out of the interacting systems that support life on this tiny blue planet. It is, first and foremost, an expression of a particular type of ecological knowledge that is testable and authoritative. We know from the scientific study of ecosystems and the basic laws of thermodynamics that much of what we call the "environment" is in fact a stunningly intricate system of cycles and flows that regulate the life-support conditions for millions of species and ecological communities. Ecoliteracy begins with knowledge of this interdependence and how it sustains the biosphere.

The goal of ecoliteracy has arguably become the ultimate aim of the Enlightenment, combining a strong emphasis on integrative systems-thinking with deep respect for the authority of science. Conventional environmental wisdom in the West holds that people who are educated about ecosystems and their interactions with human social systems will follow scientific reasoning to its inevitable conclusion: protect the environment! But the climate change debate, along with public debates about many other global environmental crises (biodiversity loss, ocean acidification, etc.), is confounding the conventional wisdom.

Debates about these environment challenges are not fundamentally contests between the educated and the ignorant. A growing body of research concludes that polarization of views about climate disruption and other complex risks (e.g., nuclear power plants) actually increases with improvement in scientific literacy and numeracy. Among people who identify with strong individualism and rank human importance by power, wealth, or other factors—so-called hierarchical individualists—concern about climate risks varies inversely with scientific knowledge. More education leads to a reduction in environmental concern.[4]

These findings suggest that certain groups use education more to justify pre-existing worldviews than to enlighten themselves with new knowledge and ways of knowing. Many researchers conclude that this knowledge-for-justification tendency is universal and varies only by degree of application. Stanford psychologist Albert Bandura, for example, argues that human abilities to justify harmful environmental practices are so strong and pervasive that society should develop strict moral sanctions to limit their use.[5]

The selective use of knowledge to avoid self-censure or to promote group bonding is well known among social scientists. Scholars refer to this phenomenon by many different names, including motivated reasoning, confirmation bias, and cultural cognition. Combining this idea with long-studied phenomena of "groupthink" and cognitive dissonance theory, researchers have woven together a persuasive but unflattering account of human reason and its self-serving uses. The importance of these research findings for the environmental science community in general, and for climate scientists in particular, is in understanding how to communicate better and to present scientific findings in tradeoff terms when they somehow threaten the dominant values and institutions of the status quo. Equally important are the insights made by scientists themselves, as human beings who remain vulnerable to these self-serving tendencies.[6]

The degree to which science-based notions of ecoliteracy are influential appears to depend on the types of environmental problems to which they are applied. Broadly speaking, there are three basic types of major environmental problems:

Familiar problems: These are straightforward and usually solvable with enough political will because they share three characteristics: the environmental science behind them is essentially settled and accepted by the public; the proposed solutions have been demonstrated and are considered "best practice" for the time; and there are politically powerful interests that will benefit from the solutions. Examples include "end-of-pipe" pollution, lost biodiversity, and human population growth.

Frontier problems: These problems invariably defy quick action because their solutions require new knowledge in science, policy analysis, and management for effective design. They involve large areas of ignorance, risk, and uncertainty, not because their causes are overwhelmingly complex but because they are novel or exploding in scale (i.e., reaching a tipping point) and, until recently, obscure or accepted (neglected) as tolerable "externalities." Examples include toxics in food and water, ocean acidification, and lost ecosystem services.

Foresight problems: This class of problems, sometimes termed "wicked," places almost impossible demands on human forecasting and policy analysis. Foresight problems are mired in ambiguity, ignorance, contradiction, and chronic indeterminacy. They require a "system-of-systems" level of understanding that identifies complex interdependencies and apparent contradictions in system behavior. Their solutions must be adaptive and evolving because these interdependencies and apparent contradictions inevitably distort the scientific understanding of their complex behavior. Such problems are easily framed and rationalized by all sides in a dispute in incompatible ways that permit no "solution" to emerge, while allowing all parties to claim, with some supporting evidence, that they are being reasonable. Examples include climate disruption, genetically modified organisms (GMOs), and a variety of as-yet "unknown unknowns."

Foresight problems can be viewed as frontier problems with a distinctive twist: their novelty or uncertainty is accompanied by a scale of complexity and long-term risk that, for many people, makes denial and disbelief preferable to action and planning. As such, foresight problems pose severe tests for democratic governance, particularly in light of the presumed knowledge deficiencies that hinder informed public deliberation.

All three types of problems are consequences in some way of unsustainable lifestyles and values; however, a relatively small number of people accounts for a large portion of the observed problems or impacts. Ecoliteracy, conceived solely as ecological science, can conveniently skirt these moral and political issues, but when it is treated in the broader context of environmental education, the issues of personal responsibility and social equity become inescapable. Because it is fundamentally misleading and self-defeating to treat ecoliteracy as science alone, it is preferable to adopt the broader view that any ecoliteracy worth having will in-

Kevin Vanden

A march in Brussels against Monsanto and its development of GMOs.

clude ethical, cultural, and political dimensions. The aim of environmental education should be a transdisciplinary form of ecoliteracy that includes experiential learning, knowledge of personal and social responsibility, and understanding of the roles of governance and communication in moving from knowledge to action.

Although most of the pioneers of ecoliteracy, such as David Orr and Fritjof Capra, understood from the outset the need to integrate environmental knowledge with political and ethical action, the typical understanding of ecoliteracy remains bounded by the science of ecology. Deep ecologists have often been outspoken in challenging this singular scientific focus. But others, who presume that formal knowledge leads inevitably to action, need no support outside of science to justify their calls for more "STEM" education (science, technology, engineering, and math). When they discover that even climate scientists tend to leave large carbon footprints, they are likely to dismiss the finding as the last vestige of behavioral momentum—habits that prove hard to break. In their view, knowledge will soon triumph and force consistent action to control personal carbon emissions.[7]

Revising the conventional notion of ecoliteracy—and models of environmental education in general—seems fully in keeping with lessons learned from adaptive management of ecosystems. Moreover, just as approaches to general education have had to be reconceived in an era of fast-changing information and communication technology, so too may ecoliteracy need rethinking in order to respond effectively to the three types of environmental problems mentioned previously. Widespread ecoliteracy, for instance, would probably have a major positive influence on action needed for "familiar" problems, and would perhaps provide a significant push for some "frontier" problems. But could it contribute much to the solution of "foresight" problems, which appear to challenge global environmental governance most urgently and divisively?

If all 7.2 billion of us were somehow given generous access to environmental education, would it make a major difference in the measurable outcomes for climate disruption, extinction rates, global freshwater availability, and so forth? The answer from social scientists appears to be a resounding "No!" First, effective environmental education tends to threaten many dominant, deeply held values and worldviews, thus relegating it to suspect status among those seeking only a selective exposure (if any at all) to ecological knowledge. Learning that helps us avoid environmentally induced illness will be treated differently than learning that challenges our freedom to have as many children as we want, consume at high levels, or drive fuel-inefficient cars.

Second, many people perceive environmental education to be deeply contaminated by values claims and frequent exaggeration. Even if free and convenient, such education will be rejected by a large percentage of the population on grounds that it undermines their ideals of personal liberty, or perhaps their ideal of unfettered market economies.

Third, and most important, learning and the knowledge that it produces lead to positive action only under very limited conditions. Knowing that change is needed is clearly not enough to motivate it in most human behavior. Individuals must have a sense of urgency and personal control over prospective outcomes and goal achievement ("self-efficacy") before they will commit to meaningful action or new behaviors.[8]

Obstacles to Learning and Action

A major barrier to public mobilization on climate and other global environmental issues is the psychological distance involved in moving from abstract environmental data (e.g., global mean temperature) to more immediate concerns about how local impacts, such as climate disruption of drought cycles in a particular area, may affect one's personal prosperity or family security.[9]

But there is an even more important kind of distancing that helps to explain the failure to promote ecoliteracy when and where it is most needed. As the boundaries of the natural world recede in the face of rapid human development, people who are disconnected from nature have less motivation to learn more about it. The consequences are especially important for children, as suggested by recent book titles, such as *Last Child in the Woods* and *Free-Range Kids*. The psychological distance separating the urbanized places where most humans reside from the shrinking remnants of natural landscape has never been greater. As a consequence, the opportunity to connect emotionally and physically with nature and wildlife has declined steadily. And implicit in this decline is an accompanying loss of attachment to natural places and wild habitat, or what is sometimes understood as lost bioregional identity.[10]

Precisely how much this growing separation diminishes human concern about the environment is unknown, but it is clear that people are more likely to protect the things they love and actively internalize. Distancing from nature may have some of the same emotionally debilitating effects as distancing from other people. This separation becomes even more significant in issues of climate change, where the most dramatic impacts are taking place in the Arctic and other remote areas that few people ever visit or monitor.[11]

The obstacles to clear thinking about these kinds of threats extend far beyond psychological distancing. Research on climate change communication has identified dozens of factors that serve to hinder or derail public support for timely action on climate risks, from poor framing of the issues to social media's role in diverting people's attention elsewhere. (See Table 4–1.)[12]

There are, of course, many other defense mechanisms and elaborate rationalizations that protect individuals and groups from painful assaults on their cherished values and behaviors. Most are considered deeply irrational and even dangerous by many scientists. Carl Sagan, the celebrated astrophysicist, devised a "baloney detection kit" to aid scientists in exposing the irrationality of anti-science arguments and pseudo-science views. But his kit fails to come to grips with the messy reality emerging within the science of psychology: that humans, rather than being stalwarts of reason, are more often irrational, neurochemically influenced lovers of self-serving community narratives—shared stories that reinforce our core values and cultural identities, and by extension our social and political behavior. Our brain chemistry tends to favor the suppression of critical reason in favor of emotions that support and defend the views and values we hold dear.[13]

As Skeptics Society founder Michael Shermer writes: "We form our beliefs for a variety of subjective, personal, emotional, and psychological reasons in the context of environments created by family, friends, colleagues, culture, and society at large; after forming our beliefs we then defend, justify, and rationalize them with a host of intellectual reasons, cogent arguments, and rational explanations. Beliefs come first, explanations for beliefs follow." Shermer goes on to assert that most people will simply disregard or rationalize away claims that contradict their beliefs. Science is commonly thought to be the antidote to such delusion. But some neuroscientists suggest that scientific reasoning and objectivity are unachievable ideals, given recent discoveries about the emotional dynamics of the human brain.[14]

From Knowledge to Behavior

The preoccupation in science with objective knowledge begs the question of whether more knowledge is the key to solving or managing impending global environmental crises. We may all be sailing on what William Ophuls calls an "antiecological Titanic." But the iceberg in this case has less to do with the

Table 4–1. Factors Contributing to Eco-Complacency and Disbelief

Distortion Factor	Example
psychological distance	human separation from the natural world
technological insulation	"turn on your solar-powered air conditioners"
"organization of denial"	disinformation campaigns and issue framing that calls attention to scientific uncertainty or conspiracy
rejection of counterintuitive information	an increase of 0.01 percent of atmospheric greenhouse gases can have a profound effect on Earth's climate
behavioral momentum	driving habits that rule human behavior even when they are dangerous or waste gasoline
absence of worrisome price signals	zero or low market price of carbon
invisible cause-and-effect relationships	you can't see carbon dioxide or methane
lack of place attachment	"we'll move somewhere that benefits from climate change"
deferred gratification for action	politicians who take aggressive climate actions incur upfront costs and will likely be out of office or dead before many of the benefits of their actions can be measured
lack of self-efficacy	"one individual can't make any dent in this problem!"
complexity	wicked problems like climate change may not have a solution
motivated reasoning	devotion to free-market libertarianism requires rejection of nonmarket-based climate policies
faiths that justify the status quo or offer salvation from climate collapse	in the Bible, God promises Noah never to allow another global flood (e.g., catastrophic sea-level rise)
techno-fix ideologies	geoengineering solutions for climate, such as fertilizing the ocean with iron
optimism bias	dismissing personal risks or discounting future threats
sunk costs	investments of dollars, dreams, and time in support of the status quo
discredence and denial	distrust of science and/or government
diverted attention	spending eight hours daily watching YouTube and other venues of "screen time"
affective image associations	emotional attachment to ideals used to define one's worldview
cultural learning theory	clinging to beliefs about abortion, homosexuality, and nationalism that serve to define one's group identity and, by extension, one's position on seemingly unrelated issues such as climate change
poor issue framing	treating green opportunities for renewable energy as threats or lifestyle sacrifices

knowledge that lies beneath the surface than with the looming ambivalence, even hostility, toward meager actions to prevent or reduce the threat. Much attention in environmental education and risk communication has been devoted to the "knowledge deficit" theory of social change when the real issue appears to be a *behavior* deficit. Even if ecoliteracy as we now define it could somehow pass through the filters of the human psyche undiminished and undistorted, would it be enough to change the course of our "Titanic?"[15]

The behavior deficit seems to exist with or without added ecological knowledge. Is this gap between knowledge and action inevitable? Or is it the result of the way that most of us acquire knowledge? In other words, is failure among the ecoliterate to take action commensurate with the perceived threat a sign of education's limits, or might it be a failure to instill knowledge about action—personal, political, and social—into learning?

How we define and teach ecoliteracy poses interesting challenges for today's educators. For the most part, we teach in indoor classrooms, not in nature. We usually avoid endorsing social or political action because prescription in education is frowned upon, or viewed as politically partisan and fraught with abuses of social engineering. We teach students that knowledge is power, but the exercise of power (i.e., action) is usually treated as a dirty process best left to sausage makers and unscrupulous politicians. Evaluating student action or behavior change is generally shunned as being too difficult and controversial, leading to a professional preference for assessing conventional learning objectives and knowledge performance. Not surprisingly, the effect of such preferences on ecoliteracy usually means that a student's knowledge of, say, the carbon cycle will count for much more, educationally, than their personal efforts to reduce carbon emissions.

Granting that ecoliteracy across vast segments of the public remains appallingly low, it is nevertheless unclear whether ignorance in this case is more of a cause or an effect (i.e., psychological defense mechanism) of growing environmental threats. Failure to act in a timely and sufficient manner cannot be explained by deficiencies in education alone.

From Behavior to Governance

Poor ecoliteracy remains a sign of crisis in education. But it is also evidence of a deepening crisis in governance. Good governance in the twenty-first century requires stewardship of planetary life-support systems and ecosystem services, accountability, transparency, informed public opinion, leadership skills in nonviolent conflict resolution, and, especially, sustainable conceptions of economic prosperity and wealth. Governance reform can be aided tremendously by applying principles of ecology and biomimicry in political design and policy formation. Examples range from biomimicry's emphasis on decentralized and distributed systems operation to economic development strategies based on ecosystem succession models (e.g., Eugene Odum's 1969 article "The Strategy of Ecosystem Development").[16]

In almost every case, ecoliteracy calls for governance that is based on life-cycle planning, regenerative design, adaptive management, and policies aimed at resilience and sustainability. It promises to improve both the structure and content of governance. But ecoliteracy will not be enough. In some countries, such as the United States, effective governance reform will also require new standards of principled compromise, campaign finance reform, and increased civility among politicians and partisan citizen activists.[17]

The most vexing problems in contemporary governance stem from converging sets of technological, social, economic, and political pressures to bypass, distort, or even dispense with democratic deliberation. Global environmental problems tend to amplify this trend. Technology is enabling political polarization and segmentation into single-minded camps through the selective use of narrowcasting and social media.

In the United States, declining trust in many large institutions and growing contempt for political compromise has resulted in social pressure to opt out of public deliberation. Meanwhile, the economic pressures to cede additional power to Wall Street and large corporations have led to highly undemocratic systems of campaign finance and political influence, not to mention a shrinking middle class that is increasingly unable to carry the burden of participatory democracy. Finally, there are muted but growing political pressures to replace a barely functioning democracy with something closer to technocratic oligarchy in order to deal decisively and swiftly with urgent domestic and international challenges, such as climate change. None of these trends bodes well for democracies in which legitimacy is regarded as sacrosanct, or at least as important as producing good policy results.[18]

The challenge for ecoliteracy in our time is to join the power of science and the joy of emotional attachment to nature with the indispensable role of governance in connecting the worlds of thought, feeling, and action for the purpose of sustaining the web of life. It is only by integrating these three objectives—in much the same way that sustainability integrates environmental, economic, and equity aims—that we can create a coherent community narrative about the interacting risks of climate change and other global environmental threats.

Responding to these threats in a timely manner is likely to require informal networks of governmental and nongovernmental organizations, using strategies that go far beyond conventional policy making or market-based incentives. Such forms of governance will probably be "glocal"—a mix of global and local—and depend on the empowerment of communities and networks of business, faith-based organizations, universities, civic groups, and many others, all of which share responsibility for addressing urgent foresight problems. Glocal foresight requires a well-educated civil society with polycentric "islands of governance" (e.g., collaborative networks ranging from the Davos Economic Forum to the World Social Forum) linked across a sea of political and economic self interest. Arguably, the principal aim of ecoliteracy would be to help disparate peoples and cultures grasp why and how their environmental self-interest, rightly understood, requires new foresight capacity and reform of governance.[19]

Increasingly, governance is about empowering collaboration that produces an expanded sense of what is possible, along with practical strategies to achieve it. The strategies arise ideally from democratic deliberation that involves community-based systems of trust and verification of claims. The choice of strategy depends on how problems are framed and narratives are constructed. Table 4–2 summarizes the broad response strategies that can be used to solve or reduce the three types of global environmental problems outlined earlier, with examples from the climate debate. Because no one strategy can be expected to make much of a dent in foresight problems, however, the ultimate collaborative challenge in governance may be in deciding the strategic mix and modalities of solutions for these very complex global problems.

Table 4–2. The Governance Tool Kit		
Approach	**Solution Strategy**	**Example from Climate Debate**
Market	change price	set a price for carbon
Science & Technology	change technology	replace coal with solar energy
Education	change (add) knowledge	promote climate literacy
Environmental Ethics	change values	accept intrinsic value of ecosystems
Religion	change spiritual beliefs	accept stewardship obligations
Policy	change policy/law	adopt post-Kyoto climate treaties
Politics	change distribution of power	Citizens United vs. F. E. C. case
Advertising	change perception/demand	climate disinformation campaigns
Geography	change place (move away)	move poleward and upward (elevation)
Adaptation	change to accept a bad situation	psychologically adjust to extreme weather
Redefine Problem	change problem features	see climate issues as migration issues
Triage	change who/what receives aid	protect only corals that have thermal resilience
Ecoliteracy	change human-nature connection	connect humans with the carbon cycle

Real-world strategies employ multiple approaches, as in the German approach of using carbon pricing (market) and feed-in tariffs (policy), along with information about renewable energy (advertising and education) and related governmental and grassroots campaigns (politics), to replace coal-fired power plants with photovoltaics or wind (technology).

If this analysis is correct, conventional ecoliteracy is too narrow a foundation on which to build an effective community narrative about climate change, biodiversity crises, and the many other global challenges confronting us. No exclusively science-based concept of ecoliteracy will be sufficient for this task. Ecoliteracy will need to accommodate the traditional knowledge derived from nature-based attachment to place. Moreover, it will need to incorporate explicit social and economic concerns within an action framework that joins ecoliteracy with political literacy about governance—in particular, deliberative, democratic forms of governance.

By recasting ecoliteracy within a larger sustainability framework, the integration of ecological knowledge with community-based concerns about social justice and economic vitality can be greatly advanced. Ultimately, if understanding the relationship between governance and sustainability can become a priority in public education, along with knowledge of the basic principles that govern ecosystems, many societies may be able to overcome some of the disbelief and suspicions that currently polarize large segments of their populations. Overcoming this polarization will require both intellectual and emotional intelligence about our common origin in the great web of life and our common future in sustaining it.

CHAPTER 5

Digitization and Sustainability

Richard Worthington

When the first Earth Day was celebrated on April 22, 1970, the assemblage of ideas, artifacts, and practices that is now known as the Internet was a research and development program of the Advanced Research Projects Agency (ARPA) in the U.S. Department of Defense. At the time, ARPANET, as it was called, connected a few dozen researchers at eight corporate and university sites around the country. Outside this limited circle, few could imagine what lay in store, but the ensuing tsunami of digital devices and systems that has since washed over society is arguably the most significant sociotechnical development of the intervening decades.[1]

Environmental advocates of the 1970s often viewed large, complex technological systems such as nuclear power or industrialized agriculture as threats to both the ecosphere and democratic self-governance. Yet critics rarely applied such concern about the political characteristics of big technological systems to digital systems when business and government began to use these in the 1980s. Instead, information and communications technology (ICT)* was often thought to bode well for environmental improvements such as the dematerialization of production, stronger democratic accountability of private and public decision makers to environmental goals, and collaboration in environmental initiatives at the grassroots level and across vast distances.[2]

These potentials associated with ICT have since borne fruit, but for the most part only in isolated cases or through nascent initiatives that have done little either to rein in ecologically damaging production or to contain the consolidation of power in the hands of global elites. These mixed outcomes suggest the value of exploring more deeply the role of digital systems in

---

* In this chapter, ICT is used interchangeably with terms such as "digital technologies," "digital systems," and "Internet." Such technologies are all electronic systems for processing, storing, and transmitting information in binary bit form. As used here, ICT also includes the people and institutions that create, manage, and use the hardware and software involved.

Richard Worthington is a professor of politics at Pomona College. This chapter was completed with the research assistance of Annie Niehaus.

environmental governance. Such inquiry can highlight important opportunities and risks now facing humanity in light of the dramatic digitization of our technological infrastructures and world since the 1970s.[3]

A rich framework for interpreting this sprawling topic is provided by the idea that "technology is legislation." This argument was first presented in a doctoral dissertation more than 40 years ago, when ARPANET was still a fledgling project and the invention of the term "Internet" lay more than a decade into the future. In setting the terms of everyday routines as well as societal possibilities, however, today's digital systems rule more clearly and consequentially than most laws.[4]

Consider the controversy caused in the United States in late 2013 by a poorly functioning website that was created to help citizens sign up for health insurance coverage made possible by policy reforms. It turns out that, despite the availability of other means of accessing the new insurance program (telephone, post, and government offices), the website mentioned only the online option on its home page. No doubt the administration of President Barack Obama deserved much of the criticism it received for a rollout that was, in many respects, clueless. Yet practically no one (critics included) noted that other means of learning about and signing up for the program were available, a condition that persisted even after Obama himself pointed to the alternatives in a national speech. Here, a technological mindset "legislates" behavior by constraining virtually everyone's consideration of the tools available for accomplishing an important task to the most "sophisticated" of them, even when that tool is not working and alternatives are readily available. Laws rarely exact such compliance.[5]

If digital technology is a form of legislation, then what are its rules for environmental governance, and how can they be navigated, applied, resisted, or changed? Exploring the prospects and pitfalls of environmental governance in a digital society raises several questions:

- Has digitization contributed to more-sustainable production systems?
- How are digitization and democracy connected, and what have been the outcomes for sustainability? Specifically, does digitization promote a governing system in which ordinary people have meaningful input into the decisions affecting their lives?
- What role has digitization played in the allocation of resources available for sustainability?

None of these questions can be answered with great confidence, because ICT has rapidly permeated virtually every aspect of society, yet the transition to a digital society is probably only in a beginning phase. Sorting out cause and effect is no easy task under these circumstances. Nevertheless, the pervasive character and powerful potential of digital systems make inquiry into them an urgent matter, and there are enough patterns in experience to date to at least identify the questions that should be considered as the governance of sustainability becomes an increasingly practical issue.

Sustainable Production

The dawning digital age spawned auspicious terms such as "the paperless office" and "telecommuting," and rapidly evolving applications that were scarcely imagined at the time, such as Skype and cloud computing, have delivered on these visions in compelling ways. Yet experience to date—for example, with dematerialized production systems that lighten environmental burdens by reducing daily commutes or cross-country flights for work, or that transmit enormous volumes of documents without using paper—has yielded limited results.

Close attention to the things measured and inferences made are critical in sifting through the studies and data on this topic. Since 1950, for example, the amount of energy required to produce a dollar of output in the United States has declined steadily. (See Table 5–1.) On a per capita basis, energy consumption grew 51 percent between 1950 and 1980—an average annual increase of 1.4 percent—and then declined for a sustained period during the 1980s. Although subsequent decades have seen both increases and declines in per capita energy usage, the decline between 2000 and 2010 is the largest since 1950, suggesting that ICT, which was being applied more aggressively to energy efficiency during this period, played a role in the shift.[6]

Table 5–1. Growth in U.S. Energy Usage, 1950–2010				
	1950–80	1980–90	1990–2000	2000–10
Energy consumption in first year of decade (quadrillion Btu)	34.6	78.1	84.5	98.8
Growth of energy consumption per capita*	1.38%	-0.14%	0.33%	-1.03%
Growth of energy consumption per dollar of GDP*	-1.12%	-2.10%	-1.75%	-1.72%

*Compound average growth rate
Source: See endnote 6.

A definitive conclusion is not possible, however, because ICT is only one of many contributors to energy conservation. Two of the main sources of efficiency gains during 2000–10—the insulation of buildings and more-efficient appliances—have little to do with ICT. More significantly, the contrast between the modest efficiency gains in energy use per person and the steep declines in energy use per dollar of output reflect in part a "rebound

123net

Energy-consuming data center in Southfield, Michigan.

effect," in which the efficiency gains of, say, telecommuting are offset by increased consumption afforded by the savings, such as taking an overseas vacation.[7]

The bottom line is that U.S. reductions in per capita energy use since 1980 have been modest, and total energy consumption in the United States has increased due to population growth. Meanwhile, for the entire world, both per capita *and* total energy consumption have continued to increase in recent decades. Digital technologies may have made these increases less than would otherwise have been the case, but there is no conclusive evidence for that claim. Regardless of ICT's role, society today remains on the same consumptive path that has created the ecological crisis.[8]

The distinctive role of political events in energy usage is especially clear when viewed at a global level. According to the World Bank, 43 countries underwent a decrease in energy use per capita between 1991 and 2010; however, 25 of them were either former Soviet republics or countries like Poland and Cuba that were in the Soviet sphere of influence. The collapse of the Soviet Union, which ended Soviet sponsorship of satellites via the provision of petroleum products at concessional prices, was a geopolitical event that accounts for most of the countries in which per capita energy usage declined. Although some countries, such as Cuba, responded to this shock in inventive ways that minimized the negative impacts of reduced energy consumption and helped transition to a more sustainable system, most simply curtailed energy use rather than using energy more efficiently, with the negative impacts on quality of life that accompany such unplanned and abrupt changes.[9]

Studies that directly address the actual contributions of ICT to environmental benefits have yielded ambiguous results, in part because of uncertain data, but also because capturing this connection poses considerable challenges. Numerous inquiries offer *projections* of future savings to be had through the application of digital technologies, although many of these studies are sponsored by global corporations in the ICT industry. In a recent analysis of the 11 most prominent studies projecting future ICT contributions to greenhouse gas reductions, 10 were sponsored by the ICT industry. The four studies published between 1999 and 2004 project significant savings but also describe how minimal or negative impacts might occur. The six studies published between 2005 and 2008, on the other hand, all project highly positive scenarios of ICT contributions to energy efficiency, thus aligning unambiguously with the sponsors' interests in ICT-driven sustainability strategies.[10]

Digital Democracy

The limited contributions of ICT to dematerialization, alongside the strong presence of corporate interests that are embedded in research about it, make the democratic accomplishments and prospects of digital systems an urgent issue. If the Internet can enhance democracy, it becomes more likely that common rather than special interests can shape new modes of governance for sustainability. For example, effective democracy can help ensure that the ICT industry's particular vision for sustainability can be complemented or (if appropriate) resisted by informed citizens who provide meaningful input into relevant decisions.

Some observers have argued that democratic governments are not up to the task of instituting the profound social and economic changes required to avert ubiquitous ecological collapse. These critics instead have advocated more authoritarian approaches along the lines of William Ophuls' evocatively titled 1973 article "Leviathan or Oblivion?" Most, however, look to improved democratic governance as a critical, but challenging, requirement for transition to sustainability. In the words of political scientist David Orr, "strong democracy may be our best hope for governance in the long emergency [of sustained ecological disruptions], but it will not develop, persist and flourish without significant changes."[11]

Among digital enthusiasts, ICT is routinely depicted as *the* key to creating new democratic forms and practices that can persist and flourish. Activist and digital observer David Bollier, for example, has described in depth how "a kaleidoscopic swarm of commoners besieged by oppressive copyright laws, empowered by digital technologies, and possessed of a vision for a more open, democratic society" has pointed toward radically new governance practices that can bypass and ultimately replace today's sclerotic institutions.[12]

The exemplar of and inspiration for these developments is the creative commons, which exchanges software and other content with limited restrictions on use. But Bollier argues that the invention of "a new species of citizenship" may create a long-term power shift in society away from unaccountable monopolies and bureaucracies and toward creative and democratic self-governance. The creative commons emerged from the largely unplanned activities of a motley array of hackers, bloggers, tech entrepreneurs, professors, and others, leaving the impression that it is at most a side show in a much larger project of digitization. Bollier argues, however, that "in truth, each is participating in social practices that are incrementally and collectively bringing into being a new sort of democratic polity."[13]

Most research on this issue is more equivocal than Bollier's account. Bruce Bimber, in his book *Information and American Democracy: Technology in the Evolution of Political Power*, closely analyzes numerous cases of digital activism and mobilization and provides aggregate quantitative data on the characteristics and political engagement of Internet users. Bimber cites a successful campaign by the Libertarian Party and other political actors that are marginal on the national scene to oppose an administrative rule proposed by the Federal Deposit Insurance Corporation (FDIC) that would enhance government access to private financial records, with the aim of cracking down on money laundering. Catalyzed by the Libertarians and others, some 250,000 citizens submitted statements during a public comment period, virtually all of them criticizing the proposed rule. In the face of this protest, the FDIC withdrew its proposal.[14]

Several observations emerge from Bimber's case studies as well as from quantitative analysis and research on shifts in political communication regimes over the course of U.S. history. In some cases, these phenomena clearly enhance democratic self-governance. For instance, the decreased cost of communication afforded by digitization has made it possible for groups to become engaged that had previously lacked the resources to participate in campaigns and policy development. The FDIC episode and many other cases provide examples where this opening to marginal players like the Libertarian Party has changed policy in directions that appear to enjoy wide public support.

In addition, large political groups, such as the Environmental Defense Fund or the World Wildlife Fund in the environmental arena, have used the Internet to extend their reach beyond traditional membership lists, and have integrated the voices of these citizens into their conventional lobbying activities. These same organizations are able increasingly to form coalitions with both large and small groups by using the power of digital communication, data sharing, and analysis.

Yet Bimber's study points to other digitally influenced developments that either result in very limited democratic gains or suggest that digitization exacerbates undemocratic tendencies in U.S. politics:

- Digital strategies increased mobilization costs for large groups, because these were added to and integrated with conventional lobbying approaches. In the environmental arena, this can widen existing disparities between large and small groups, or in the framework of environmental justice advocates, between "mainstream" and "grassroots" organizations.
- The use of digital systems has not increased the number of politically engaged citizens, although those already engaged have more informa-

tion and increased opportunities to use it. This, too, widens the information and engagement gap. At one end of the spectrum are a relatively few highly informed and active citizens, whose information sources are more biased toward their views than was the case before the advent of digital systems. At the other end are the vast majority of citizens, who have relatively little information or interest in politics, and whose views are subject to the messages emanating from an increasingly concentrated mass media.

- The actions of political groups are increasingly event-driven and rapid-response in nature, which may have occasioned a shift toward short-term planning.
- The Internet has not been as effective a means of attracting and directing citizen attention as the traditional mass media; however, the evidence of widespread citizen concern about an issue generated by digital activism has, in some cases, been used to win media coverage.
- Digital communication and coordination cannot substitute for the personal relationships among political elites that are central in effective lobbying. As with media attention, however, in some cases, digital activism has shown the salience of an issue to the citizenry and thus helped place it on elite agendas.
- Digital mobilizations, such as the Libertarian campaign, cannot sustain influence through the policy cycle that includes agenda-setting, passing legislation, and ensuring that the legislation is properly implemented. Conventional lobbying is required for this type of success.
- The Internet spawns enormous volumes of "cheap talk," i.e., Internet petitions and similar communications to elected officials and other elites, which are ignored by the latter. This wastes time and erodes the quality of political communications.

Taken together, these research results show that the Internet has expanded citizen access to policy makers, with some positive results. The wider effects on power and democracy, however, are minimal at most, and the novelty of digital activism makes predictions unreliable. Bimber's study was published in 2003, but research since then has not changed this outlook in any substantial way.

Several more recent studies are consistent with Bimber's results on three important points. First, the Internet has facilitated many collaborative actions by people, and of those that have addressed public issues, some have attained results. Second, the means of taking public action have changed because of digitization, although there is no evidence that the overall structure of power has changed appreciably. Finally, the changes in activism associated with digitization continue at a dizzying pace, so predictions about democracy, whether optimistic, pessimistic, or something in between, are unreliable.[15]

Funding Sustainability

In a course on technology policy that I taught a few years ago, a very capable and diligent student proclaimed in class one day that use of the Internet is free, and the benefits to the user are virtually unlimited because of the many applications already available and yet to be invented. When I pointed out that someone had paid for the laptop and smartphone in front of him, the subscription for the latter, and the roughly 4 percent of his $50,000 annual tuition, room, and board at Pomona College that is dedicated to information technology infrastructure and services, he readily acknowledged the point that the cost of his digital usage was significant but was embedded in capital and operating expenses in a way that obscured his awareness of them.

Policy discourse and social commentary on the Internet are similarly inattentive to the costs of digitization, a noteworthy omission because these costs are not insignificant. Since the late 1990s, ICT has accounted for about one-third of private investment in the U.S. economy. (See Table 5–2.) Meanwhile, the sums required to attain sustainability far outstrip actual investments. The capital costs for ICT may therefore be an impediment to environmental balance.[16]

Table 5–2. U.S. Total Investment versus ICT Investment, 1992–2012					
	1992	1997	2002	2007	2012
	billion U.S. dollars				
Total investment*	728.5	1,142.8	1,350.1	1,909.2	1,951.9
Total ICT investment*	224.0	371.7	455.1	619.6	642.1
ICT as share of total investment	30.7%	32.5%	33.7%	32.5%	32.9%

*Private, nonresidential investments
Source: See endnote 16.

One recent study concluded that the investment required to align U.S. greenhouse gas emissions with a global goal limiting the rise in atmospheric temperature to 2 degrees Celsius (which many scientists think can avert a severe disruption of industrial society) would have to increase by $25.6 billion, to an annual total of $52.5 billion. This does not include the cost of adapting to climate change (e.g., building sea walls, resettling populations away from expanded flood zones) or investments for other critical elements in the balance between people and nature, such as contributions to biodiversity preservation in developing countries. One might argue that these increased capital costs can be financed from economic growth, including the considerable profits yielded by the ICT industry, but this approach might well continue to produce the rebound effect that limits the benefits of greener production by increasing overall output.[17]

A society that maximizes digitization with little attention to its consequences can expect two additional phenomena that are already in evidence. The first is the increased enclosure of individuals in communication silos that reinforce a narrow sense of self, where commercial priorities are embedded in a radically expanded assemblage of structures and messages. The second is polarization in wealth and income that is produced by the high-tech strategy of economic development. Noting the high incomes that prevail in the "knowledge sector" of the economy that employs relatively few people, economist James Galbraith has concluded that "the effect of a redistribution toward the K-sector must truly be a massive funneling of income from the many to the few."[18]

The polarization of income and wealth observed worldwide in recent decades, some of which is caused by the growth of high-tech industries such as ICT, undermines the social resilience that is essential for sustainability. More-equal societies have fewer poor people; in such societies, it costs less to transfer income to the poor because there are fewer of them. Likewise, more-equal societies have fewer rich people, which mitigates against the rich "opting out" of society through privatization of public goods and exaggerated individualism. And finally, more-equal societies have lower private debt, which is an important means of transfer from the poor to the rich. The money not transferred to the rich through debt payments can be used instead for public investments, such as in environmental preservation, good education, and the arts.[19]

At its core, investment capital is a measure of a society's freedom, representing the resources to address urgent issues and to enhance prosperity. The large claim that digital industries have on these resources in a global society that faces severe ecological disruption warrants closer attention than it has received to date.

Looking Ahead

What prospects and constraints does digitization harbor for environmental governance? One significant fact is that profound changes have already occurred. The most compelling reason to use ICT in production and politics for sustainability, then, stems not from any inherently beneficial or effective properties, but from the fact that there is little choice in the matter.

In this respect, the challenge bears similarities to the issue of mass transit in Los Angeles: having built an entire metropolis around the automobile and sprawl, there is no sensible way to shift to a fixed rail system, because there are no concentrated commercial, industrial, and residential settle-ments among which to fix the rails. Buses, on the other hand, can travel on the dispersed network of roads and highways that sprawl throughout the metropolis, thus delivering economic, environmental, and social benefits by putting the problematic infrastructure to a more sensible use. Digital sys-tems surely harbor more positive prospects and are less rigid than Los Ange-les's transport system, but the common element in both is that creativity is required to nudge an existing infrastructure in better directions.[20]

This means that a strategic orientation toward the Internet is critical. It is possible that, on balance, digital systems have added to ecological destruc-tion and sociopolitical polarization, although no unambiguous answers are available on this matter, nor are there likely to be any in the future. But the possibility that this hypothesis is just shy of outrageous, and not clearly out-rageous, suggests that proposals to deploy digital systems for environmental purposes should be greeted with skepticism, and should only proceed if the skepticism is taken seriously.

A good example is Ecoinformatics, an effort to integrate data on bio-diversity so as to better understand what is actually happening to natural systems, to determine the most effective projects for preserving them, and to support the decisions of people who manage those projects. Megaprojects like this routinely disappoint, and often fail outright. Yet the scale and com-plexity of the transition toward sustainability poses a trenchant question: how can such a massive and expensive task be successful in the absence of a technologically sophisticated means of bringing intellectual order to what is known about the problem, establishing priorities, monitoring efforts, and supporting experts and citizens alike in the implementation of the transi-tion? Debates over these types of issues should become an ongoing feature of governing for sustainability.[21]

Finally, if a meaningful public discourse about public and private invest-ment ever takes place (and if it does not, sustainability will not happen), it will have to address more than simply diverting some of the trillions of dollars invested in ICT in recent decades toward sustainability. The enor-mous profitability of ICT will have to be addressed as well. Can sustain-ability compete with ICT in this regard? If not, how will it be politically and economically possible to adjust investment patterns?

In sum, there is little choice about engaging digital systems in environ-mental governance, but naïve attachment to them will perpetuate distorted patterns of investment and other features of the socioeconomic model that has generated the environmental crisis. Critical engagement, careful strate-gizing, and most of all a commitment to profound change are preconditions for using these systems for different ends.

CHAPTER 6

Living in the Anthropocene: Business as Usual, or Compassionate Retreat?

Peter G. Brown and Jeremy J. Schmidt

Human activity is changing the earth at a global scale. Atmospheric carbon dioxide reached 400 parts per million (ppm) in 2013, and there are no policies in place to prevent it from passing 450 ppm. This makes it highly unlikely that the 2009 Copenhagen agreement to limit warming to 2 degrees Celsius will be achieved, and there are many reasons to believe that this goal itself is too high to be safe. Projected sea-level rise will encroach on many of the world's urban centers and agricultural lands, while shifts in regional weather patterns are leading to additional concerns about food, water, political insecurity, and massive migrations of climate refugees. All of this occurs in a world where already-high rates of species extinctions are set to rise dramatically due to climate change.[1]

We have entered the Anthropocene, the geologic era in which humans are a primary driver of the evolution of planetary systems. In this context, geoengineering is seen as an increasingly credible option for climate mitigation and as a way to buy the time needed to pursue more permanent solutions. But geoengineering is controversial because it brings serious risks. Moreover, it sets in contrast two governance philosophies that differ profoundly in how we conceive of the human-Earth relationship.[2]

Geoengineering

Geoengineering is the "intentional large-scale manipulation of the environment, particularly manipulation that is intended to reduce undesired anthropogenic climate change." This manipulation can take several forms, such as removing carbon dioxide (CO_2) from the atmosphere and pumping it deep underground, or genetically altering plant leaves to increase the amount of sunlight reflected back into space. Each technique comes with its own costs, benefits, and risks.[3]

For instance, dispersing sulfates high into Earth's atmosphere would be relatively inexpensive and would mimic the cooling effect produced when

Peter G. Brown is a professor at the McGill School of Environment and the coauthor of *Right Relationship: Building a Whole Earth Economy* with Geoffrey Garver. **Jeremy J. Schmidt** is a post-doctoral fellow in social anthropology at Harvard University. He runs The Anthropo.Scene blog at www.jeremyjschmidt.com.

volcanoes release similar particulates that reflect solar radiation back into space. But it comes with potential costs, starting with hazier skies. More serious is the chance that other events, such as natural volcanic eruptions, could compound cooling effects in undesirable ways. Perhaps topping the list is the possibility that adding sulfates *en masse* to the atmosphere will directly affect other biogeochemical systems, such as the oceans and soils that are intricately and intractably linked with the climate system. And even if atmospheric temperatures were stabilized using this method, it would do nothing to stabilize CO_2 concentrations, which already are making the oceans increasingly acidic and hostile to life.[4]

Of course, humans have been altering the earth to our advantage for thousands of years, but the planetary scale and influence of our impact is unprecedented. From this vantage point, we have a choice between two paths. As theologian Thomas Berry describes them, we may proceed from a *technozoic* mindset and continue to digest the biosphere, excavate the lithosphere, and dispose of our waste in the land, sea, and air, not to mention our own bodies. Or we can take the *ecozoic* path and seek a mutually enhancing human-Earth relationship. In this kind of relationship we would seek to restore and repair the earth's life-support systems, and to emulate, respect, and enable those societies characterized by respectful reciprocity with the sources of their being. The question before us would no longer be understood as one of weighing marginal costs, benefits, and risks. On the contrary, it is a profound moral and political question about the human-Earth relationship itself.[5]

Governance in the Anthropocene

Underlying the Anthropocene is the idea that, just as we distinguish previous geological periods from one another based on sedimentary layers, so too are humans laying down markers of their domination of the planet that will stand the test of geologic time. One such marker is the radioactive fallout from nuclear explosions in the twentieth century. Others might be the layers of ocean sediment accumulating as the result of acidification, higher carbon concentrations in the atmosphere deposited in accumulating glaciers and ice sheets, and extensive erosion and redeposition from global land cover change.[6]

The defining characteristic of the Anthropocene is human dominance of Earth systems. Previously, the earth was seen as consisting of natural biomes—areas of tundra, taiga, savannah, desert, etc.—that humans affected in ways that sometimes permanently disturbed them, such as when agricultural practices led to desertification. But what is new in the Anthropocene is that there are no longer any "natural biomes" at all. Rather, humans have changed planetary land-cover patterns to such an extent that there are only "anthromes." Even where there has been no direct intervention, such as deforestation or plowing, the biomes are changing in profound, if less readily observable ways under pressure from temperature changes, acid rain, higher levels of CO_2, and invasive species. The assumption that there is a background set of stable conditions—"nature"—is no longer tenable.[7]

For this reason, we must abandon the assumption that Earth systems, while variable, fluctuate within an overall envelope of stability. Even commonplace ideas like "renewable water" assume that there are perennial stocks and annual flows of water that remain relatively consistent over time. But once the background conditions of the hydrologic cycle are altered by human impacts on the climate system, we must fundamentally rethink such ideas because the stable basis that once seemed "natural" can no longer be assumed. Grappling with this new normal requires revolutionary changes to the way we think, and goes far beyond the technical problems of climate change. In this way, our entrance to the Anthropocene should also compel us to revisit the governance norms, or rules of right conduct, that have led us into this new, unstable era.[8]

There are two principal ways to think about what norms should guide the use of geoengineering, which roughly parallel Berry's technozoic and ecozoic paths. A "management first" approach would seek to optimize the climate for human well-being. Here, the goal would be to respond prudently to the climate crisis given the political stalemates and ineffective regulations that have thus far prevented meaningful action. An "ethics first" approach, by contrast, would seek to reposition politics and environmental regulations using norms that see humans as interdependent parts of Earth systems. From this perspective, reducing human impacts on planetary systems is the first step in recognizing our interdependence on Earth systems.[9]

These two approaches are not mutually exclusive, but they present very different understandings of the human place on Earth and in the universe. Within the former, geoengineering represents the most recent iteration of what has been called the "emancipation project." The project began several millennia ago with agriculture and the attempt to free ourselves from life as hunter/gathers. Paleoclimatologist William Ruddiman has argued that this was also the beginning of human-induced climate change.[10]

landrovermena

A Land Rover expedition passes abandoned ships in the dried-up Aral Sea bed.

The goal of the emancipation project is threefold: (1) to emancipate ourselves from nature, (2) to emancipate ourselves from obligations to the earth's lesser peoples, principally the non-agriculturalists, and (3) to emancipate ourselves from our natural selves—in other words, to conform the self to the project of emancipation from nature and the domination of others. Since World War II, this project has focused on continuous economic growth and has required the creation of a global supply system and the substitution of consumers for citizens. This is the lens through which geoengineering should be viewed.[11]

Technozoic approaches to understanding human-environment relationships have led us into the Anthropocene. Although the Anthropocene has been building for centuries, it was hastened and magnified by what has been described as the "Great Acceleration"—the massive increases in the extraction and use of natural resources to produce material wealth in the latter half of the twentieth century. These efforts were supported by ideas about individuals as rational and autonomous persons who were free to make individual economic choices and whose collective democratic choices established their political sovereignty. But throughout the twentieth century, we also learned precisely the opposite: individuals and human communities are part of ecological systems. We must ask whether technozoic practices and the ideas that support them fit with ecological understandings of humans as interdependent beings whose collective choices at once shape and depend on Earth's life-support systems.[12]

The technozoic approach is undercutting its own future by destroying and destabilizing Earth's life-support systems. As Columbia University political theorist Timothy Mitchell has shown, fossil fuel energy, assumed to be inexhaustible, was thought to free the economy from material limits in the mid-twentieth century, and was used to support modern forms of democracy premised on indefinite growth. So much trust is now placed in economic models that computer programs currently drive global financial markets at rates faster than humans can respond to. When the economy falters, democracies that depend on this techno-political model declare that its institutions are "too big to fail." But the assumption that "the economy" operates free of material constraints is false and has led to the degradation and destabilization of many of Earth's life-support systems. To maintain the technozoic project, a new language of "distributing risk" and determining "sacrifice zones" has arisen to describe how it is eroding its own foundation.[13]

By contrast, an ecozoic approach rejects all three dimensions of the emancipation project. The human relationship to nature must be one of respect and reciprocity with the goal of a mutually enhancing human-Earth relationship. It requires that we recognize that many of those who do not accept the dominant model of "development" often have insights on how to live peacefully and respectfully on Earth. Respecting these alternatives requires a return to the idea of the person as a responsible citizen, not a consumer.[14]

Hence, ecozoic approaches reject management techniques that seek to control natural or social diversity because they misunderstand ecological relationships among humans and between humans and Earth systems. They look instead for alternative ways of organizing human societies and human-environment interactions wherein humans are members, not masters, in the community of life. This does not mean that we never use the planet, but it does mean that the earth and the rest of life on it are not conceived of as reservoirs for human gratification—a world of so-called natural resources.[15]

The entire technozoic form of life must be considered when making decisions about geoengineering because these planetary-scale decisions will affect other cultures, their rights to distinctive ways of life, and potentially all life on Earth, both now and into the future. So we must carefully weigh whether continuing in a business-as-usual, technozoic form of life should be maintained. We argue it should not, and offer the ecozoic as an alternative.

Geoengineering: Managing First

The case for geoengineering is rooted in a technozoic approach. Although he is currently advocating a moratorium on the practice, a leading researcher in geoengineering is Harvard University's David Keith, whose recent book, *A Case for Climate Engineering*, dismisses the idea that we should not manipulate the climate system on the grounds that we have been intervening technologically in the environment for years. For proponents of geoengineering, such as Stewart Brand, technical solutions are made necessary by governance failures on climate mitigation, the long-term expense of dealing with non-technical carbon removal (such as through biomass accumulation), the need for immediate political solutions to impending conflicts, and the costs of a low-carbon world. Brand claims that any of these provide necessary and sufficient warrant for disciplining the earth through what he calls "planet craft"—reconfiguring nature's climate system through technical means.[16]

But these arguments are not convincing. Recall, for instance, that geoengineering is supposed to be something novel, yet when it comes to linking science to policy we see that it is only a variation on the technozoic mindset that legitimates interventions without questioning the worldview generating the very problems that technology is supposed to solve. In this way, it is a form of addiction. This creates three issues that the proponents of geoengineering fail to address.

First, there is the social inequity produced by climate change. Here we see that, because climate policy is also about broader democratic concerns, we must use something *other than science* to determine what sorts of risks

Sharada Prasad

Monsoon rains in Lalitpur, Kathmandu Valley, Nepal.

we are willing to accept and to create systems of decision making that allow those bearing the risks to consent to them. This is a fundamental concept of environmental justice.[17]

The second problem is that climate is the result of complex interactions among multiple subsystems, such as ocean cycles and land cover. So taking a single, one-dimensional action, such as altering the earth's radiation balance, by no means guarantees enhanced climate control. For instance, a surprising effect of atmospheric aerosols, such as those emitted by coal-fired power plants, is that they reduced global warming's effects on the hydrologic cycle. Precipitation was expected to intensify with warming in the twentieth century due to the increased amount of moisture that warm air can hold, but the aerosols reflected sunlight and thus curbed that effect to an extent. The intensity of the hydrologic cycle is now beginning to increase, however, as aerosol reduction policies enacted years ago take effect. The point here is that climate policies are not projected onto a natural, unperturbed backdrop but onto a complex set of systems that are already heavily influenced by human activity in ways that we do not fully understand.[18]

Third, there is the problem of overconfidence. In *Earthmasters*, ethicist and philosopher Clive Hamilton argues that geoengineering wrongly "plays God" with the climate because "the grander schemes to regulate the climate trespass in a domain properly beyond the human.... [W]e want to supplant the gods in order to counter the mess we have made as faulty humans." Hamilton's arguments expose the assumption that even if management could increase climate control—which is far from clear—this does not establish that *we have the capacity to do the managing*. Numerous studies show how "command and control" resource management fails to be either socially democratic or ecologically sound even at small scales, such as a watershed or forest.[19]

So ramping up to a planetary scale through geoengineering is deeply misguided. Hamilton believes that this kind of faith in our ability to control complex systems is evidence of "epistemic hubris"—a false and dangerous belief in our own brilliance and power—that actually invites calamity. The plain fact is that we are not nearly smart enough to manage the complex and inherently indeterminant systems that make life on Earth possible. As Wes Jackson has pointed out, "ignorance is our strong suit."[20]

Ethics First: Compassionate Retreat

In contrast to the management-first approach, an ethics-first approach begins by identifying the mistaken assumptions about human-environment relationships that brought on the climate crisis, and indeed the Anthropocene itself. It then seeks new rules, or a rediscovery of old ones, for right conduct that can make an ecozoic view operational as a platform for reducing impact on Earth systems in ways that are fair and just. We envision this as a compassionate retreat—a step back from conquering nature, the "lesser" people, and ourselves. It is a step toward developing rules, institutions, and practices that mend the torn fabrics of Earth's life-support systems in which we live; an opportunity to relearn the ancient wisdom still held by the world's traditional peoples and, in so doing, to free ourselves from a self-inflicted tyranny.

The ecozoic view sees the universe as a communion of subjects, not a collection of objects. It departs from the technozoic view of humans as the only active agents in the cosmos and as ecologically and morally independent. It argues that there are many degrees and kinds of agency on the planet. And it acknowledges that many other human cultures exist in traditional communities all around the world that share this belief in treating the natural world with respect and reciprocity. These offer alternate rules, practices, and institutions as management models. An ecozoic view recognizes the unforeseen consequences of conquering nature, which is the potential to undo the fabric of evolution by simplifying complex systems in support of only one form of life: that of the conqueror. It counters simplification with an emphasis on diversity, redundancy, and respect for the alternate aims and ends of other agents. It supports democratic norms conducive to life within the Anthropocene.[21]

Compassionate retreat is a way to make an ecozoic view operational. It has three key elements. First, it is mindful of scientific uncertainties regarding Earth systems and the potentially detrimental effects of acting on limited knowledge. It does not see scientific uncertainty as something to be overcome, but as something that is intrinsic to the way that we know the world around us. Because of this, it suggests a disposition toward the human-environment relationship that is tempered by humility and respect for other forces that also have important effects on the complex systems that coproduce the climate. As a result, contests over geoengineering are seen not as technical decisions, but as social decisions about how we want to live and steward the planet in an era in which technology is not simplifying our lives, but rather making them more complicated.[22]

Second, compassionate retreat acknowledges that the current balance of power favors a minority of the world's wealthiest over the vast majority of others, who also bear a disproportionate burden of negative climate effects. This is unjust. Compassionate retreat acknowledges that there is no neutral arena in which to adjudicate the competing demands that face us from climate challenges, which means that the very idea of geoengineering could only legitimately go forward with the consent of those affected by it—and even then, only if it would be likely to improve those who are worse off, and if it respects alternate forms of cultural organization, political decision making, and choice.[23]

Third, and perhaps most important, compassionate retreat implies rethinking the assumptions that brought us the climate crisis and the Anthropocene. It does not begin with the assumption that humans have rightful dominion over Earth. It rejects the idea that nature exists as a set of stable background conditions that Earth systems will revert to if we simply stop perturbing them. That is unlikely, given that humans have fundamentally altered planetary systems, such as the climate, that may take thousands of years to return to pre-industrial patterns, if they return at all.

Compassionate retreat requires recognizing that our existing legal systems are not grounded in empirical knowledge about the interdependence of humans or their communities, but assume that humans are independent agents rather than fully embedded in the earth's energy and material flows and in sociocultural systems of meaning. In fact, we have created a number of such systems—across economics, finance, law, governance, ethics, religion—that legitimate and foster a technozoic relationship with life and the world. But none of these has been critically rethought in light of the fresh understandings brought to us by the scientific revolutions of the last 200 years, nor by our radically changed circumstances. They are like orphans: their intellectual parents have died but they live on in teaching and practice. As a consequence, our mental maps of the world are not maps of where we are.[24]

Compassionate retreat is a way to think about the transition from the technozoic to the ecozoic. In practical terms, it requires a movement away from economic growth for those whose needs are already saturated and the release of ecological space for those who lack the minimums required by justice; in the already rich countries, *degrowth* should become a goal of macroeconomic policy. It aims for a human population that can live within the human share of the earth's energy and material budgets. And it seeks to rethink the political, ethical, and governance "orphans" that have hastened our headlong rush into the Anthropocene. We must urgently redirect investment away from a fossil fuel-based economy to renewable energy sources—

ideally diverting the vast sums spent on the military to these ends; coupled with an immediate end to inappropriate subsidies of the fossil fuel industry.

In reaching and living in the ecozoic, atonement must be a central virtue—it forms one dimension of a moral foundation for healing the earth. Although, taken in any literal sense, ecological restoration to a previous state is impossible, ecological function and the restoration of flourishing life are well within reach. And this provides another incentive for a return to the ecozoic: the joy of the return of life. Rather than throwing yet more money to the banks, fiscal and monetary policy should be tied directly to the regeneration of Earth's life-support capacities. In this vein, we need to develop economic and governance institutions that, as ecological economists have shown, are not premised on indefinite growth and yet produce the ability to live well and justly on a healing planet.[25]

Engaging in geoengineering so that we can continue business as usual, or buy time to fix a failing governance system, reveals a paucity of imagination and a technozoic conception of the human place on Earth and in the universe for which there is no evidence. By contrast, compassionate retreat offers a way to frame the tasks before us in the context of the ecozoic. It honors all members of life's commonwealth with whom we share heritage and destiny. It suggests that we cultivate an open disposition toward alternate cultural ways of life whose members live on Earth with respect and reciprocity.

Our task within the Anthropocene is to re-learn what it means to be a citizen; not just of our earthly community, but of the universe. And it raises sharp questions about whether geoengineering is the latest version of the Faustian bargain struck by a wealthy minority who have brought life's commonwealth to an unwanted and undeserved, yet fateful, choice.

CHAPTER 7

Governing People as Members of the Earth Community

Cormac Cullinan

The governance systems of today's dominant consumerist cultures are facilitating, rather than preventing, degradation of the natural systems that support life and are the foundation of human well-being. By defining all of nature (other than humans) as property, legal systems enable people and corporations to exploit and trade aspects of nature as if they were slaves. Economic systems reward those who extract natural resources and accumulate assets handsomely, and society rewards the financially wealthy with power and status.

Contemporary governance systems are creating incentives for and legitimizing human behaviors that are harmful to the common good. Climate change and the many other "environmental crises" that confront us are the symptoms of this failure of governance. The crucial questions are: why are our governance systems failing, and what can be done about it?

Governance systems reflect a community's or a society's collective view about what it is, what it believes in, and what it wishes to become. Most governance systems today reflect the narcissistic belief that humans are exceptional beings who are superior to the rest of nature and who are not subject to its laws in the same way as other beings. The evidence, however, does not support the proposition that humans differ fundamentally from other species or that it is possible for us to transcend and escape the ordering principles that we observe throughout the universe. On the contrary, the more we discover, the more apparent it is that everything that exists is interrelated and forms a single reality that is ordered on the basis of consistent, universal principles.

Most contemporary governance systems do not account for the fact that they are established within a preexisting system of natural order that is binding on us all. To the contrary, they assume that these universal principles are not relevant to the design and functioning of our legal, political, and economic systems. Consequently, governance systems often function in ways

Cormac Cullinan is an environmental attorney in South Africa, a research associate of the University of Cape Town, and the author of *Wild Law: A Manifesto for Earth Justice.*

that run counter to nature and that cannot be sustained. The overexploitation of a fish stock, for example, may be promoted by the political system, authorized by the legal system, and incentivized by the economic system, but all of these systems are powerless to prevent the ultimate collapse of the fish stock, which the laws of nature dictate. Sustainability depends on governance systems that ensure that people understand and comply with the laws of nature. The penalties that nature imposes for failing to do so are severe and nonnegotiable.

Most contemporary governance systems reflect the fundamental belief of consumerist societies that "more is better," as well as the aspiration to enhance human well-being by amassing ever more material wealth and the technological power to transcend the limitations of nature. Consequently, these systems have been designed to facilitate human appropriation of ever increasing amounts of "natural resources" and "ecosystem services" to fuel infinitely increasing gross domestic product (GDP). Despite the logical absurdity of the goal of achieving infinite GDP growth and the abundant evidence that achieving it would require confounding natural principles of dynamic balance, this model informs most collective decision making.

Earth Jurisprudence

Earth jurisprudence, in contrast, is a philosophy or approach to governance that embraces the reality that humans are an integral part of the whole living community that we call "Earth," and that, in order to flourish, we must govern ourselves in ways that accord with the laws of that community. Logically, if humans are part of a larger ordered universe (or Earth) without which we cannot exist, it must follow that we cannot flourish over any extended period of time unless the systems that we establish to govern human behavior are consistent with those that order the system as whole.

A subsystem based on principles that are entirely incompatible with those on which the whole system is based cannot persist for long. More importantly, if the forces that give us life and that enhance our well-being flow through the web of relationships that connect all that has come into being, then alienating humans from nature and establishing mutually antagonistic or competitive relationships between humans and Earth is entirely self-defeating.

The governance systems of industrial and consumerist societies are designed to legitimize and facilitate the exploitation of Earth on the basis that this is the best way of ensuring human well-being. *Earth jurisprudence*, on the other hand, seeks to guide humans to pursue their own well-being by contributing to the health, beauty, and integrity of the Earth communities within which they live.

One of the central premises of the Earth jurisprudence approach is that long-term human well-being and survival (as with other species) depends on the extent to which we are adapted to our habitat. Thus, the primary goal of human governance systems should be to ensure that humans behave in a manner that enables them to thrive without degrading the Earth community that is essential to life. This means that human governance systems should be aligned with, rather than run counter to, the principles that we observe governing the ever-renewing communities of life.

From this perspective, the purpose of governance is to enhance our fitness to survive (in the Darwinian sense) by progressively fine-tuning our adaptation to Earth. This will require drawing on the best available understanding of how the universe orders itself to inspire the design of congruent human governance systems that regulate people in accordance with the reality that we are embedded within Earth communities. The purpose of legal, economic, and political systems would be to promote behavior that contributes to the ongoing evolution of a healthy Earth community and to discourage behavior that is "anti-social" from the perspective of the community of life.

Earth jurisprudence reflects a worldview that is profoundly different from the materialist worldview of most contemporary cultures. It therefore poses a significant challenge to contemporary governance systems. Earth jurisprudence is not simply another theory jostling with other jurisprudential approaches for attention with an agreed framework of reference. It challenges the framework itself and calls into question the very purpose of governance as currently understood. It is an aspect of a wider cultural shift in our understanding of the universe and our place within it.

Transforming Governance

An essential first step is creating a vision of what a governance system for an ecologically viable human society in the twenty-first century might look like. However, because governance systems are so fundamental to a society and reflect its worldview, values, and aspirations, transforming a governance system requires transforming the society, and *vice versa*. Fundamental social change may be catalyzed by a small group of determined people and inspired by new ideas, but history teaches us that achieving lasting structural changes in society requires the combined actions of many people acting in concert. How much impact the idea of Earth-centric governance will have is likely to be determined by whether it is adopted by enough people who are sufficiently organized to be able to sustain collective action over a long period of time. Change must be both fundamental and rapid because of the speed with which phenomena such as climate change are closing the window of opportunity.

Historically, widespread and fundamental changes in societal values have occurred within relatively short periods of time, but usually by means that we would not wish to emulate. For example, religious and cultural values have been changed by conquering armies and expanding empires (such as the spread of Islam after the seventh century), the disintegration of governance systems (the collapse of the Soviet Union), and traumatic events (the Black Death plague, which killed 30–50 percent of Europe's population). Yet significant changes in societal values and practices also have been achieved by social movements such as the U.S. civil rights movement.[1]

The Secretary of the Bhutan Ministry of Information & Communications (left) inspects a low-cost XO computer.

The prospect of conquest or empire-building spreading ecocentric values during the twenty-first century appears remote. Disasters (such as the predicted avian flu pandemics or climate change-related natural disasters) may well play a role in changing values; however, trauma-driven change usually involves massive loss of human life and often causes negative changes in values, such as the persecution and killing of minority groups identified as scapegoats for the plague in Europe. A fear-based response is unlikely to increase empathy and to shift values toward the more inclusive and cooperative values that are central to an integral Earth-oriented approach.

It is also unlikely that such a transformation will be led by national governments, international institutions, businesses, or religious organizations, although all may play a role.

So far, only a few governments (Bhutan, Bolivia, Ecuador) have shown an appetite for fundamentally reconceptualizing their governance systems in order to steer their country toward ecological sustainability. Even in Ecuador and Bolivia, which recognize the rights of "Mother Earth" and profess a commitment to living well in harmony with nature, the governments have continued to authorize mining and infrastructural development projects that are difficult to reconcile with that commitment. Furthermore, the "Law of Mother Earth and Integral Development for Living Well," promulgated by Bolivian President Evo Morales in October 2012, reflects a retreat from key elements of the philosophy that informed the declaratory "short law" adopted by the Bolivian congress in December 2010.[2]

At the international level, the ability of the United Nations or similar institutions to drive any such transformation will be limited until member states adopt these ideas on a broad scale. Moreover, experience shows that such organizations have consistently failed to act decisively and effectively in response to phenomena such as climate change and the loss of biological diversity.

Corporations, too, are unlikely to drive the transformation of governance systems. Corporate laws and internal corporate governance systems create significant practical obstacles to corporate leadership in this area. Although addressing climate change would be in the commercial interests of some companies (such as insurers), most of the largest global companies derive the bulk of their income from the exploitation of oil, coal, gas, and minerals and are likely to oppose governance systems that would inevitably lead to restrictions on the activities of extractive industries. (See Chapter 20.)

This means that if any widespread shift in values and worldviews—and significant reorientation of governance systems—is to occur, it is most likely to be driven by civil society organizations. There is now some evidence that this is beginning to happen.

Progress to Date

Numerous articles and books have outlined the philosophy and broad framework of what a governance system could look like that recognized and protected the rights of the whole Earth community. These ideas continue to spread. "Wild law" conferences are held annually in England, Scotland, and Australia. Organizations that are members of the Global Alliance for the Rights of Nature are actively involved in promoting and developing these concepts in Australia, Italy, Ecuador, the United Kingdom, and the United States, among other countries.[3]

The idea of shifting the purpose of governance systems to ensure that humans live harmoniously within an Earth community in which all members have legal rights is no longer unthinkable. As this approach infuses civil society organizations around the world, and as the use of language regarding the rights of nature and Mother Earth increases, it is beginning to shift the global discourse about governance.

Existing social movements have increasingly taken up these ideas since April 2010, when the 35,000-strong World People's Conference on Climate Change and the Rights of Mother Earth convened in Cochabamba, Bolivia, to proclaim a Universal Declaration of the Rights of Mother Earth (UDRME). The declaration recognizes that Earth is an indivisible, living community of interrelated and interdependent beings with inherent rights, and defines fundamental human duties to other beings and to Mother Earth as a whole. The reasons for its adoption are reflected in the resulting People's Agreement: "In an interdependent system in which human beings are only one component, it is not possible to recognize rights only of the human part without provoking an imbalance in the system as a whole. To guarantee human rights and to restore harmony with nature, it is necessary to effectively recognize and apply the rights of Mother Earth."[4]

The People's Agreement adopted at Cochabamba has created a common manifesto for many civil society organizations throughout the world. Since 2010, for example, the peasant's organization La Via Campesina has made statements showing that it regards mobilizing to defend the rights of Mother Earth as an integral part of strategies to defend the rights of exploited groups such as peasants and women. Faith communities' and indigenous peoples are adopting this language in public declarations, and the UDRME has sparked numerous other initiatives, including one to develop a global Children's Charter for the Rights for Mother Earth.[5]

Initially, some indigenous people's organizations had concerns about using non-indigenous concepts such as "rights" to express indigenous perspectives, and about whether advocating rights for nature might undermine efforts to enhance the rights of indigenous peoples. Indigenous peoples from South America's Andean region have helped to allay these concerns, and indigenous organizations in North America speak increasingly of the importance of defending the rights of Mother Earth. Indigenous peoples' activists in Africa and Australia also are beginning to explore the relevance of this discourse to their culture and political struggles.

The language of the rights of nature and Mother Earth is penetrating international discourse as well. The United Nations General Assembly has convened several discussions on "living in harmony with Nature," and references to the rights of nature are found in several reports of the UN Secretary-General as well as in both the official declaration from the 2012 World Summit on Sustainable Development ("Rio+20") and the declaration of the parallel People's Summit.[6]

The most significant example of the application of these ideas to date is Ecuador. In September 2008, a referendum of the people of Ecuador approved the adoption of a constitution that explicitly recognizes that nature, or *Pachamama* (Mother Earth), has legal enforceable rights that every Ecuadorian person must respect and that the state has a legal duty to uphold. (See Box 7–1.) Significantly, this recognition of the rights of beings other than humans is characterized as part of a wider project of building a new social order in which citizens will seek to achieve well-being in harmony with nature.[7]

Bolivia has since adopted a law recognizing the rights of nature as well. Both countries are now grappling with how to reconcile the socioeconomic demands of their electorates, the ambitions of extractive industries, and the rights of Mother Earth, with mixed results. In Ecuador, for example, a lawsuit in the name of the Vilcabamba River was successful and the provincial government was ordered to rectify damage caused by the tipping of soil and

---

Box 7–1. Extracts from the Constitution of Ecuador

"We, the sovereign people of Ecuador… celebrating Nature, the Pachamama [Mother Earth] of which we are part and which is vital to our existence, … decided to build a new order of cohabitation for citizens, in its diversity and in harmony of nature, to achieve *el buen vivir*, *sumak kawsay* [well-being]" (Preamble).

"*El buen vivir* requires that individuals, communities, peoples and nationalities shall effectively enjoy their rights, and exercise responsibilities within the framework of inter-culturality, respect for their diversity and harmonious cohabitation with Nature" (article 275).

Individuals and communities have the right to benefit from the environment in order to enjoy *buen vivir* (articles 73 and 74).

"Nature or Pachamama, where life is reproduced and exists, has the right to exist, persist, maintain and regenerate its vital cycles, structures, functions and its processes in evolution" and empowers every person or community to demand the recognition of these rights before public bodies (article 72).

All Ecuadorian women and men must respect the rights of nature, preserve a healthy environment and use national resources in a rational, viable and sustainable manner (article 83(6)).

The state must –

1. guarantee the rights of Nature as well as of individuals and groups, (article 277(1));
2. promote forms of production which will ensure quality of life for the people and discourage those which threaten those rights or those of nature (article 319);
3. guarantee a sustainable model of development which protects biodiversity and the natural capacity of ecosystems to regenerate (article 395(4));
4. apply any ambiguous legal provisions relating to the environment in the way most favorable to the protection of nature (article 395(4)).

Source: See endnote 7.

---

earth from a road-widening project into the river. Subsequent litigation to prevent major mining projects has been unsuccessful, however.[8]

In the United States, a quiet grassroots revolution among local communities continues to gather momentum. Since 2006, when the Pennsylvania-based Community Environmental Legal Defense Fund first helped the Borough of Tamaqua pass a local ordinance recognizing the rights of nature, scores of local communities (and even cities like Pittsburgh) have claimed their right of self-determination by enacting local legislation that protects the health of local ecosystems. This legislation recognizes that local ecosystems have a right to thrive and flourish that must take precedence over corporate interests and rights.[9]

New Zealand provides one of the most interesting examples of how indigenous understandings of the interrelation between human well-being and nature can influence the development of legal systems. In 2012, after protracted litigation, the government signed an agreement with the Whanganui iwi, a Maori tribe with strong cultural ties to the Whanganui River, acknowledging that the river would be recognized as a legal person, called *Te Awa Tupua*. The agreement recognizes the Whanganui River as an indivisible and living whole, from the mountains to the sea, incorporating its tributaries and all of its physical and metaphysical elements.[10]

The agreement provides for the appointment of two persons (*Pou*) of high standing to play a guardianship role, one appointed by the Crown and the other appointed collectively by all iwi with interests in the Whanganui River. Significantly, the guardians are regarded as being accountable to the river and not to their appointors. In the coming years, all parties with an interest in the river—including iwi, central and local government, commercial and recreational users, and other community groups—will collaborate to develop a "whole of river" strategy for the river's management and use.[11]

Winds of Change

Several factors are combining to create a climate that is more conducive for civil society organizations to take up ecocentric governance ideas. First, acceptance of the need for fundamental changes to our governance systems is growing. Dissatisfaction among many people is rising as their overall well-being declines in response to population growth, the intensifying impacts of climate change and other forms of environmental damage, the rising cost of extracting "natural resources," the growing concentration of wealth, and slowing economic growth. It is increasingly apparent that existing international and national governance systems are incapable of responding effectively to these challenges.

Second, public faith in the development models and solutions that governments and the international community have proposed to address these challenges effectively is declining. For example, the civil society organizations participating in the 2012 Rio+20 conference rejected in its entirety the summit's main declaration, *The Future We Want*, which proposed a "green economy" based on commoditizing and trading ecosystem services. This rejection demonstrated that most civil society organizations do not believe that the significant challenges of the twenty-first century can be addressed by employing the same market-oriented thinking that created them, and exposed the gulf between the aspirations of civil society and those of governments and business.[12] (See Chapters 13 and 15.)

Third, the increasing discourse around the rights of nature and Mother Earth is helping to break down the debilitating barriers between social justice organizations and environmental organizations. For centuries, movements for social change have articulated their concerns in the language of rights, justice, and freedom. Yet until now, few environmental activists used these terms because the law defines nature as a collection of objects that are by definition incapable of holding rights or volition. Climate change activists and the UN Human Rights Commission have made significant progress in shifting climate change discourse from a predominantly scientific, technological, and economic debate to one about human rights (although not yet about rights of nature).[13]

On the other hand, local communities (particularly indigenous peoples) increasingly are using the language of rights to reassert their worldview that Earth is animate and sacred. In India, the Dongria Kondh tribespeople, who recognize that their livelihoods and well-being are dependent upon the Niyamgiri Hills, met with determined opposition a project by Vedanta Resources to establish an open-pit bauxite mine in their territory. In April 2013, after the tribespeople framed their efforts as protecting the rights of the hills as a sacred natural person, the Supreme Court of India upheld the religious and cultural rights of the most-affected villages to prevent the mining.[14]

Fourth, as natural resources become scarcer, corporations are intensifying their attempts to exploit areas that local and indigenous communities value highly or regard as sacred. Because existing governance systems are designed to facilitate activities such as mining and because the tax revenues from extractive activity encourage governments to authorize it, these communities are increasingly exploring other means to protect their interests. For example, the rising worldwide use of hydraulic fracturing ("fracking") to extract oil and natural gas from subterranean shale rock has intensified conflicts between local communities and large corporations (usually supported by governments). In the United States, many communities have responded by adopting local ordinances and charters that assert community rights of self-determination, recognize the rights of nature, and, in some cases, provide that if corporations infringe those rights then their status as separate legal persons holding legal rights will no longer be recognized.[15]

Prospects

Over less than a decade, the idea of expanding notions of governance to recognize the rights of all aspects of Earth to self-expression—and imposing legally enforceable duties on humans and corporations to respect those rights—has gained a significant foothold in global discourse. The adoption of an Earth jurisprudence approach is no longer unthinkable or laughable, and countries like Ecuador and Bolivia are grappling with how to implement it in practice. People around the globe are now using this language to express their understanding that their well-being, rights, and freedoms cannot be maintained without recognizing and protecting the rights and freedoms of all beings that collectively constitute the Earth community that enfolds and sustains us all.

At present, the societal forces that wish to maintain current approaches to governance remain formidable, and for many people the prospects of shifting to governance based on Earth jurisprudence seems implausible. Yet because industrialized civilization is demonstrably unsustainable in its current form, fundamental change is inevitable. Only the nature and direction of that change remains to be determined. The factors that appear to be driving the move toward a more integral approach to governance are all strengthening, and they will likely become increasingly powerful in the future. At the same time, the solutions proposed from within the industrialized worldview (such as increased commodification of ecosystem services and better technology) lack conviction and would exacerbate many problems, such as the yawning chasm between the very wealthy and everyone else.

If ecocentric approaches to governance are to gather force during the next few decades, activists and opinion makers that are currently active in a wide range of realms—animal rights, human and civil rights, indigenous peoples' rights, conservation, environment and climate change, youth, faith, labor, and women—will need to recognize that they would all benefit by collaborating on a common agenda. This agenda would recognize the rights and freedoms of all beings as the basis for a new form of society and as a means of counteracting the property-based powers of corporations and the public institutions that advance their interests.

Global society may already be in the early stages of a tectonic shift in thinking that will fundamentally change the terrain on which the future direction of society will be contested. If the trends discussed above continue, concepts like living well in harmony with nature will begin to replace economic growth as the primary goal of societies, and impairments of ecosystem functioning will be regarded as unlawful unless there are exceptional circumstances justifying a temporary infringement. Communities will see their primary allegiances as being to the places that they love and that sustain them, rather than to political parties or nation states, and will assert their rights to self-determination increasingly strongly. We may then begin to use our creativity in ways that are consciously aligned with the wild creativity that animates the unfolding of the universe, and our civilizations may recover a sense of belonging and purpose.

CHAPTER 8

Listening to the Voices of Young and Future Generations

Antoine Ebel and Tatiana Rinke

When conquering the North American continent in the seventeenth century, few European colonizers recognized that the native populations they encountered were organized in state-like groups, ruled by ethical and moral values. One of these ethical principles remains visionary even today: the Seventh Generation principle of the Iroquois peoples, which states that any action or decision should take into account its consequences for up to seven generations to come. Colonizers likely did not understand this then, and, it seems, we do not understand it now. Judging by our current course of development, we are, as a species, incapable of preserving the ecological well-being of one or two generations down the road, let alone seven.[1]

Fortunately, issues of intergenerational equity and governance have gained significant traction at the global level and have a growing presence in national and international texts and preambles. Several organizations, such as the World Future Council, have made it their mission to make intergenerational equity a reality. Related declarations, commissions, and policy recommendations are multiplying. But have they helped to improve the future prospects of the young and unborn?[2]

Future Rights: From the Page to the Court Room

At the national level, several countries have embedded future generations and intergenerational governance into their constitutions, including Bolivia, Ecuador, Germany, Kenya, Norway, and South Africa. The Norwegian constitution, for example, states in its article 110(b) that "natural resources should be managed on the basis of comprehensive long-term consideration whereby this right will be safeguarded for future generations as well." This language is in line with the spirit of the 1987 Brundtland Commission report *Our Common Future*, which popularized the concept of sustainable development. The report eloquently summarized the moral bias at the heart of intergenerational injustice in terms that still ring true today: "[W]e borrow

Antoine Ebel and **Tatiana Rinke** are members of CliMates (www.studentclimates.org), a student-led, international think-and-do-tank striving to research and implement innovative solutions to climate change.

environmental capital from future generations with no intention or prospect of repaying…. We act as we do because we can get away with it: future generations do not vote; they have no political or financial power; they cannot challenge our decisions."[3]

Carving intergenerational governance and solidarity into law can appear as a relatively easy fix to this bias—but time is a thorny subject for legal deliberation. International law traditionally has been spatially oriented: many court rulings relate to the spaces we occupy and the borders we define, but few legal decisions focus on past generations, and almost none on upcoming ones. Even though the first reference made to "future generations" in a lawsuit dates to 1893, the practical applications of this concept are an exception, not the rule. In one noteworthy case in 1993, *Minors Oposa v. Secretary of State for the Department of Environmental and Natural Resources*, the Supreme Court of the Philippines examined the complaint of a group of children opposing deforestation. In response to the plaintiffs' claim that ongoing logging of their country affected not only living generations, but also future generations, the court ruled that the children were indeed allowed to "represent their yet unborn posterity."[4]

In another relevant case heard by the International Court of Justice in 2010, *Argentina v. Uruguay*, Argentina was concerned about pollution from a pulp mill that Uruguay had built on the river separating the two countries. The court ruled in favor of Uruguay, but a judge, Cançado Trindade, wrote a dissent noting that "the acknowledgment of inter-generational equity forms part of conventional wisdom in International Environmental Law" and that "inter-generational equity has significantly been kept in mind by both contending parties."[5]

This case reflects the current state of intergenerational equity: the concept is influential enough to be mentioned in one of the world's most important courts, but too weak to be the prevalent basis of significant rulings. For all of the current efforts to make intergenerational equity more than words on a page, there is still no legal instrument designed to bind states legally to the principle of protecting the environment for future generations. Moreover, counting on unelected officials to steer our societies toward more temporal justice cannot suffice; instruments and actors that bring future interests to the heart of democratic debates have an important role to play as well.

Representing Tomorrow, Today

As the Brundtland Commission report aptly noted, future generations cannot take their frustration to the streets or voice their concerns in parliamentary hearings. This raises the question of who will speak for them, with the legitimacy to do so. As an alternative to including future generations in law, various countries—such as Canada, Finland, Germany, Hungary, Israel,

New Zealand, Norway, and Wales—have taken a step forward in intergenerational governance by creating specific institutions for it. The creation of such an institution within the United Nations system is currently under discussion. (See Box 8–1.)[6]

Box 8–1. Representing Future Interests Within the United Nations

The need for a solid institutional infrastructure to address intergenerational concerns within the United Nations (UN) system has led to various proposals in recent years—many of which surfaced in the lead-up to the Rio+20 conference in 2012. One was to appoint a Special Envoy to serve as a global independent advocate for the welfare of future generations, different from the Secretary-General's Special Envoy on Youth. Weaker proposals suggested addressing intergenerational solidarity and the needs of future generations as a recurring agenda item in the UN's High-Level Political Forum, or through interagency coordination within the UN system.

Ahead of Rio+20, one proposal received strong support from civil society and many countries: the establishment of a High Commissioner for Future Generations. This proposal stated that the Rio+20 outcome document should commit countries to a clearly defined process for establishing such a commissioner, which should be an independent office within the UN, funded from the regular UN budget. This idea was ambitious, given that only two similar Commissioner positions exist today—one for Refugees and the other for Human Rights.

The proposal was eventually cut from deliberation as several countries blocked the initiative, and the outcome document ended simply with a paragraph inviting the UN Secretary-General to present a report on intergenerational solidarity and the needs of future generations. The report was issued in August 2013 at the 68th UN General Assembly and offers an important set of recommendations to promote this agenda. The question now is how to move from stating broad principles to ensuring their implementation. Progress is occurring, but with very small steps, leading many to wonder whether this critical position will be created before it is too late to be relevant.

—Mirna Ines Fernández
CliMates member, Bolivia
Source: See endnote 6.

In 2001, Israel was the first state to establish a Commission for Future Generations, strong with investigative and advisory powers. The non-political entity could voice its opinion on any legislative text to pass through the Knesset, the nation's congress, which amounted to an informal veto power on any law considered harmful to the interests of future generations. Despite its ambitious mandate—or because of it—the Commission did not last; Israel did not renew its mandate five years later, deeming it costly and ineffective.[7]

Other countries have attempted to create an ombudsperson (or mediator) position at the service of future generations. In 2008, Hungary created such a position that implied significant independence—including the ability to sanction public institutions—and that explicitly mandated frequent interactions with regular citizens. In 2012, however, the function was merged into a broader Commissioner for Fundamental Rights. It is still too early to tell whether this move will effectively take intergenerational concerns off the political agenda, but it certainly makes the initiative less unique.[8]

These promising experiences have revealed their own limitations. Ombudspersons or commissioners today rarely hold enough political clout to be more than a needle of big-picture thinking in a haystack of short-termism. Changes also are needed with regard to how we produce and consume, how we define prosperity and progress, and how much we are willing to sacrifice for it. The issue of intergenerational governance is every bit as economic as it is political, and it must be tackled from both sides.

Putting "Long-term Goggles" on Business

The last few decades have blurred the lines of decision-making power between national governments and private companies, particularly multinational ones—often to the advantage of the latter. This, in turn, has raised civil society's expectations regarding the sometimes questionable behavior of large corporations; consider, for instance, the fact that just 90 companies have contributed a staggering 63 percent of human-caused greenhouse gas emissions since the beginning of the Industrial Revolution. Because economic actors are chiefly responsible for pushing future generations toward the brink of ecological catastrophe, the contribution of the business community will be crucial in keeping the young and unborn safe.[9]

Yet companies seem even more geared toward short-termism than political institutions. The average tenure of a chief executive officer is under four years; most CEOs rely on hedge funds and stock options for a significant part of their salary, and are required to submit quarterly earnings expectations to shareholders. In the face of such powerful incentives working against the interests of future generations, it is difficult to see where the solutions could arise.[10]

According to a Business for Social Responsibility report, because corporate social responsibility policies are too vulnerable to short-term financial stresses, corporations need permanent structures embedded within them to represent long-term interests. So-called "Futures Councils" would be composed of company employees and executives, alongside independent experts. They would issue yearly reports on whether their corporation is "operating in a manner compatible with sustainable development," offer timelines and targets, and even make policy recommendations. While the idea is worth considering, it is difficult to see how it would avoid the shortcomings identified in similar experiments, such as a lack of weight in the decision-making balance, dependence on the good will of power holders, and so on. A closer look at economic decision making tends to suggest that the roots of short-termism run deeper.[11]

The Devil in the Details of Economic Thinking

Mainstream economic theory began taking seriously the notion of "natural capital" only in the 1990s; even then, it was perceived as one form of capital among others, such as technology or knowledge. The fundamental hypothesis made by economic thought leaders such as Robert Solow was that the different forms of societal capital could be substituted for one another. In terms of intergenerational justice, it meant that a generation could use up ecosystem services and natural resources, as long as it made new and equivalent means of production available to the next generation.[12]

In terms of "real life," this perspective implied two things: first, that a technical fix would be available for all major instances of resource exhaustion and environmental destruction; and second, that future generations would consent to this substitution. A look at ecological trends worldwide makes it clear how influential these ideas remain today. However, initiatives such as the creation of sovereign wealth funds dedicated to generating wealth for future generations point to a slow change in mentality. (See Box 8–2.)[13]

Another key factor of economic short-termism is the discount rate. Economic analysis uses discount rates to express a preference for the present: future costs or benefits are *discounted* to show that they mean less in present terms. This rate will, for instance, determine whether long-term infrastructure projects get the go-ahead, or help put an "appropriate" price tag on carbon emissions. In short, the discount rate will determine what is cost effective and what is not. Yet this apparently neutral tool contains moral judgments: the higher the discount rate, the less importance we give to the economic prosperity of future generations—and the more we tilt toward instant gratification.[14]

British economist Nicholas Stern drew considerable criticism when he chose a very low discount rate in his famous 2006 review of climate change, which concluded that climate-friendly investment was very profitable in the long run. Opponents argued that this ethical stance—weighing present and future needs equally—was out of place and led to spending too much, too early on climate action and restraining economic growth. Why sacrifice present economic prospects, critics asked, when future generations will presumably be more prosperous and blessed with better and cheaper technologies to fight climate change? Such debates remind us that the pillars of business-as-usual thinking remain firmly in place: "growth equals well-being" and "environmental action equals economic loss." We are finding out, at great cost, that treating nature like any other form of capital—as a discountable factor in the equation—is immoral and misguided, but this truth has yet to find its way into mainstream economics.[15]

Is the current development model the only one that works? Is it still

Box 8–2. Sovereign Wealth Funds: The Financial Arm of Intergenerational Governance?

Intergenerational justice has emerged as a concept underpinning several sovereign wealth funds around the world. If governed in the right way, such funds can provide a financial resource for future generations. Yet the policy settings that underwrite sovereign wealth funds vary globally, influencing the purpose of the funds, their success, and what they are ultimately used for.

Generally, sovereign wealth funds are state-owned funds that invest in financial assets. They are often established from a balance-of-payments surplus and channeled into investments. The Sovereign Wealth Fund Institute estimated that in 2013, the total size of sovereign wealth funds reached $5.8 trillion. The funds thus can make a sizeable contribution to help future generations cope with potentially devastating environmental damage. It is important, however, that the governance settings that support them ensure that they are managed ethically and in favor of intergenerational justice.

The nature of these funds can vary depend-ing on their source of funds, their administration, and their financial structure. Resource-rich countries often use sovereign wealth funds to manage profits from resource extraction; some dedicate these profits to generating wealth for future generations, such as the oil-funded Norwegian Government Pension Fund (see Chapter 17) or the Kuwait Investment Authority.

Out of coherence with the principle of intergenerational justice, there is increasing public pressure on governments with sovereign wealth funds to invest in "ethical" activities. Australia's Future Fund, for example, no longer invests in tobacco or munitions projects. An increasing proportion of funds, such as the pension fund in Norway, acknowledge explicitly that their interests are linked to sustainable development, and have adopted specific ethical guidelines for shareholder investments.

—*Elizabeth Buchan*
CliMates member, Australia
Source: See endnote 13.

worth pursuing, even if it means closing the doors to sustainable prosperity for a substantial part of the world's population? Too many decision makers, whether political or economic, make choices every day that effectively answer "yes" to these two questions. To achieve sustainability, we will need more voices able to articulate a resounding "NO!" Fortunately, such voices are multiplying, particularly among young people—to the point where listing all youth-initiated efforts for the planet would require a full publication of its own. In the area of climate change in particular—with its far-reaching consequences and intrinsically temporal dimension—young people are being inspired in droves to take action, with many increasingly making it the defining battle of their generation.[16]

A New Phase in the Climate Movement

In June 2013, close to 500 youth climate leaders representing more than 130 nationalities gathered in Istanbul, Turkey, in an unprecedented effort to scale up the climate movement worldwide. Participants in the so-called Global Power Shift shared campaigning skills and techniques, built a common understanding of the challenges at hand, and committed to organize their own "Power Shifts" once they returned home. Power Shifts provide training and strategizing opportunities for budding activists and have helped local climate movements bloom even in the more unlikely reaches of the planet, from India to Kyrgyzstan.[17]

The dynamic at work in initiatives like the Global Power Shift is exceptional given that, until recently, the youth climate movement devoted most of its energy to influencing international climate negotiations. Although the deliberations under the United Nations Framework Convention on Climate Change (UNFCCC) are highly technical and often disappointing, young people identified the talks early on as an important venue to urge their leaders to take action on climate change. Beginning with the Montreal conference in 2005, youth activists have organized an annual Conference of Youth ahead of the UNFCCC's two-week formal negotiations, gathering hundreds of young people to meet, strategize, and build a stronger voice in the talks.[18]

At first, young climate activists hailed mostly from industrialized countries, a reality that is not surprising, given the cost of attending the negotiations. Although representation is still far from balanced, developing-country youth started joining in increasing numbers. The movement structured itself in national and sometimes regional coalitions. Mobilization peaked in 2009, around the Copenhagen conference, where more than 1,000 young people made the trip, while many more supported them from home. Sadly, much like its "grown-up" counterpart, the youth climate movement suffered a terrible blow as the hopes of a "fair, ambitious, and binding" global deal vanished, replaced by an empty four-page declaration.

The youth climate movement has yet to fully recover from "Hopenhagen." Hundreds of young people continue to follow the negotiations cycle, but with much greater cynicism and impatience. By forming an activism space independent from all official political processes, the Global Power Shift is thus unique. It has opted for a change in narrative, identifying the fossil fuel industry as the clear "villain" of the climate story.[19]

One of the most successful youth-led efforts for climate—which embodies well this change in narrative—has been the divestment campaign, an effort to make institutional investors such as university endowment funds remove fossil fuel shares from their portfolios. Although many divestment campaigns have been run since the 1980s (anti-apartheid, tobacco, arms trade, etc.), evidence suggests that the "fossil-free" declination is the fastest growing of all. The success of the movement owes considerably to the gravity of the perils that young people are facing. People under the age of 30 have never been through a month that was colder than the twentieth-century average, and they might well live to see a brutal shift in the state of the world's ecosystems. But initiatives like divestment would not have taken off so quickly were it not for the under-30 generation's unique qualities and opportunities.[20]

Trusting the Millennial Generation

With more than 3 billion people worldwide under the age of 24, nine in ten of whom come from developing countries, the so-called Millennials are the largest youth generation to date. They are also the most educated and the best positioned to seize the movement-building and communication opportunities found on the Internet. (See Chapter 5.) Although the spread of English as the world's *lingua franca* has its own drawbacks and limitations, it allows for much easier cross-cultural collaboration than ever before.[21]

All of these advantages can, and must, be put at the service of sustainability—yet there is still a long way to go. With its many imperfections, the youth climate movement is representative of this. Mutual incomprehension persists between those who are still active within the climate talks and those who have decided to wage an even more challenging battle against the fossil fuel industry. Despite vigorous efforts, the movement is not enlarging to the developing world as quickly as it should. And while young people are effectively making the ethical case for fossil-free societies, they have yet to be equally convincing when showing the many advantages of fossil-free lives. Nevertheless, the world's youth could become a key game changer in favor of environmental progress, if they manage to strike a balance between uncompromising denunciation and a more solutions-oriented approach, to create stronger connections across borders, and to make better use of their diversity of methods and targets.

Yet for all their great qualities and passion, young generations can go only so far in standing up for themselves, their descendants, and the planet. By definition, they are resource-constrained and have little access to media or political power. Legal texts, ombudspersons, futures councils, or ethically sound discount rates can help give them more weight in the balance, but these measures will be effective only if the overall perception of the young and unborn changes as well. For now, humanity's behavior brings to mind, in many ways, Groucho Marx's famous quip: "Why should I care about future generations? What have they ever done for me?" It is striking to observe how much even the best-intentioned people like to speak in favor of the young, but how little they really stop to listen to what young people might have to say.[22]

When Nelson Mandela passed away in December 2013, many of his most inspirational quotes resurfaced on social media. It cannot be a coincidence that one in particular spread so quickly among young environmentalists: "Sometimes, it falls upon a generation to be great. You can be that generation." In the face of growing ecological threats, but also unprecedented opportunities for change, the Millennials have little choice but to be great. In rapidly increasing numbers, they are trying to make the best of the difficult hand that they have been dealt. They will do it more, and they can do it better. But they should receive all the help and appreciation they need.[23]

CHAPTER 9

Advancing Ecological Stewardship Via the Commons and Human Rights

David Bollier and Burns Weston

Climate change and other ecological challenges grow faster and larger with each passing day. Yet little has been done to address the immense legal and political factors that lie at the heart of this unprecedented crisis. Much effort is spent on remedying specific environmental harms, typically after the fact. Even proactive legislation and regulation, on the rare occasions that they are enacted, tend to be piecemeal and incremental, and irregularly enforced.

The basic challenge is that our very conception of the problem is too limited. Nearly all current policy "solutions" insist upon a worldview that subordinates the environment to economic prosperity. They take for granted most prevailing, but outmoded, conceptions of economics, national sovereignty, and law, both domestic and international. They focus on technical fixes and business-friendly interventions—more-efficient technologies, "smarter" environmental policies, emissions trading schemes, etc.—that are inherently limited in what they can achieve.

If serious and enduring progress in protecting natural ecosystems is to be made, however, we must address the core pathologies that dominate contemporary politics and culture: the governance structures and logic of the state, intergovernmental organization, the structure of corporate enterprise, globalized commerce, mainstream economic thought, and the mores and laws that underwrite all of these realms.

This is a daunting agenda, to be sure. But after decades of failed environmental policies and the imminent catastrophes that climate change will inflict on us and our children, it is time to face up to the systemic roots of our predicament. We must imagine and implement a new set of legal and political initiatives that shift ecological governance away from the prevailing framework of neoliberal economics and policy—characterized by an ideological commitment to free trade, deregulation, privatization, and reducing democratic oversight of economic activity—to one that is based on commons- and rights-based ecological governance.

David Bollier is an author, activist, and independent scholar of the commons. He is cofounder of the Commons Strategies Group and blogs at Bollier.org. **Burns Weston** is the Bessie Dutton Murray Distinguished Professor of Law Emeritus and senior scholar of the Center for Human Rights at the University of Iowa. Bollier and Weston jointly direct the Commons Law Project.

We call this paradigm "green governance." We believe that the rigorous application of a reconceptualized human right to a clean and healthy environment—achieved through a growing "commons sector" that blends productive activity and governance—is the most promising, feasible way forward. It can help us slip off the shackles of neoliberal economic policy and its relentless growth and destruction of nature while promoting environmental stewardship that can meet everyone's basic needs and respect the more-than-human world.[1]

Making the Transition to a New Paradigm

It is our premise that human societies will not succeed in overcoming our myriad eco-crises through better "green technology" or economic reforms alone. We must pioneer new types of governance that allow and encourage people to move from anthropocentrism to biocentrism—recognizing the value and interconnectedness of all living things—and thereby develop qualitatively different types of relationships both with nature and with each other.

A political economy that valorizes growth and material development as the precondition for virtually everything else is ultimately a dead end—literally. Achieving a clean, healthy, and ecologically balanced environment requires that we cultivate a practical paradigm of governance that is based on, first, an ethic of respect for nature, sufficiency, interdependence, shared responsibility, and fairness among all human beings; and, second, a logic of integrated global and local citizenship that insists upon transparency and accountability in all activities affecting the integrity of the environment.

We believe that commons- and rights-based ecological governance—green governance—can fulfill this ethic and logic. Properly done, it can move us beyond the neoliberal alliance of state and market that is chiefly responsible for the current model of governance that fails to address environmental degradation adequately.

A basic difficulty is that the price system falls short in its ability to represent notions of value that are subtle, qualitative, long-term, and complicated. These are, however, precisely the attributes of natural systems. The price system has trouble taking account of qualitatively different types of value on their own terms, most notably the carrying capacity of natural systems and their inherent usage limits. Exchange value is the primary, if not the exclusive, concern of conventional economics, with gross domestic product (GDP) serving as the respected, though crude, measure of our society's health and progress.

Conversely, anything that has no price or cannot be traded in the market is regarded (for policy-making purposes) as having subordinate or no value. This orientation has led to a systematic blindness to the actual costs of economic growth. As Ida Kubiszewski, Robert Costanza, and a team of other ecological economists showed in a 2013 paper that uses an alternative metric—the Genuine Progress Indicator (GPI)—to expose the built-in fallacies of GDP, the externalized costs of economic growth around the world have outweighed the benefits since 1978.[2]

Moreover, it is an open secret that various industry lobbies have captured if not corrupted the legislative process worldwide, and that regulatory mechanisms, for all their necessary functions, are essentially incapable of fulfilling their prescribed mandates, let alone pioneering new standards of environmental stewardship. Regulation has become ever-more insulated from citizen influence and accountability, while scientific expertise and technical proceduralism increasingly have become the exclusive determinants of who may credibly participate in the process.

Still, it will not be easy to make the transition from the current approach to ecological governance to commons- and rights-based ecological governance. A system of green governance requires serious reconsideration of some of the most basic premises of our economic, political, and legal orders, and of our culture as well. It requires that we enlarge our understanding of "value" in economic thought to account for nature and social well-being; that we expand our sense of human rights and how they can serve strategic as well as moral purposes; that we liberate ourselves from the limitations of state-centric models of legal process; and that we honor the power of non-market participation, local context, and social diversity in structuring economic activity and addressing environmental problems.

Fortunately, some robust and encouraging developments are now beginning to flourish on the periphery of the mainstream political economy. These include insurgent schools of thought in economics, ecological management, and human rights aided by fledgling grassroots movements. From the Occupy movement to Istanbul's Gezi Park; from Cairo's Tahrir Square to São Paulo's plazas; and from the streets of Athens and Madrid to diverse Internet communities such as Anonymous and the Pirate Party in Germany and even such right-leaning agitators as the Tea Party in the United States, the pulse of citizen protest against "the system" is quickening.

Although disparate and irregularly connected, these various protests, each in its own way, seek to address the many serious deficiencies of centralized governments (corruption, lack of transparency, incompetence, anti-democratic policies) and concentrated markets (externalized costs, fraud, wealth inequality, the ethos of "development" as the overriding goal of the economy). As *The Economist* noted in a June 2013 cover story: "A wave of anger is sweeping the cities of the world. Politicians beware."[3]

Taken together with other paradigm-shifting governance movements—among online activists, subsistence farmers, indigenous peoples, "alter-

Paul Townsend

In 1913, the new Village Hall on Oldlands Common, Gloucestershire, England.

globalization" activists, the Slow Food movement, relocalization projects, and more—we see in green governance the contours of a new paradigm of ecological governance, a non-market mode for communities to manage resources and govern themselves based on the theory and practice of the commons.

A commons is a regime for managing common-pool resources that eschews individual property rights and state control. It relies instead on common-property arrangements that tend to be self-organized and enforced in complex, idiosyncratic social ways, and generally is governed by what we call *vernacular law*—the "unofficial" norms, institutions, and procedures that evolve from commoner practice and decision—to manage shared community resources, typically democratically. State law and action may set the parameters within which vernacular law operates, but it does not directly control how a given commons is organized and managed.

In this way, the commons operates in a quasi-sovereign manner, similar to the market but largely escaping the centralized mandates of the state and the logic of market exchange while mobilizing decentralized participation "on the ground." Broadly conceived, the commons could become an important vehicle for assuring a right to environment at the local, regional, national, and global levels. This, of course, would require innovative legal and policy norms, institutions, and procedures that could recognize and support commons as a matter of law.

The commons represents an advance over existing approaches to ecological governance because it gives us practical, democratic ways of naming and protecting value that the market is incapable of doing. The commons represents an advance over the regulatory state or "self-regulating" market because it gives us a vocabulary for talking about the proper limits of market activity—and for enforcing those limits. To talk about the commons is to force a conversation about "market externalities" that often are shunted to the periphery of economic theory, law, politics, and policy making. People who rely on their commons to manage resources needed for their everyday lives have a keen practical interest in such questions as: How can appropriate limits be set on the market exploitation of nature? What legal principles, institutions, and procedures can help manage a shared resource fairly and sustainably over time, sensitive to the ecological rights of future as well as present generations?

Despite its venerable history as a paradigm of resource management and governance, the commons has been largely ignored by economists and policy makers as a serious alternative to the prevailing modes of control and regulation. There are many reasons for this fact, but clearly one of them is the commons' lack of grounding in state law sufficient to give it popular force and effect. We believe, however, that the legal as well as moral claims of human rights can be powerful agents for enabling and operationalizing the new paradigm we propose. This is a central theme of our book *Green Governance* and the purpose of a key proposal that we include within it: "A Universal Covenant Affirming the Human Right to Commons- and Rights-based Governance of Earth's Natural Wealth and Resources."[4]

The Human Right to a Clean and Healthy Environment

Human rights signal a public order of human dignity, for which environmental well-being is a prerequisite. They trump most other legal obligations, being juridically more elevated than commonplace "standards," "laws," or mere policy choices, and carry with them a state of entitlement on the part of the rights-holder, facilitating legal and political empowerment. To assert human rights is to challenge and make demands upon state sovereignty as well as on the parochial agendas of private elites.

For these and other reasons, the human right to a clean and healthy environment can be a powerful tool for imagining and securing a system of ecological governance in the common interest. Yet, despite many efforts championing the right worldwide—whether understood as an entitlement derived from other substantive rights, as a substantive right autonomous unto itself, or as a cluster of procedural entitlements—its standing in the current state sovereignty system is essentially limited in official recognition and jurisdictional reach.

It is for good reason, therefore, that in recent years, two attractive, alternative approaches have emerged, the first focusing on the environmental rights of future generations, the second on the rights of nature. Both approaches go beyond the narrow anthropocentrism of existing law. And politically, each reflects a deep frustration with the environmental community's conventional terms of advocacy and the formal legal order's deep commitments to neoliberalism.

But neither of these approaches persuades sufficiently in our view. The first, although firm in legal theory, is handicapped by a culture of modernity that prioritizes the present and thus relies heavily on moral appeal for its acceptance. The second, granting nature legal standing, trains on altering mainly the procedural playing field. Moreover, each remains in the grip of the existing state-market regulatory system that is, fundamentally, responsible for most of the environmental damage that threatens our collective future.

We thus believe that it is important to reimagine and establish the human right to environment in a form and substance that is different from existing and proposed incarnations. We propose a human right to commons- and rights-based ecological governance that would constitute a foundational or "meta" right and, where necessary, take precedence over other rights, notwithstanding the problematic of "a hierarchy at odds with the assertion that all rights are equal and indivisible." It would not privilege any right or cluster of rights (present or future), but instead would embody the inclusive spirit of Article 28 of the Universal Declaration of Human Rights, which says that, "Everyone is entitled to a social and international order in which [all] the rights and freedoms set forth in this Declaration can be fully realized."[5]

The human right that we envision embraces structural and procedural issues equally with normative ones, and thereby better integrates with people's everyday experiences and practices within society and as producers. And most importantly, unlike the limited right to environment that is now espoused and practiced, it is anchored in a well-defined, rich history of both substantive and procedural justice. As such, it makes way for a sensibly collaborative and democratic alternative to the current ecologically dysfunctional regulatory system at the heart of our worldwide environmental crisis.

The concept of green governance is based on active stakeholder engagement and innovation, and so fosters more responsive forms of ecological practice, law, and accountability. It is not only credible and necessary, but also politically attractive because it offers a feasible alternative to coercive, top-down approaches that may or may not command popular support.

True, barring some game-changing ecological disaster, investors, corporations, and their political allies will continue to resist innovative legal gambits for both historical and philosophical reasons. Furthermore, most people accept the existing system as a given and therefore do not seek to transition away from it. But absent an overhaul of the current regulatory framework or a radical shift from it, no alternative to commons- and rights-based ecological governance is likely to prove sufficient over the long term. Let us be blunt: neither the state nor the market has been very successful at setting limits on market activity because neither wants to.

The Commons as a Model for Ecological Governance

The paradigm of green governance is compelling because it comprises at once a basis rich in legal tradition that extends back centuries, an attractive cultural discourse that can organize and personally energize people, and a widespread participatory social practice that, at this very moment, is producing practical results in projects big and small, local and transnational.

The history of legal recognition of the commons, and thus the commoners' right to the environment, goes back centuries and even millennia. Pharaoh Akhenaten established nature reserves in Egypt in 1370 BC, and forestry conservation laws were in effect in Europe as early as 1700. Hugo Grotius, often called the father of international law, argued in his famous treatise of 1609, *Mare Liberum*, that the seas must be free for navigation and fishing because the law of nature prohibits ownership of things that appear "to have been created by nature for common things." Antarctica has been managed as a stable, durable intergovernmental commons since the ratification of the Antarctic Treaty in 1959, enabling scientists to cooperate in major international research projects without the threat of military conflict over territorial claims. And the Outer Space Treaty of 1967, although yet to be seriously tested, declares outer space, the moon and other celestial bodies to be the "province of all mankind" and "not subject to national appropriation."[6]

Commons have been durable transcultural institutions for ensuring that people can have direct access to, and use of, natural resources, or that government can act as a formal trustee on behalf of the public interest. Commons regimes have acted as a kind of counterpoint to the dominant systems of power because managing a forest, fishery, or marshland as a commons addresses certain basic human wants and needs that endure: the need to meet one's subsistence needs through cooperative uses of shared resources; the expectation of basic fairness and respectful treatment; and the right to a clean, healthy environment to meet (non-market) household and personal needs.

In this sense, the various historical fragments of what may be called "commons law" (not to be confused with the common law) constitute a legal tradition that can advance human and environmental rights. They reflect the elemental moral consensus that all the creations of nature and society that we inherit from previous generations must be protected and held in trust for future generations. Also, they acknowledge the functional necessity of vernacular law.

As Trent Schroyer puts it: The vernacular space is the sensibility and rootedness that emerges from shaping one's own space within the commons associations of local-regional reciprocity. It is the way in which local life has been conducted throughout most of history and even today in a significant proportion of subsistence- and communitarian-oriented communities.... It is also central to those places and spaces where people are struggling to achieve regeneration and social restorations against the forces of economic globalization.[7]

In our time, the state and market are seen as the only credible or significant forces for governance. In large part, this is due to the state and market having developed a powerful alliance for economic growth, technological progress, and top-down governance—one that has systematically destroyed and marginalized the commons so that its viability has been generally overlooked.

But it is due also to the "tragedy of the commons" parable, popularized by ecologist Garrett Hardin and used often to denigrate the commons and to celebrate libertarian, private-property regimes as the best way to manage natural resources. Even though, according to the International Association for the Study of the Commons, an estimated 2 billion people depend on forests, fisheries, water, wildlife, and other natural-resource commons for their everyday needs, the commons paradigm is essentially invisible to mainstream policy making. It has so many diverse manifestations and so many different types of commoners that economists and policy elites have more or less ignored it.[8]

Yet the commons is an eminently practical and versatile mode of governance for ecological resources, among many other forms of shared wealth. By helping us get beyond the false choices of "state versus market" and "public versus private," the commons points the way toward a diverse array of working governance models that operate at appropriate scales, recognize the realities of ecological limits, and enlist people to become active stewards (and beneficiaries) of commons that matter to them.

Imagining a New Architecture of Law and Policy to Support the Ecological Commons

For a shift to this paradigm to take place, public laws and policies must formally recognize and support the countless commons that now exist and the new ones that must be created. They could legally recognize, for example, the subsistence and indigenous commons that depend on customary practices and rights that presently have scant standing in official law. Governments also could take steps to facilitate such commons as land trusts, cooperatives, and online peer networks that monitor ecological resources ("participatory sensing" communities of water quality; crowdsourcing the spotting of birds, butterflies, and endangered species). They could create new sorts of stakeholder trusts to manage shared assets similar to the Alaska Permanent Fund, or enter into state-commons partnerships that contract with self-organized collectives of commoners, as some Italian municipalities have done.

By such means, government, working with civil society, could facilitate the rise of a commons sector—an eclectic array of commons-based institutions, projects, social practices, and values that advance the policy of collective action. Care must be taken, however, to ensure that state involvement in supporting commons does not stifle their moral or operational integrity, because the quasi-autonomy of commons in making and enforcing their own rules and managing a shared resource is critical to their effectiveness. Extending legal recognition and financial support to the commons is entirely warranted given the expansive legal and financial privileges that governments have provided to corporations for generations. State support for commons could unleash tremendous energy and creativity for improved stewardship of our planet.

If the commons is going to achieve its promise as a governance template, however, there must be a suitable architecture of law and public policy to support and guide it. Innovations in law and policy are needed in three distinct fields:

1) General internal governance principles and policies that can guide the development and management of commons.

Nobel Laureate Elinor Ostrom's eight core design principles, first published in 1990, remain the most solid foundation for understanding the internal governance of commons as a general paradigm. In a book-length study published in 2010, Amy Poteete, Marco Janssen, and Ostrom summarize and elaborate on the key factors enabling self-organized groups to develop collective solutions to common-pool resource problems at small-to-medium scales. Among the most important are the following: (1) reliable information is available about the immediate and long-term costs and benefits of actions; (2) the individuals involved see the resources as important for their own achievements and have a long-term time horizon; (3) gaining a reputation for being a trustworthy reciprocator is important to those involved; (4) individuals can communicate with at least some of the others involved; (5) informal monitoring and sanctioning is feasible and considered appropriate; and (6) social capital and leadership exist, related to previous successes in solving joint problems.[9]

Ostrom has noted that "extensive empirical research on collective action…has repeatedly identified a necessary central core of trust and reciprocity among those involved that is associated with successful levels of collective action." In addition, "when participants fear they are being 'suckers' for taking costly actions while others enjoy a free ride," it enhances the need for monitoring to root out deception and fraud.[10]

If any commons is to cultivate trust and reciprocity and therefore enhance its chances of stable collective management, its constitutive and oper-

Westport Wiki

The Taylortown Salt Marsh, preserved by the Aspetuck Land Trust of Westport, CT.

ational rules must be seen as fair and respectful. To that end, ecological commons must embody the values of human dignity as expressed in, optimally, the Universal Declaration of Human Rights and the nine core international human rights conventions that have evolved from it and are applicable. As this suggests, both human rights and nature's rights are implicit in ecological commons governance.

2) Macro-principles and policies—laws, institutions, and procedures— that the state and market can embrace as ways to facilitate the development of a quasi-autonomous sector of commons and "peer governance."

For larger-scale common-pool resources—national, regional, global— the government must play a more active role in establishing and overseeing commons. It may have an indispensable role to play in instances where a resource cannot easily be divided into parcels (the atmosphere, oceanic fisheries) or where the resource generates large economic benefits relative to the surrounding economy, e.g., petroleum. In such cases, it makes sense for the government to intervene and devise appropriate management systems. *State trustee commons* typically manage hard and soft minerals, timber, and other natural resources on public lands, national parks, and wilderness areas, rivers, lakes, and other bodies of water; government-sponsored research; and civil infrastructure, among other things.

In such circumstances, however, a structural tension exists between commoners and the state-market alliance because governments have strong economic incentives to forge deep political alliances with the market and thus promote an agenda of privatization, commoditization, and globalization despite the adverse consequences for ecosystems and commoners. Any successful regime of commons law must recognize this reality and take aggressive action to ensure that the government does not betray its trust obligations, particularly by colluding with market players in acts of enclosure. This would require that commons-based entities have legal rights of action and access to the courts, which would be empowered to defend the rights of commoners as they have done over the centuries when upholding the rights of commoners enumerated in the Magna Carta.

The overall goal must be to reconceptualize the neoliberal state and market as a "triarchy" with the commons—the *state/market/commons*—to realign authority and provisioning in new, more beneficial ways. The state would maintain its commitments to representative governance and management of public property just as private enterprise would continue to own capital to produce saleable goods and services in the market sector. But governments must shift their focus to become "Partner States," as Michel Bauwens puts it, so that they are not just catering to the interests of capital and markets, but also to the diverse constellations of commons that can and do serve public needs.[11]

3) Catalytic legal strategies that civil society and distinct communities of commoners, governments, and international intergovernmental bodies can pursue to validate, protect, and support ecological commons.

Perhaps the most significant challenge in advancing commons governance is a widespread indifference or hostility to most collectives, at least in Western societies. Accordingly, commoners must use ingenious innovations to make their commons legally cognizable and protected. Since legal regimes vary immensely worldwide, our proposals should be understood as general approaches that necessarily will have to be modified and refined for any given jurisdiction. Still, numerous legal and activist interventions could help to advance commons governance in select areas:

- *Devising ingenious adaptations of private contract and property law is a potentially fruitful way to protect commons.* The basic idea is to use conventional bodies of law that serve private property interests, but to invert their purposes to serve collective or public rather than individual or private interests alone. The most famous example may be the General Public License, or GPL, which copyright owners can attach to software to ensure that the code and any subsequent modifications of it will be forever accessible to anyone for use. This model has been applied to such diverse shared resources as data, scientific knowledge, bioengineered products, and copyrighted works, helping to create a commons of shareable material.[12]

- *Eco-minded trusts that serve the interests of indigenous peoples and poorer countries can emulate private-law workarounds to property and contract law to create new commons.* For example, the Global Innovation Commons, a massive international database of lapsed patents assembled by the Virginia-based enterprise M-CAM, enables anyone to manufacture, modify, and share ecologically significant technologies. Companies and governments in marginalized developing countries, for example, could use these patent-free technologies to develop their own fuel-efficient vehicles and energy systems based on solar, tidal, and wind power.[13]

- *The "stakeholder trust" can be used to manage and lease ecological resources on behalf of commoners, with revenues being distributed directly to commoners.* A well-known model is the Alaska Permanent Fund, which collects oil royalties from state lands on behalf of the state's households. Some activists have proposed an Earth Atmospheric Trust to achieve similar results from the auctioning of rights to emit carbon. And activists in the state of Vermont have proposed stakeholder trusts for common assets within the state, such as water, minerals, rocks, and wind.[14]

- *Some of the most innovative work in developing ecological commons (and knowledge commons that work in synergy with them) is emerging in local and regional circumstances.* The reason is simple: the scale of such commons makes participation more feasible and the rewards more evident. Salient examples are being pioneered by the "re-localization movement" in the United States and the United Kingdom, and by the Transition Town movement in more than 300 communities worldwide, where local citizens seek to reinvent local economies and lifestyles in anticipation of the impending disruptions associated with climate change and peak oil.[15]

- *Federal and provincial governments have a role to play in supporting commons formation and expansion.* Their commerce departments typically host conferences, assist small businesses, promote exports, and so on. Why not provide analogous support for commons? Governments could also help build translocal structures that could facilitate local and subnational commons, such as Community Supported Agriculture (CSA) initiatives and the Slow Food movement—thereby amplifying their impact.

- *The public trust doctrine of environmental law can and should be expanded to apply to a far broader array of natural resources, including protection of Earth's atmosphere.* (See Box 9–1.) It is an important way to ensure that governments act as conscientious trustees of our common ecological wealth. Using various digital networking technologies can render administrative processes more transparent, participatory, and accountable—or, indeed, managed as commons. For example, government wikis and "crowdsourcing" platforms can help enlist citizen-experts to participate in policy making and enforcement. "Participatory sensing" of water quality and other environmental factors can be decentralized to citizens with a stake in those resources. Networks of commoners whose work is mediated by online platforms can report reliably on their performance as stewards of resources through automatic online mechanisms.[16]

Moving Forward

It might be claimed that green governance is a utopian enterprise. But the reality is that the current pursuit of ever-expanding consumption on a global scale is the utopian, totalistic dream. It cannot fulfill its mythological vision of human progress through ubiquitous market activity. It simply demands more than nature can deliver, and it inflicts too much social inequity and disruption in the process. The first step toward sanity requires that we recognize our myriad ecological crises as symptoms of an unsustainable cultural, socioeconomic, and political worldview.

In our book, *Green Governance*, we outline a variety of legal tools and initiatives that can be helpful in spreading a vision of commons- and rights-based ecological governance. Moving forward, however, requires bridging the divide among activists between "intellectual dialogue" and "movement building." There is an urgent need for intensive mutual collaboration between creative thinkers and activists in co-developing bracing new forms

Box 9–1. Litigating for the Public Trust

When legislation fails to address social or environmental wrongs, litigation may be the only recourse, and the environmental movement has a long history of asking the courts to address various ills or failures to uphold environmental law. The current threat of atmospheric degradation runs afoul of state and federal constitutional and statutory protections, as well as the government's common-law obligations to protect public resources for the benefit of present and future generations. (See Chapter 8.) But the U.S. legislative and executive branches have been paralyzed by politics and unwilling to pass laws that will help stabilize the climate.

That is why I sued the U.S. government, on May 4, 2011. The lawsuit (*Alec L., et al. v. Gina McCarthy, et al., USCA Case #13-5192, D.C. Circuit*) demands that six major federal agencies develop a comprehensive climate recovery plan to reduce U.S. carbon dioxide (CO_2) emissions and protect the atmosphere. In addition to the federal case, over the past three years, young people have filed legal actions in all 50 states and around the world.

The cases are based on a legal theory called atmospheric trust litigation (ATL). ATL is grounded in commons law (which has existed since Roman times and is reflected in such codes as the Magna Carta) and the public trust doctrine, under which the state serves as a trustee for rights and resources held in common by all people. According to Mary Christina Wood, the legal scholar at the University of Oregon who developed ATL, a trustee has "an active duty of vigilance to prevent decay or waste to the asset." ATL holds that these assets—including rivers, groundwater, the seashore, and in this case, the atmosphere—cannot be privatized or substantially impaired because they belong to everyone equally, including those yet unborn. As representatives of the youngest generation and generations to come, the plaintiffs in all these suits are the beneficiaries of this atmospheric trust, and the government has a fiduciary duty to protect the atmosphere on behalf of our generation.

The public trust doctrine has been used successfully in the past to defend the commons from destruction by private interests. As the U.S. Supreme Court itself stated in *Geer v. Connecticut*, "the ownership of the sovereign authority is in trust for all the people of the state; and hence, by implication, it is the duty of the legislature to enact such laws as will best preserve the subject of the trust, and secure its beneficial use in the future to the people of the state."

Examples of the success of the public trust doctrine include the case of the pollution and diversion of water from Mono Lake in California, argued as *National Audubon Society v. Superior Court*, in which the court held that the public trust doctrine restricts the amount of water that can be withdrawn from navigable waterways. Also, in a landmark trust case, *Illinois Central Railroad v. Illinois*, the U.S. Supreme Court declared that the state can no more abdicate its trust over property in which the whole people are interested." And in December 2013, the Pennsylvania Supreme Court invoked the public trust in a landmark decision to overturn a state statute that promoted fracking.

There are many other examples of successful public trust cases that address situations of localized environmental damage, but never the atmosphere as a whole. Our case asks the courts to establish a comprehensive climate recovery plan that brings atmospheric CO_2 levels down to no more than 350 parts per million by 2100, which is what the world's leading climate scientists say is necessary to stabilize the earth's climate. To achieve this, fossil fuel emissions must decline at a rate of at least 6 percent per year, beginning now, and aggressive reforestation must be promoted throughout this century.

continued on next page

Box 9–1. continued

This is what our lawsuit is demanding.

The first hearings before the court were dominated by lawyers representing the fossil fuel industry, who intervened in the case and moved for dismissal (which the district court eventually granted), claiming that our complaint did not allege that the defendants violated any specific federal law or constitutional provision. We have filed an appeal, however, which is currently pending in the U.S. Court of Appeals in the D.C. Circuit in Washington, D.C. Several critical *amicus*

curiae briefs were filed in support of our case by top U.S. scientists, national security and legal experts, local government officials, and leaders on behalf of indigenous, faith, and human rights communities. The federal appeal will be heard in spring or summer of 2014, with a decision expected before the year ends.

—Alec Loorz, 19
Founder, Kids vs Global Warming,
iMatterYouth.org
Source: See endnote 16.

of ecological governance that transcend and transform existing frames of legal-political governance and economic and policy analysis. New ideas do not self-actualize, and policy advocacy may not have ready access to bold new ideas.

Moving to green governance will entail many novel complexities and imponderable challenges. Yet there is little doubt that we must re-imagine the roles of the state and market, and imagine alternative futures that fortify the commons sector. We must gird ourselves for the ambitious task of mobilizing new energies and commitments, deconstructing archaic institutions while building new ones, devising new public policies and legal initiatives, and cultivating new understandings of the environment, economics, human rights, governance, and commons.

CHAPTER 10

Looking Backward (Not Forward) to Environmental Justice

Aaron Sachs

On September 4, 1882, at 3:00 p.m., Thomas Edison was in J. P. Morgan's offices on Wall Street—literally inside the mahogany walls. And when he closed a switch shortly after the clock struck three, hundreds of his incandescent bulbs lit up simultaneously in a five-block radius. It seemed like a miracle to the gathered crowd—like magic. People started murmuring, "They're on!" The bulbs stayed on as evening fell, and everyone in lower Manhattan noticed how different they were from the smelly, unsteady gas lamps that they were used to. The next day, the *New York Times* reported that the "light was soft, mellow, and graceful to the eye. It seemed almost like writing by daylight to have a light without a particle of flicker and with scarcely any heat to make the head ache."[1]

What the crowd did not see, of course, were the six steam generators a few blocks away on Pearl Street, each the size of an elephant (they were nicknamed Jumbos, after the famous pachyderm who starred in P. T. Barnum's circus). To power the generators, men had to shovel loads of coal into large furnaces (and, of course, no Manhattanites had seen the coal being scraped from the mountains), and those furnaces boiled water to form steam, which in turn rotated turbines to create electrical energy. To connect the generators to the light bulbs, a work crew had dug some 30 kilometers of tunnels that were lined with brick and laid with copper wire, and then they had connected smaller wires from these main channels into sockets in the walls of various Wall Street buildings. So Edison's light bulbs lit up lower Manhattan—but simultaneously consigned labor and environmental harm to the shadows. More than 130 years later, many of us flick switches every day without recognizing that we are committing acts of violence.[2]

Society changed drastically when people started believing that energy could be captured and harnessed at any time of day or night at virtually no cost. Most old cultures have fables teaching that you can never get something for nothing. We live in a young culture, but it is old enough to have

Aaron Sachs teaches history, American studies, and environmental studies at Cornell University and is the author of, most recently, *Arcadian America: The Death and Life of an Environmental Tradition*.

1885 view of Broadway in New York City, showing the recent advent of telephone, telegraph, and power lines.

a history. And while I understand the environmental movement's tendency to invoke the future in making its arguments, I think the past might be even more relevant and, well, illuminating. Since modernity itself lives in the future tense, the very act of retrospection, so often dismissed as nostalgic, represents the potential for a radical reorientation. To make the effort of connecting the present back to the past is to start the process of reconnecting ourselves to the true sources of our energy—and to each other.

That's why, in part, six years after Edison's demonstration, the American socialist Edward Bellamy called his utopian novel *Looking Backward*. It famously looked ahead to the year 2000, but the real point was to imagine how future U.S. societies would evaluate the massive changes (such as electrification) that occurred in the late nineteenth century. And Bellamy was also looking back, with longing, to the seeming simplicity of the century's earlier decades, before the young republic had moved so definitively from agrarianism to industrialism, from rural homesteading to urbanization, from a culture attuned to cycles to an embrace of linear progress.[3]

The whole structure of Bellamy's society was shifting to accommodate the needs of Big Business. In 1883, the major railroad companies imposed standardized time zones across the nation, and in an 1886 Supreme Court decision, the Southern Pacific Railroad won the rights and protections of personhood for all corporations. But the Robber Barons' dreams of orderly commerce and steady profits unleashed a new chaos on the land. The Gilded Age economy was exploding and collapsing every few years, forests were being decimated, labor struggles were growing more violent, Native Americans were fighting to retain their land, newly emancipated African Americans were clinging to their hard-won rights—and society in general was changing so rapidly that people found it difficult to take stock. So Bellamy catapulted his main character into the future and gave him the leisure to gaze back at the dawn of corporate capitalism. The main thing he saw was a growing inequality, a sense of utter disconnection between certain groups of people—as embodied, perhaps, by the gap between the Wall Street bankers in their well-lit offices and the Pearl Street laborers in their blackened basements.[4]

Today, despite assumptions of hyperconnectivity, such gaps have only widened. When policymakers discuss projects like the Keystone XL pipeline, their emphasis tends to be almost exclusively on the question of how best to secure our supply of energy. They rarely talk about how development of the Canadian tar sands has already devastated several First Nation communities and the environmental resources on which they depend. How can the public take a responsible position on the pipeline if most of us don't even know what an open-pit mining operation looks like?

■

The fundamental problem with a future-oriented environmentalism is that we can't actually predict the future, so our pronouncements become vague, and we revert to fear-mongering and the abstraction of "saving the planet." For whom? For what version of "civilization?" The planet is going to be fine; it will support some form of life no matter what we do to it. But if we try to look backward at our present moment, it becomes clear that this era bears striking resemblances to Edison's and Bellamy's: we are in a period of rapid transformation that generates both excitement and anxiety, characterized by all the temporal, spatial, and interpersonal discontinuities of modernization, including a radically unjust distribution of ecological resources and services.

Some of us live in comfortable, climate-controlled homes, with little idea of how exactly our comfort is delivered to us or at what cost; others have no access to clean drinking water. Some of us eat far more than we need to survive; others suffer from severe malnutrition. The people living in relative ease number in the millions; those barely surviving in the billions.[5]

Awareness of these kinds of disparities bubbled to the surface of the mainstream environmental movement in the early 1990s. As intellectuals reckoned with the legacies of imperialism, and eco-activists started to acknowledge their own privilege, ethics came to the fore, and the idea of environmental justice took hold. Certain grassroots organizations, which had sprung up in the 1980s to fight proposed incinerators and dumps in poor, minority neighborhoods, increasingly demanded more information, transparency, and inclusion in the decision-making process. Meanwhile, scholars published numerous compendia examining how the most vulnerable communities got exposed to the most toxic chemicals, breathed the most polluted air, drank the dirtiest water, and had the least access to green spaces.[6]

For a few years, environmental rhetoric began to suggest the ethical need for the most rapacious consumers to make sacrifices for the sake of the underprivileged around the world; ethics, as a field of study, urges us to take responsibility, to examine the rippling impact of our actions. In 1994, President Bill Clinton issued Executive Order 12898 on environmental justice, designed explicitly to prevent the classic practice of building hazardous waste facilities in the communities that are least well equipped to mount a protest. At the time, some of us hoped that the increasing difficulty of siting such facilities would force manufacturers to find ways of producing less hazardous material in the first place.[7]

Unfortunately, Clinton and his collaborators were simultaneously making it easier for companies to shift their dirtiest operations to other countries. And, more importantly, by the late 1990s, global climate change was starting to dominate environmental discussions, and this new focus led to invocations of planetary crisis and impending doom. Environmental justice went out of fashion as quickly as it had come in.

Now, whenever we despair over the coming storms, floods, and heat waves, whenever we worry publicly about the environmental conditions that our grandchildren will face, we risk coming across as insensitive to the terrible injustices of today's world. Those who look ahead in this way are usually well-intentioned people desperate to shock the public into becoming more politically active, and their assumption is that concern for one's own descendants will be a motivating force. But successful social movements in history have been based on the immediacy of ethics, not on weather forecasts. The ethical purchase of climatology is shaky at best; historical analysis has significantly more to offer.[8]

History suggests, for instance, that we need to understand ourselves as outliers—that the era of fossil fuels has been a truly exceptional one. Never before have so many people lived in such ease, able to focus on consumption and comfort—and never before have we seen such levels of poverty, exploitation, pollution, and certain kinds of violence. Consumer society rests not just on oil and machines but on degrading labor, and on degraded environments where vulnerable populations are losing their homes and livelihoods. Climate change is killing people and creating refugees right now, today, in many parts of the world, and the groups that are most affected have had little to do with creating the conditions they are facing.

This injustice will be perfectly clear to future generations, just as today we all recognize the evil of the slave trade. And isn't it intriguing, as the incisive journalist Andrew Nikiforuk has remarked, that the defenses of today's social order mounted by conservative ideologues—that our energy system and military industrial complex employ millions of people, make us all happy, and allow us to live more secure lives, with more time to further the aims of civilization—sound a lot like the justifications of plantation owners in the pre-Civil War American South? History is not prescriptive, but, like the study of ethics, it forces us to consider our role in social processes.[9]

Most of us are further removed from the physical reality of injustice than the average slaveholder was, but we are ethically obliged to shorten that distance. We need to understand, for instance, the impact that our dependence on coal-fired power plants has on the Appalachian region—just as antebellum northerners should have known how their sugar, cotton, and tobacco were produced. Perhaps most importantly, if we're going to focus on climate change, then we ought to contribute as many resources to adaptation (adjustment to the realities of a changing climate) as to mitigation (the effort to halt its progress), given the burden that is already being borne by so many poor people, especially in the developing world.[10]

As climate scientists always note, it is virtually impossible to pin a given environmental event on something as complex as the climate. But clear patterns have emerged in the last two decades to suggest that climatic factors in certain parts of the world are wreaking more havoc than ever before. Even in some temperate, relatively well-off communities, people have noticed that "hundred-year floods" are happening every few years. Generally speaking, however, conditions are worst in the highest and lowest latitudes.

Inuit hunters, who have a slim margin of error even in the best of times, can no longer rely on their knowledge of animals' migration networks, given how warmer temperatures have changed local ecosystems. And the 2 billion people living in dryland environments, 90 percent of whom are in the developing world, have seen tried-and-true farming techniques starting to fail, year after year. There is also a clear correlation now between "below-normal" rainfall and violent conflict, in both agricultural and pastoral communities. In some relatively arid regions, like East Africa, average rainfall has increased, but instead of arriving regularly and gently, it comes in short, explosive bursts and then disappears, resulting in periods of flooding and erosion followed by drought. Meanwhile, coastal peoples (especially those living in delta regions) have experienced much greater instability, with higher-frequency and higher-intensity storms creating hundreds of thousands of refugees. Such migrants are sometimes characterized as "burdens" on the more "stable" communities designated to take them in, but from the perspective of a climate refugee in the developing world, the burden and injustice travel in the opposite direction: they derive from greenhouse gas emissions caused by the industrial world's investments in towering buildings, elaborate transport systems, and massive manufacturing operations.[11]

Of course, environmentalists and other activists have been trying for decades to get wealthy citizens of the so-called Global North to care about injustices in the impoverished regions of the so-called Global South, with limited success. So it makes sense that U.S. climate change campaigners have started to invoke threats to future generations of Americans rather than reminding people of current water scarcity in Africa. Unfortunately, though, this strategy seems to be floundering as well, in a culture that is so committed to technological optimism and so unaccustomed to confronting the need to sacrifice.

Why not, at the very least, encourage Americans to help the world's neediest—by contributing to efforts to secure water supplies and sanitation systems in the coastal cities of the developing world, or to grow drought-tolerant crops, or to bolster public health regimes and even public insurance initiatives? Behaving ethically—working for justice—even tends to make people happy, in contrast to the American lifestyle of overconsumption, which is more than likely to leave people endlessly dissatisfied.[12]

■

I have no perfect, clear strategy for encouraging people to sacrifice in order to address climate change and environmental injustices; nor do I claim to have made sufficient sacrifices myself. But I have come to believe that historical and ethical approaches, with the sense of connection and investment that those approaches tend to generate, could yield better results than any of the strategies we are currently emphasizing. Perhaps the key lesson of history is that all change is contingent, and nothing is inevitable—and that alone is good grounds for hope. And to ponder ethics is to have faith that individual values and decisions matter deeply, in part because they sometimes cohere into social values. Think of the civil rights movement: it turned out to be deeply significant when one woman refused to move to the back of the bus.[13]

I was intrigued to learn recently that climate activist Bill McKibben has decided, based on his study of history (especially abolitionism and civil rights), that the best thing we can do is to demonize the oil industry, because social movements have traditionally needed an enemy, and environmentalists are not going to get anywhere telling Americans to beat up on themselves for their overconsumption. He has a point: the petroleum lobby is powerful and insidious, and we need to fight hard for deep, structural changes in the economy that will lay most of the burden on those who hold the most power. In my environmental history courses, I always tell my students that what I really want them to do is to march on Washington and demand a carbon tax rather than just shop at our local farmers market. But in truth I want them to do both. We need spaces like farmers markets to help foster political action. Moreover, bringing down the oil companies and replacing them with solar power companies would not erase our complicity in mass-market consumerism or our addiction to energy. Photovoltaic cells might seem like a clean, green technology, but we still don't have a non-toxic way of producing them, which means that the solar industry, like every other energy industry, is leaving communities polluted all around the world.[14]

Justice will not be served until privileged people—my students and I are good examples—reassess their needs and sacrifice some of their privileges. Such people tend to be reluctant to undertake this kind of serious self-evaluation. Even the best-intentioned young environmentalists, who often emphasize governance and "efficacy," tend to scoff at my insistence that they read Thoreau: given the enormity of our problems, what does it matter if one more hermit goes off the grid? But the point of working one's way through *Walden* and Thoreau's other writings is not so much to dwell on his specific actions in the woods as to analyze his way of thinking and his resistance to certain elements of the status quo, to engage with his New England spirit of self-reliance and civil disobedience.

Or perhaps the point is to consider the way in which Thoreau may have inspired Gandhi and Martin Luther King, Jr., who in turn led broad-based social movements that succeeded thanks in part to a determined desire simply to make things right. Sometimes acknowledging the sacrifices of our forebears can spur us toward making our own sacrifices. Or, as economist and philosopher John Broome has argued in his book *Climate Matters: Ethics in a Warming World*, perhaps the point is to remember that justice requires each of us, first and foremost, to do no harm. Thoreau's famous refusal to pay his poll tax stemmed from a fear that the government would use his money to wage war in Mexico and extend slavery southward.[15]

Unfortunately, our current level of consumption in industrial countries—especially of fossil fuels—is directly harmful to billions of people, although it remains difficult to track the harm. Edison, the Wizard of Menlo Park, magically disconnected us from the consequences of our actions. This reality is overwhelming and distressing to recognize, so another of the most important contributions of the environmental movement might be to seek ways of boosting everyone's morale. It's probably time for some climate change knock-knock jokes (perhaps involving Jumbo the Elephant, or some of the nicer things that happen in the dark). But, again, it's also time to do more sustained historical thinking—to remember that cheap, highly concentrated power has been with us for only a short time, and that human societies did find ways of thriving even before fossil fuels were dominant, back in the era of nighttime darkness and wood and walking. In fact, working one's body (Thoreau built his cabin himself using discarded materials) is another well-established way of bolstering one's mood and resilience, so another important and multifaceted goal might be to recapture an older, more positive vision of work. Recent neuroscience research suggests that some forms of modern depression (in the industrial world) are linked to the fewer opportunities we have to use our bodies to accomplish necessary tasks. On cold days, some of my neighbors go outside and chop wood to fuel the stoves that warm their houses, and they find it deeply satisfying; last winter, my gas-powered furnace stopped working early one morning, and there was nothing my wife and I could do to keep our children warm except take them elsewhere. I found that I was still angry at the world even after I had paid the repairman. (In industrial countries, the production of fuelwood is also associated with much less social damage than the production of fossil fuels.)[16]

Energy just means the capacity to do work. That sounds humble and ordinary, but a society's energy system goes pretty far in setting assumptions about what is possible and normal in that society. A family in colonial America with several healthy children and one hired hand and a team of oxen—in other words a middle-class or upper-middle-class family—had about three horsepower, and needed to convert that into its equivalent in food and fuel. Today, a typical middle-class suburban family has about 100 times as much power available. So, on the one hand, we can now envision accomplishing much more, which makes it hard to imagine going back to an earlier standard of living: it has become normal to travel great distances, to eat food that has traveled great distances, to dream of curing cancer and ending poverty. But, on the other hand, we now actually *work* much less on average, although of course some hard labor still has to be performed. The basic work that keeps us alive is done mostly by fossil fuels rather than bodies, which is a great advantage in some ways, because it frees us to do more interesting and useful things. But it can also be seen as a disadvantage, because by buying into a system where we do much less bodily work, we have also bought into higher rates of depression, heart disease, obesity, and general alienation (not to mention all the social and environmental harm caused by fossil fuel extraction). Here we are with far more energy at our disposal, yet how often do we note that we're feeling "low energy?" That's not something people said in colonial times.[17]

Modernity has been liberating and exhilarating in all kinds of ways—I am often grateful for my furnace, not to mention my electric lights—but also brutally damaging and horrifically unjust. Computers, planes, modern surgical techniques, antibiotics, electricity—these are marvels. But we rarely consider what they actually cost in suffering and destruction, because that cost is hidden in the shadows. We take antibiotics to make ourselves feel better without knowing how they actually work or how they were made or tested, but hey, they work—why not take more? Some antibiotics are of course critical; others, as we are now learning through hard experience, may ultimately do more harm than good. Pressing the button to turn up the heat is so easy—for those who can afford it—that it becomes nearly impossible to know whether we actually need to turn up the heat or just want to turn up the heat. (The advertising industry, which arose in the late nineteenth century as corporate capitalism's handmaiden, also helped to expand our needs.) What if thermostats were decorated with pictures of open-pit mines?

What if we had to ride a stationary bike for five minutes for each extra degree of warmth? My favorite household device is the crank-driven flashlight that my wife and I hand to our seven-year-old son every evening at 8 o'clock: if he wants to stay up reading (he always does), he has to turn the crank.[18]

Most Americans are now wired into an elaborate energy system that we have little control over and cannot hope to understand in any thorough way. But we can understand some of its history. In the twentieth century, for various cul-

Los Angeles freeways, 2009.

tural and economic reasons, Americans became addicted to cars (Europe and Asia kept their emphasis on trains) and sprawling suburban developments (Europe and Asia have more densely packed populations and housing stock that is much easier to heat efficiently). And now we use 40 percent more energy than Germany in per capita terms, twice as much as Sweden (where it's pretty darn cold), and three times as much as Japan or Italy. Those are all places with a high quality of life.[19]

Especially given the cost of our energy consumption to so many less-privileged people, do we really want to be this dependent on our machines and on a shaky power grid and on a volatile, leaky supply of oil and natural gas? Perhaps, for those of us with sufficient energy, it might be time to see if we can replace a certain amount of fossil fuel consumption with human power, to see if we can do our work on a more human scale. Bike to the office, use a push mower, join or start a community garden, slow down, ease up. Sometimes it might feel like a sacrifice, and sometimes it might actually be fun. Who isn't interested in avoiding traffic jams? Who would object to seeing more constellations in the night sky? Wouldn't it be cause for celebration if we could show that we were doing less damage to vulnerable communities?

We could try to get our carbon emissions as close to zero as possible—because it is our duty to do no harm—and then, as John Broome suggests, we could direct our money in ways that would offset whatever emissions we couldn't eliminate. We could embrace smaller, more local economies (with much shorter and simpler supply chains), and we could generally try to live, as Bill McKibben has eloquently proposed, more "lightly, carefully, gracefully." It is not a matter of insisting on a joyless efficiency, but perhaps of following the example set by people like Thoreau, or, as the cultural critic

Lewis Mumford once proposed, that of the Benedictine monks: "Rewarding work they kept for themselves: manuscript copying, illumination, carving. Unrewarding work they turned over to the machine: grinding, pounding, sawing. In that original discrimination they showed their intellectual superiority to many of our own contemporaries, who seek to transfer both forms of work to the machine, even if the resultant life proves to be mindless and meaningless." History reminds us that there are always choices, and that communities have flourished in many different contexts.[20]

I don't intend to romanticize physical labor or to glorify the specific version of society developed in Thoreau's day, when much of the hard work was done by exploited peoples and when many hardy men and women, including Thoreau himself, died of diseases like tuberculosis. Yet a town like Thoreau's Concord had much to recommend it: there were no slaves; wage earners at mills and small factories could earn a decent living; many people were independent farmers or artisans; there was a thriving intellectual culture, with a strong undercurrent of utopianism, indicating a commitment to work for change; and the meadows were surrounded by a "border of wild wood," as Thoreau put it. On the other hand, there was little ethnic diversity and most local Native Americans had been killed or driven away (there were also too many dams on the river, and people's tax dollars and consumer purchases sometimes supported militarism and slavery). No place has ever been perfect. But isn't it time to admit that society has not in fact become more and more perfect through the ages—to acknowledge that there will always be work and someone will always have to do it, and to dispense with the modern platitude that "we can't go back?" We can at least go slightly *backward*, and there are excellent reasons to do so—the most compelling of which being, perhaps, that the more work we do for ourselves, the more just our society will become.[21]

Walking to the farmers market instead of driving to the grocery store is not going to halt climate change or eliminate environmental injustices. But maybe, as you try to tell the celery root from the rutabaga, you'll feel a bit more connected to the way people used to live. Maybe, looking backward amid the buzz of public exchange, you'll recognize more fully how each individual is implicated in social structures, and thus how structural change depends on the public airing and coordination of seemingly personal decisions. Maybe you'll be inspired to engage in an act of civil disobedience, to protest corporate irresponsibility or government inertia, following in the footsteps of the Bellamy Clubs that formed in the years after *Looking Backward* was published and that helped the Populist Party re-insert the issue of inequality into American politics. Maybe, lugging your vegetables back to your house or apartment or dorm room, you'll feel a jolt of energy.[22]

The Too-Polite Revolution: Understanding the Failure to Pass U.S. Climate Legislation

Petra Bartosiewicz and Marissa Miley

Passage of an economy-wide cap on greenhouse gas emissions has been one of the great, unrealized ambitions of the environmental movement of this generation. With the effects of climate change already in our midst, and environmental catastrophe very much a threat in this century, curbing human-caused emissions of carbon dioxide (CO_2) has become imperative. To this end, over the past two decades the U.S. environmental community has mounted a series of increasingly well-funded and organized efforts toward adopting federal legislation to cap and reduce greenhouse gas emissions. But such a comprehensive measure has proved elusive. In the past decade, more than 20 bills have been proposed in Congress to create a federal market-based carbon emissions cap; not one of them has become law.[1]

The 2008 presidential election was supposed to change all that. Although not a time-tested environmental ally, Barack Obama named clean energy among his top domestic policy priorities and called for a graduated cap on CO_2 emissions while on the campaign trail. "No business will be allowed to emit any greenhouse gases for free," he pledged. Obama, moreover, was a skilled organizer with the largest grassroots base of any president in history. "For the first time in decades, a President will enter office at the spearhead of a social movement he created," noted *Time* magazine in January 2009.[2]

With a Democratic majority in both the House and Senate under a Democratic president for the first time in 14 years, a coalition of national green groups, backed by deep-pocketed funders, mobilized for what they believed was a historic opportunity to address climate change. The policy vehicle that the green groups put their efforts behind was a cap-and-trade system similar to one already in effect in the European Union. Under such a program, the government places an economy-wide cap on greenhouse gas emissions and ratchets it down over a specified period of time. Individual polluters are issued emissions permits, which can then be traded in a market exchange with other polluters. According to its proponents, cap and trade thus employs

Petra Bartosiewicz is a freelance journalist who lives in Brooklyn, New York, and is currently a Knight-Wallace journalism fellow at the University of Michigan. **Marissa Miley** is a Boston-based freelance journalist and author, and a correspondent at GlobalPost.

Roy Luck

Dow Chemical plant on the Mississippi River just upstream from New Orleans.

financial incentives for companies to move toward more-efficient, lower-carbon energy solutions.[3]

It was, of course, no secret that any kind of carbon emissions regulation in the United States would provoke vehement protest from major polluters in the oil, gas, and electric industries. Since the early 1990s, these corporations have spent more than $3 billion in total lobbying dollars on Capitol Hill, in part to ensure that similar proposals do not get very far. Opponents of climate legislation also have flexed their muscle in the international arena. Because of industry pressure, the United States never ratified the 1997 Kyoto Protocol. It is the only signatory not to have sanctioned what is the most significant global climate agreement to date.[4]

With this in mind, in this most recent legislative campaign the green groups resolved to bring industry to the table. In 2007, major environmental organizations and corporations came together under the banner of the U.S. Climate Action Partnership (USCAP). By the end of 2008, the coalition comprised nearly three dozen members, among them the country's most influential environmental advocacy groups, led by the Environmental Defense Fund (EDF), the Natural Resources Defense Council (NRDC), and the Pew Center on Global Climate Change, along with a corporate membership that included some of the country's biggest polluters: General Electric, Dow Chemical, Alcoa, ConocoPhillips, BP, Shell, and DuPont.[5]

The environmental groups aimed to broker a deal with traditional adversaries and to show lawmakers on Capitol Hill that there was industry support for carbon regulation. The green groups were banking on the political power of the major corporations to sway members of Congress, especially those from states where coal was produced or consumed, to support a climate bill. The corporations, meanwhile, had watched rising public awareness of climate change and believed that comprehensive carbon regulation was imminent. Naturally, the businesses wanted a hand in shaping whatever federal legislation might be crafted. "You're either at the table or on the menu," said Michael Parr, senior manager of government affairs at DuPont, one of the founding USCAP companies.[6]

But despite passage of a cap-and-trade bill in the House of Representatives in June 2009—itself a historic achievement—no such legislation ever made it to a floor vote in the Senate during that Congress. By mid-2010, after several attempts at crafting a bill had failed, the campaign was officially declared over. Given the backlash felt by House members who had cast a tough vote only to see it come to naught, the Senate's inability to pass even a compromise bill effectively killed prospects for any comprehensive carbon cap in the near future and perhaps longer.

The cap-and-trade campaign was driven by the choice of the country's leading environmental organizations to place their faith in a strategy that required them to make deep concessions to the country's biggest polluters. Although significant external factors contributed to the bill's failure—among them a souring economy, a sharp, rightward shift in the Republican Party base, and the president's choice of health care as the major legislative priority of his first term—the green groups made tactical errors that diminished their chance of success.

At the core of this failed campaign was the green groups' belief that any victory they achieved would be modest and incremental. Repeated failed attempts at passing carbon cap legislation had primed the green groups to seek a compromise from the start. The resulting cap-and-trade proposal was brokered among a small group of stakeholders and was largely without broad-based, grassroots support. The diminished role of the grassroots in the climate campaign was no anomaly. Rather, it reflects a fundamental structural disconnect in the environmental community between the big Washington, D.C.-based green groups that have a predominantly inside-the-Beltway approach and the panoply of local, state, and regional environmental groups that focus on coalition building and citizen engagement.

In keeping with this disconnect, the bulk of the money that financed the cap-and-trade campaign came from a small cadre of wealthy hedge fund owners and foundations headquartered primarily in California. This underscored the green groups' reliance on a few large stakeholders rather than on a wide array of on-the-ground supporters. These major funders pooled their resources and coordinated their strategies leading up to the climate campaign. While this may have been done with the intention of marshaling their finances toward a singular goal, it also had the effect of drawing advocacy groups to a preordained mission, rather than trusting the groups to use their ingenuity and expertise to seek out solutions on their own.

The Promise of Cap and Trade

Recognizing the danger presented by climate change, beginning in the early 1990s European nations developed policies aimed at curbing carbon emissions, with the first cap-and-trade system taking effect in the European Union in 2005. But in the United States, the world's second-largest carbon polluter behind China, calls for a carbon or energy tax have been fiercely opposed. The fossil fuel-burning companies that contribute the majority of U.S. human-generated carbon emissions, along with the nation's coal and oil producers, have formed one of the most powerful political lobbying blocks in Washington. Addressing climate change has become all the more contentious with increasing partisanship in Congress, with Democrats generally supporting climate action and Republicans making resistance a central tenet of their party's ideology.[7]

The idea of a market-based emissions cap is itself nothing new in the United States. The model first gained currency in environmental policy circles in the 1980s, when it was implemented to phase out lead in gasoline in lieu of "command and control" approaches, such as having the Environmental Protection Agency (EPA) mandate the reductions itself. In 1990, President George H. W. Bush made good on a campaign promise to swing-state environmentalists to pass amendments to the Clean Air Act that significantly reduced sulfur dioxide emissions from coal-fired power plants—thereby curbing acid rain, which had been a growing problem in the United States and Canada.[8]

In the years leading up to USCAP, green groups such as EDF, which was pivotal in helping draft the acid rain legislation, advocated for a cap-and-trade model to curb greenhouse gas emissions both domestically and on the international level. Regional cap-and-trade programs for climate change mitigation were successfully proposed in northeastern states in 2003 and in California in 2006. In the two years before Obama's election, no fewer than 10 pieces of federal economy-wide carbon cap-and-trade legislation were presented in the House and Senate.[9]

The benefit of cap and trade for climate policy, according to its proponents, was that it was an attractive model for all stakeholders. The green groups liked that it placed an actual cap on carbon, something that had never been done before. The corporations liked that it created a single market-based policy that would trump EPA regulation of greenhouse gases—bureaucratic oversight that was subject to change from administration to administration—and preempt states from implementing their own carbon policies. Republican leaders whom the USCAP coalition hoped to sway to its side could vote for it because it was, in its purest form, a market-based solution.[10]

Most importantly, cap and trade did not appear to be a tax, something that the green groups had long regarded as a nonstarter in gaining the support of lawmakers. This was a lesson learned painfully through the failure of a 1993 energy tax—known as the "Btu tax" after the British thermal unit, the measure of energy it proposed to regulate—which was met with such hostility that while it passed the House it was considered a factor in the defeat of 28 of its Democratic champions in the 1994 elections.[11]

In January 2009, five days before Obama's inauguration, USCAP issued a blueprint for action, calling for a cap-and-trade system with up to an 80 percent reduction in 2005 U.S. greenhouse gas emissions by 2050. (As ambitious as the plan seemed, the proposed reduction goal still fell short of scientists' recommendations for averting catastrophic climate change.) Agreeing on even the most top-level points—how stringent the cap should be and who would receive the bulk of the pollution allowances—had required thousands of hours of negotiations. Despite the historic nature of the coalition, reaction to the blueprint was critical, especially from groups like the nation's leading wind and solar companies, which were never invited to the table. "When you look at the companies that were in USCAP, they were not interested in regulating carbon," said Jigar Shah, founder of solar services company SunEdison. "They were interested in a huge amount of wealth being transferred to their companies in exchange for their vote on climate change."[12]

Many in the environmental community also expressed doubts. "The time to negotiate with industry is when you've had major successes beating industry back and you're holding really strong hammers," said Kierán Suckling, executive director of the Center for Biological Diversity. "These folks sat down with industry when they weren't threatened."[13]

From Earth Day to Inside the Beltway

The 20 million Americans who participated in the first Earth Day in 1970—considered by historians as a watershed moment in the modern environmental movement—proved that a vast public constituency was concerned about the environment. Over the next decade, the environmental movement became a political force as a new crop of environmental advocacy groups and law firms, such as EDF and NRDC, successfully sued industry and the government to enforce the nearly two dozen federal environmental acts that were signed into law, including the Clean Air Act of 1970, the Clean Water Act of 1972, and the Superfund act of 1980 to clean up toxic sites.[14]

Yet in contrast to most social movements in U.S. history—such as women's suffrage and civil rights initiatives that successfully mobilized the public to achieve their goals—the environmental movement has been led increasingly by organizations that pursue an inside-the-Beltway approach. Rather than marshal the power of public concerns, these groups have focused on lobbying Capitol Hill—albeit with a fraction of the resources of their well-heeled industry opponents and with severe limits on how those resources could be spent, given that their federal tax-exempt status imposes strict lobbying limitations.

In spite of this strategy, virtually no major federal legislation has passed since the 1990 Clean Air Act amendments, and some successes have even been rolled back. Many key environmental victories have, in fact, been at the state and local levels and have been spearheaded by grassroots organizations. The largely grassroots environmental justice movement, which led to stricter pollution controls across the country, for example, was catalyzed by community outrage over an epidemic of illnesses at Love Canal—an upstate New York neighborhood built on a chemical company landfill.

The divide between the big greens and the grassroots is underscored by the philanthropic community, which overwhelmingly funds national green groups rather than smaller local organizations. According to a 2012 report by the National Committee for Responsive Philanthropy, although large national organizations with revenue over $5 million comprised only 2 percent of environmental public charities in 2009, they received half of all environmental contributions and grants from foundations.[15]

Private foundations, in particular, have played a crucial role in developing and sustaining the major environmental groups over the past four decades. One analysis estimates that their support was $750,000 in 1970 and has since grown to as much as $1.9 billion in 2008, according to the Foundation Center, a nonprofit group that tracks the philanthropic sector. As these donations have grown, they have been concentrated among fewer organizations. In 2008, according to the Foundation Center, just five foundations were responsible for nearly half of all foundation giving for the environment; at the same time, more than one-third of environmental funding from all foundations went to just five recipients.[16]

A sharp rise in funding occurred in 2007, after the publication of a foundation-commissioned report, *Design to Win*, which outlined the key steps that philanthropists needed to take to combat climate change. The authors of the report, consultants at California Environmental Associates and the Stockholm Environment Institute, estimated that the philanthropic community was, at the time, devoting $210 million annually toward the fight against climate change—far less, they argued, than the philanthropic donations in the United States for health ($3.2 billion), education ($3.1 billion), and the arts ($1.5 billion). To adequately fight the global climate crisis, the report concluded, it would be necessary to invest $525–$660 million annually, of which $80–$100 million should be directed toward implementing a carbon policy, especially in the United States. The *Design to Win* authors wrote that a "cap on carbon output—and an accompanying market for emissions permits—will prompt a sea change that washes over the entire global economy."[17]

As a direct result of *Design to Win*, in 2008, the William and Flora Hewlett Foundation, David and Lucile Packard Foundation, and McKnight Foundation— among the wealthiest foundations in the country—pooled their resources and committed more than $1.1 billion over five years to launch ClimateWorks, a foundation whose primary mission was to combat dangerous climate change. Along with Hewlett, Packard, and ClimateWorks, two addi-

Courtesy of The Bullitt Foundation

Denis Hayes speaks on the first Earth Day, April 22, 1970.

tional California-based foundations, the Energy Foundation and Sea Change Foundation, invested substantially in pursuing a cap-and-trade policy. Together, the five West Coast funders formed a group of grant makers whose geographical proximity underscored their close funding relationships.

In addition to the funds from these groups, a number of wealthy individual donors made sizable contributions to the green groups at the forefront of the cap-and-trade push. Julian Robertson, Jr., an EDF board member who ran one of the most successful U.S. hedge funds in the 1990s, gave EDF more than $40 million between 2005 and 2009 for work on climate change, and the charitable trust of Robert W. Wilson, another former hedge fund manager and EDF board member, gave the green group nearly $24 million in general support between 2008 and 2010.[18]

The clustering of partnered foundations around a single issue and solution supported, in the words of one funder, a larger trend toward "lean and mean" grant making. Funders "want to make sure the money gets spent in the best way," said Ron Kroese, director of the environmental program at the McKnight Foundation. By entrusting larger sums of money to a single organization, said Kroese, foundations can keep costs down and make their donations as impactful as possible.[19]

But having a limited number of people controlling so much money can be "dangerous," said Betsy Taylor, former board president of 1Sky, a grassroots coalition campaign of hundreds of organizations seeking climate legislation. "We have a problem structurally, because Energy Foundation, Hewlett, Sea Change, ClimateWorks, they all fund each other and are all advised by a handful of people," she said. "Let's say they're all brilliant. Let's say they are the very best we could ever have. There's still a structural problem." The result of these close relationships, Taylor said, is an atmosphere of "groupthink" where money is channeled toward one shared strategy rather than distributed across a diverse number of possible options.[20]

Although the precise figure that environmental groups spent promoting cap and trade in Congress is unknown, it is clear that an unprecedented amount of money was allocated to climate action in the United States and that a significant portion of this funding, in turn, went toward the legislative campaign to place a cap on carbon emissions. According to Paul Tewes, former head of Clean Energy Works (CEW), a field and media campaign formed by the green groups to push for comprehensive climate change legislation in the Senate, at least $100 million was spent on the Senate campaign alone. Meanwhile, the leading green groups in USCAP prioritized climate issues above all other program areas in their budgets. EDF, which spent half of its program budget between 2008 and 2010 on climate issues, identified federal cap-and-trade legislation as its top priority for climate work. NRDC spent $35.8 million on its Clean Energy Future program out of $78.5 million in total program services between July 2009 and June 2010, according to its IRS filing.[21]

A 2005 report about the future of philanthropy, funded in part by the Packard Foundation, described this type of focused grant making as "high-engagement giving." Under such a model, which takes its cues from the venture capital world, funding is contingent on the achievement of measurable performances. But according to Jigar Shah, a former venture capitalist himself, the truly successful venture capital model involves something more nuanced: trusting the ingenuity of businesses and the entrepreneurs who lead them. The downside of a lockstep funding structure, said Shah, is that green groups work toward a preordained policy solution rather than coming up with ideas of their own. "These guys believed that if we actually put all of our eggs in one basket, then we have the best chance to pass something," Shah said.[22]

The Battle in Congress

When President Obama took office in January 2009, the USCAP coalition knew that their greatest challenge lay in persuading lawmakers to support their plan. In this respect, the green groups believed that the president would play a decisive role by pushing for the bill, just as George H. W. Bush had done with acid rain legislation under the 1990 Clean Air Act amendments, when he sent his White House counsel, C. Boyden Gray, to personally shepherd the bill through the Senate floor. But if Obama was sincere about addressing climate change, external events lowered the issue on his list of priorities before he was even sworn in.

By late 2008, the U.S. economy was in serious trouble, with the gross domestic product (GDP) shrinking by its worst quarterly contraction in half a century. Obama took office facing the highest unemployment level in 16 years, skyrocketing foreclosures amid the subprime mortgage crisis, and the country's banking and automotive sectors on the brink of collapse. From the start of the administration, it was also clear that climate would have to vie with health care reform for the top spot on the president's domestic policy agenda. With these competing priorities, the green groups soon realized that despite the president's encouraging speeches, he preferred to let Congress hash out the details of a bill before expending his own political capital to lobby for legislation.[23]

Even so, the green groups had reason to be optimistic. The bill was first taken up in the House of Representatives, where they had a strong ally in one of the most skilled legislators in Congress, Representative Henry A. Waxman, Democrat from California. Waxman, the newly appointed chairman of the House Energy and Commerce Committee, started to draft a bill with then Representative Edward Markey, Democrat from Massachusetts, in early 2009. But as the negotiations carried deep into the spring, some grassroots-based environmental groups grew uncomfortable at the compromises being made. The bill was being hammered out behind closed doors, with direct input from big oil and big coal interests, which, in addition to supporting Republican Party members, have given generously to key Democrats. According to Ted Glick, policy director at Chesapeake Climate Action Network, it was becoming clear that industry groups were in a win-win position: "They knew if they defeated the bill that was good, but if what passed was completely watered down, that would be good, too."[24]

When the Waxman committee bill was finally released on May 21, 2009, it numbered almost 1,000 pages and was a nearly indecipherable piece of policy making that attempted to reconcile vast and conflicting special interests through allowances and offsets and other enticements. Groups such as Friends of the Earth and Greenpeace came out against the bill, arguing that it had lost the narrative of the urgent need for the reduction in carbon emissions. "It was really a political bill; it wasn't a science bill," said John Passacantando, former executive director of Greenpeace USA. "It wasn't a bill that was going to address atmospheric CO_2. It was, how are we going to buy off the coal industry first because it's a huge player in the Democratic Party."[25]

Despite these divisions, on June 26, the Waxman-Markey bill passed in the House, marking the first time that comprehensive carbon-cap legislation passed one of the congressional chambers in a full vote. The close tally—219 to 212, with only eight Republicans voting for the bill—reflected just how bruising a vote it was for members, many of whom later said they had taken the toughest vote of their careers.[26]

The backlash against the House vote began immediately, with the most

Climate demonstrators in Washington, D.C. on March 3, 2009.

damning reaction coming from the right-wing Tea Party movement, which labeled the bill as "cap and tax" and lambasted lawmakers who voted for it as "cap and traitors." Tea Partiers had already rallied a powerful base of grassroots activists to protest Obama's economic stimulus bill and to support an anti-regulation and ultra-free-market agenda. The movement's grassroots populism was accompanied by big-money advocacy from such sources as the Koch brothers, the billionaire conservatives behind energy conglomerate Koch Industries who had long fought action on climate change. According to Greenpeace, the Koch brothers have given more than $61 million to "climate-denial front groups" since 1997, with the majority of the funds (nearly $38 million) given between 2005 and 2010.[27]

Despite the victory in the House, passage of a Senate bill proved difficult from the start. The bill needed every one of the 58 Democrats in office at the time, but not all the Democrats supported cap and trade. Following a failed legislative attempt by Senator Barbara Boxer (Democrat from California), a coalition of three senators—John Kerry (Democrat from Massachusetts), Lindsey Graham (Republican from South Carolina), and Joe Lieberman (Independent from Connecticut)—began working on a "grand bargain" for an emissions cap. Yet despite huge concessions to industry interests—including increased production of natural gas, nuclear power, and offshore oil drilling—the bill never gained enough momentum to be taken to a floor vote in the Senate.[28]

A competing bipartisan bill in December 2009 by Senators Maria Cantwell (Democrat from Washington) and Susan Collins (Republican from Maine) called the Carbon Limits and Energy for America's Renewal (CLEAR) Act similarly failed to gain traction. The CLEAR Act called for a "cap-and-dividend" strategy of auctioning 100 percent of the pollution permits under a carbon cap and pledged long-term carbon reductions similar to those of the Waxman-Markey bill. Unlike the Waxman-Markey bill, however, it offered Americans an average $1,100 annually to a family of four between 2012 and 2030 to cover the anticipated increased energy costs from a carbon cap. Even if its policy promises might have resonated with much of the American public, most of the big greens were committed to cap and trade and treated the legislation as a distraction.[29]

By spring of 2010, prospects for a Senate bill finally unraveled. On April 20, an explosion on a BP oil rig in the Gulf of Mexico led to a massive oil spill. Any chance that the disaster might have created support for climate legislation was offset by the fact that the Kerry-Graham-Lieberman bill had called for large-scale expansion of offshore drilling. Two days after the BP explosion, Senate Majority Leader Harry Reid (Democrat from Nevada), whom the green groups had hoped would help push the bill to the Senate floor, announced that he was placing climate change on the back burner in favor of immigration reform. Despite several additional attempts to pass a climate bill in subsequent months, Senate Democrats announced on July 22 that they were abandoning efforts to pursue climate change legislation before the summer recess. Reid's adviser, Chris Miller, said that while Reid was supportive of passing climate legislation, the green groups simply "didn't have a Senate strategy" that made passage of a bill realistic.[30]

The Grassroots Alternative

Marshall Ganz, a veteran grassroots organizer who worked on Obama's 2008 presidential campaign, said that real societal change "almost never comes from an insider deal." A key to Obama's 2008 victory, said Ganz, now a lecturer at Harvard University, was the strong local, state, and national leadership of Obama's more than 2,500 field directors and organizers. Civic organizations such as the green groups have become what Ganz describes as "bodiless heads"—professionally staffed, Washington-based organizations that are largely disconnected from the public they purport to represent. "To think that a deep reform of our energy policies was going to happen because somehow it was going to be negotiated in D.C., it was just ahistorical," he said.[31]

And yet, pursuing an inside game is precisely the path that the green groups chose. Had they tapped into existing mass mobilization efforts, they might have formed valuable alliances with groups such as 1Sky, a network of smaller environmental organizations that was organized well in advance of the congressional legislative battles. As early as 2007, 1Sky, which championed a strong carbon cap, had built up a grassroots base of youth, labor, and faith-based groups as well as some of the strongest regional environmental organizations in the country. Unlike the green groups, 1Sky built broad support, deploying some 2,300 field volunteers across 29 states. Gillian Caldwell, 1Sky's former campaign director, told *National Geographic News* that the climate campaign suffered from "a chronic and historic underinvestment in grassroots mobilizing."[32]

This most recent lack of investment in the grassroots was certainly not limited to climate and energy issues. A February 2012 report published by the National Committee for Responsive Philanthropy, *Cultivating the Grass-*

roots, found that between 2007 and 2009, just 15 percent of all environmental grant dollars benefited marginalized communities and 11 percent went to social justice issues—two investment areas that the report's authors identified as critical to cultivating grassroots support.[33]

Part of the reluctance among green groups to make such an investment is because such on-the-ground work is resource and time intensive, requiring commitments that go against the current trend of funding allocations in one-, two-, and three-year cycles with short-term deliverables. As Maggie Fox, president and CEO of the Climate Reality Project, formerly the Alliance for Climate Protection, put it: "Funders don't do grassroots."[34]

Although the green groups did launch a more localized media and field campaign after passage of the House bill—Clean Energy Works (CEW)—the effort came late in the game. CEW's director, Paul Tewes, a veteran Democratic operative, deployed some 200 individuals in more than two-dozen swing states such as Arkansas and Ohio to generate grassroots support for climate action and to develop intelligence on the senators and their staffs. In an effort to address the most pressing concerns of voters, CEW pushed two chief benefits of cap-and-trade legislation: better national security through energy independence and the creation of "green jobs."

But while CEW claimed that 1.9 million new jobs would result from climate legislation, even those leading the campaign recognized that the figure was merely "a number cobbled together from a number of reports," said David Di Martino, former CEW communication director. In truth, despite the fact that the White House had been clear that green jobs was an important message, the green groups never believed they were the right messengers. "We're not about job creation," NRDC president Frances Beinecke said.[35]

The lack of grassroots organization around climate is in sharp contrast to the 2010 passage of health care reform. In 2008, health care reform advocates faced similarly strong and well-funded opposition as the environmental community did, but how they organized themselves was radically different. Leading the health care push in Congress was Health Care for America Now (HCAN), a reform coalition launched in 2008 that now includes 1,000 groups representing 30 million people in all 50 states. This primary lobbying vehicle was a broad-based organization of like-minded members, including public charities, advocacy groups, physicians and nurses, and labor unions. Rather than seeking to broker a compromise solution from the start as US-CAP did, the HCAN approach was more oppositional. Even so, underscoring the difficulty of pushing through such contentious legislation, the health care bill that passed in 2010 was an enormous departure from the principles set out by HCAN.[36]

The failure of the cap-and-trade campaign, which did not have the same grassroots support as health care reform, has left the environmental community even further away from passing comprehensive climate legislation than when Barack Obama first came into office. If anything, opponents of the cap-and-trade bill have been emboldened by its failure and have mounted an assault on the EPA's scope of authority—particularly its ability to regulate greenhouse gases under the Clean Air Act. In both houses of Congress, dozens of bills have circulated to weaken the 44-year-old law. Even as 2012 went on record as the hottest year in the United States, President Obama faced a tough re-election campaign against Republican candidates who labeled climate change as "manufactured science" and a "hoax." He shelved plans to tighten Bush-era ozone standards and instead advocated for "the importance of reducing regulatory burdens and regulatory uncertainty, particularly as our economy continues to recover."[37]

The cap-and-trade defeat in 2010 was so profound that it is unclear when another attempt at passing federal legislation can be made. At the beginning of his second term, President Obama called on Congress to "pursue a bipartisan, market-based solution to climate change." In the absence of Congressional action, Obama has vowed to use his executive powers to act on climate. Notably, his administration has proposed the nation's first federal limits on power plant carbon emissions. But among green groups, there has been no apparent marshaling of resources around a different approach to climate policy, such as, for example, a carbon tax, which in recent years has been increasingly supported by politicians, economists, and think tanks along the political spectrum for promising to drastically cut the federal deficit.[38]

Recently, green groups have shown signs that they are trying to engage the public more, mobilizing successfully, for example, in defense of the EPA's authority to regulate greenhouse gases and other air pollutants through partnerships with groups such as the American Lung Association and the League of Women Voters. But it remains unclear whether the big greens will be able to build mass demand for a national climate policy, or even whether they will decide it is in their interest to do so. The most visible grassroots mobilization these days is being spearheaded not by organizations like EDF or NRDC, but by groups such as 350.org, which has successfully rallied students from more than 300 colleges and universities in a nationwide fossil fuel divestment campaign and thousands more Americans to protest the controversial expansion of the Keystone XL pipeline.[39]

"Democratic mobilization becomes the norm when would-be leaders can achieve power and influence only by drawing others into movements, associations, and political battles," Harvard sociologist Theda Skocpol writes in *Diminished Democracy*, her study of civic engagement in American life. This incentive to mobilize was largely absent in the green groups' campaign for climate legislation. Their fundamental assumption was that success lay in negotiating with industry and lawmakers directly, and not in building grassroots support.[40]

This reasoning is, of course, not without some merit. A real transformation has taken place in the civic landscape over the past four decades, Skocpol notes, from the days when politicians won office in closely fought, high-turnout elections, and American civic life was characterized by participation in far more local and community-based groups. The focus today on Washington-based advocacy and lobbying is reflected in the expansion of congressional staffers who serve as the primary conduit to elected officials— the number of these staffers has risen from 6,255 in 1960, to 10,739 in 1970, to about 20,000 in 1990. By 2000, the number was 24,000.[41]

The composition of the national green groups today, with their professional staffs and their Washington focus, reflects this shift. But given that the green groups are likely to remain vastly outspent by industry lobby groups that oppose their efforts, future campaigns will run into the same obstacles as in this most recent push for climate legislation. Tapping into the grassroots base and learning how to mobilize the public may be the only way to balance the scales. It was, after all, the rise in the public's environmental consciousness in the 1960s that led to the first Earth Day in 1970 and gave a mandate and a constituency to EDF, NRDC, and the Sierra Club, which then leveraged this energy to push for reforms.

Whatever policy approach is embraced, the path to meaningful action will require a fundamental paradigm shift. Climate is the defining issue of our generation. Yet solving this problem requires confronting market capitalist forces that are considered fundamental to the American way of life. As writer Naomi Klein astutely points out in her essay "Capitalism vs. Climate," in *The Nation*, what climate deniers understand (and the big green groups do not) is that lowering global carbon emissions to safe levels will be achieved "only by radically reordering our economic and political systems in ways antithetical to their 'free market' belief system." In this sense, writes Klein, the climate deniers have a firmer grasp of the high stakes at the core of the climate debate than "professional environmentalists" who "paint a picture of global warming Armageddon, then assure us that we can avert catastrophe by buying 'green' products and creating clever markets in pollution."[42]

In 1995, the journalist Mark Dowie observed in *Losing Ground* that for too long, mainstream environmental advocacy in the United States has taken the form of a "polite revolution," one that has been marked from the start by "polite activism" that favors an elitist and insider approach rather than aggressive grassroots and coalitional forms of activism. The failure of the legislative effort during President Obama's first term is perhaps the most definitive evidence to date that climate change will not be resolved through *politesse*.[43]

CHAPTER 12

China's Environmental Governance Challenge

Sam Geall and Isabel Hilton

On posters and banners across China's cities, the new leadership has made "Ecological Civilization" and "Beautiful China" two of its most prominent slogans. But underlying these buzzwords is a complex, unenviable, and worsening problem.

China's environmental and climate governance is at a crisis point. While China attempts to transition to a more sustainable model of development—a difficult enough process for one-fifth of the world's population—legacy political structures and associated powerful interest groups have made necessary reforms all the more difficult by restricting, rather than harnessing, the potential for citizen participation in environmental protection.

As the 2012 review from China's Ministry of Environmental Protection illustrates, creating a "Beautiful China" will be no easy task. In China's countryside, the environmental situation is "grim." The cities and waterways are not much better: in 198 cities inspected in 2012, more than 57 percent of the groundwater was rated "bad" or "extremely bad," and more than 30 percent of the country's major rivers were "polluted" or "seriously polluted." The air in 86 out of 113 key cities did not reach air quality standards. A recent study in *The Lancet* suggests that in 2010 alone, air pollution in China caused some 1.2 million premature deaths.[1]

China recently surpassed the United States to become the world's leading emitter of carbon dioxide (CO_2) by volume, accounting for 29 percent of global CO_2 emissions in 2012. In the same year, China's average CO_2 emissions per person increased by 9 percent, to 7.2 tons; this puts China's per capita emissions roughly on a level with the European Union's, according to the Netherlands Environmental Assessment Agency.[2]

For most people in China, such dire assessments will come as little surprise: the visible effects of pollution are everywhere. In early 2012, heavy smog blanketed more than 1 million square kilometers of China for several days. More recently, in October 2013, record-setting levels of smog effec-

Sam Geall is a research fellow at SPRU (Science and Technology Policy Research) at the University of Sussex in the U.K. and the executive editor of chinadialogue.net. **Isabel Hilton** is a journalist and broadcaster and the editor and CEO of chinadialogue.net.

tively shut down the major northeastern city of Harbin. According to a Pew Research Center survey, Chinese citizens' concerns about the environment rose sharply in 2013: 47 percent considered air pollution a "very big problem," up from 36 percent in 2012.[3]

Chinese government officials have stated that pollution now may be the country's single greatest cause of social unrest. Chen Jiping, formerly of the Chinese Communist Party's Committee of Political and Legislative Affairs, claimed in 2013 that the country sees an estimated 30,000 to 50,000 "mass incidents," or protests, every year. Of these, Chen said, the "major reason… is the environment, and everyone cares about it now." Other studies indicate that the frequency of environmentally related social incidents has been increasing by 30 percent every year. As Chen put it, "If you want to build a plant, and if the plant may cause cancer, how can people remain calm?"[4]

In July 2013, local authorities in southern China's Guangdong province bowed to this rising discontent when they cancelled the construction of a $6 billion uranium processing plant after hundreds of protesters took to the streets, having organized the demonstrations through social media and online messaging services. The city government continued to defend the project until the last moment, finally issuing on its website a simple one-line statement: "To respect people's desire, the Heshan government will not propose the project."[5]

The protests, and the local government's last-minute turnaround, are phenomena that increasingly worry senior government officials. Over the past several years, a succession of so-called not-in-my-backyard protests have opposed large industrial facilities and infrastructure around the country. The first such major uprising, in 2007, focused on the proposed construction of a petrochemical plant that was manufacturing paraxylene, or PX, in Xiamen in southeastern China.[6]

Since then, waves of social unrest have halted many more projects: a second PX plant in Dalian in northeastern China; a copper and molybdenum refinery in Shifang, in the west; and incinerators in Panyu (Guangzhou province) and Xierqi (Beijing)—to name only a few. The specter of urban discontent, amplified by the growth of new media and mobile computing, looms large for China's decision makers—as does the potential for such opposition to derail economic development plans and trigger even greater social unrest if economic growth were to falter.

China now has 591 million Internet users and more than 460 million mobile Internet users, according to the China Internet Network Information Center. Sina Weibo, the country's largest micro-blogging service, has more than 500 million registered users. More than ever before in the history of the People's Republic, news and opinions can be shared among the public with ease—and the environment has become a key issue of concern. In effect, new media have given voice to a generation of citizens, many of whom are becoming economically enfranchised but are frustrated by their lack of a meaningful political stake in planning and other decisions that will affect their and their children's health.[7]

Tang Hao, an academic at South China Normal University, summarized the situation in a typically insightful fashion, noting that in China, "pleasant living environments are getting harder to find—and scarcity leads to competition and conflict." But, since the country has no mechanisms in place for managing such competition, "the outcome is unruly conflict," Tang wrote.[8]

Top-down Directives for Ecological Civilization

China's rhetorical push for an "Ecological Civilization" has accompanied an ambitious raft of top-down environmental targets, regulations, and policies. These include strategies launched under the country's 12th Five-Year Plan (2011–2015), which has enshrined sustainable development as a core state policy. Among other measures, the plan includes significant investments in low-carbon energy technologies, policies to support "strategic emerging industries" (including electric cars and energy conservation), and a nationwide target for reducing carbon intensity—carbon dioxide emitted per unit of gross domestic product (GDP)—of 17 percent.[9]

The plan also includes reduction targets of 16 percent for energy intensity, 8 percent for sulfur dioxide and chemical oxygen demand (a measure of water pollution), and 10 percent for ammonia nitrogen and nitrogen oxides. It establishes a 30 percent reduction target for water intensity—water consumed per unit of industrial added value—and pledges to boost forest cover to 21.7 percent and increase forest stock by 600 million cubic meters.

A coal-fired power plant and industrial area near downtown Yangzhou.

Furthermore, the government has now designated 13 regions as "low-carbon economy" pilot zones, and in August 2013 it launched a smart-city program with nine pilot cities. Also in August, the environment ministry took the unusual step of halting new projects for two major state-run oil companies after they failed to meet pollution targets.[10]

Despite the laudable ambition of such moves from the top, however, China's authoritarian structures do not always facilitate rapid and effective policy implementation, as is commonly perceived. In reality, power in the People's Republic is highly negotiated; academic observers often refer to China's system as "fragmented authoritarianism." Policies, laws, and regulations are not only weakened through protracted bargaining among bureaucratic elites, but also frequently ignored further down the system.[11]

One environmental law expert, Peking University professor Wang Jin, argued memorably that China's "green laws are useless." Although China has many environmental laws on the books, Wang observed, their enforcement provisions are often weak, and the legal system is underdeveloped and hobbled by political interference. Chinese laws are often vague and are more akin to policy statements; many "encourage" rather than "require" specific steps to be taken. According to academic Alex Wang, this is well understood by Chinese environmental officials, who have openly acknowledged that such weaknesses result from compromises in the legislative process—compromises driven by concerns about limiting China's economic growth.[12]

China's phenomenal growth over the past three decades was unleashed in large part through the considerable devolution of power from the center, which spurred economic competition among regional government chiefs. But a notable cost of this arrangement has been an ecological race to the bottom, where collusive alliances of money and power at the local level commonly trump environmental regulations. Significantly, China's local environmental protection bureaus are funded not by the central government's Ministry of Environmental Protection, but instead by the very same local officials they are tasked with regulating.

Prominent green projects launched from the center often have turned out to be less impressive than the rhetoric accompanying them. Jiang Kejun, with the Energy Research Institute, an influential government think tank, said in 2010 that most of China's "low-carbon" city projects were not "genuine," and that many of these cities were still very much on high-carbon development pathways. Without clear and transparent regulations or effective systems for political implementation, the cities had simply "all piled in to become 'low-carbon cities' and it's been disastrous," Jiang said.[13]

The opacity of decision making and the restricted public access to implementation mechanisms adds to the difficulty of uncovering such problems, and indeed of predicting whether any given policy will be effective. In 2010, when the environmental news site chinadialogue.net commissioned an investigation into environment and health in Dongguan, a manufacturing hub in southern China, the research was made difficult by a culture of official secrecy. Researchers' requests for interviews with scientists and environmental and public health officials were constantly refused. In some cases, academics initially agreed to researchers' requests for interviews, but later were told by government officials not to speak to them. Even the proceedings of public academic conferences were deemed "confidential."[14]

Such experiences are familiar to Chinese journalists, whose ability to conduct investigations is regularly curtailed by censorship and obfuscation. In the context of rising environmental concern from networked citizens, Chinese authorities have regularly extended such censorious approaches into the realm of the Internet as well. Terms like *sanbu*, or "stroll"—a euphemism commonly used by citizens to describe a street demonstration—are often scrubbed from websites when environmental protests are expected to occur. Truthful information leaked by whistleblowers, such as information about an oil well blow-out in the Bohai Gulf in 2010, is often initially suppressed as dangerous "rumor."[15]

In September 2013, Dong Liangjie, an "environmental expert," was arrested as part of a nationwide "anti-rumor" crackdown. The cofounder of a water-purifier company with more than 3 million followers on the microblogging site Sina Weibo, he had frequently commented on environmental issues, but police said that many of his posts contained sensational or false information that exaggerated the problem of environmental pollution in China.[16]

Effective environmental governance in China is hampered further by a weak and restricted civil society. Some within China's fragmented governmental system have actively encouraged the growth of civil society, in part to help provide necessary services, such as elder care, that the post-socialist society has increasingly struggled to provide. But another reason for the push has been to supervise the implementation of environmental laws and regulations at the local level.

By 2011, China was home to an estimated 449,000 legally registered civil society organizations, many of them environmental groups. Many more groups, possibly as many as 3 million, operate as unregistered organizations, having failed to meet the requirements of a highly restrictive registration process (which requires, for example, that every independent group find a government sponsor). These groups exist in a gray zone, with no protection from prosecution or other official sanctions.[17]

China has also introduced laws and regulations that provide for public participation in environmental decision making, but, as with other environmental laws, the existence of such measures on the books is no guarantee of their effective implementation. China's Environmental Impact Assessment (EIA) Law and Administrative Licensing Law require the government to solicit public opinion on new projects. Yet even when these laws are enforced, participation is not invited at the early, scoping stage of a project when it could be used to make more informed and sustainable decisions. It is sought only after a project design has been finalized and an EIA has been completed, just before the EIA is submitted for official approval.[18]

Furthermore, the full EIA will not be disclosed for the public to read. In 2008, China adopted government transparency regulations, which led to the creation of a specific decree on the release of environmental information. This not only requires the proactive disclosure of certain types of environmental data, but also allows citizens to request information from government. But these requests are still commonly rejected, and more sensitive data—not only EIAs, but also, for example, information on the disposal of hazardous waste—is almost impossible to obtain.[19]

Because transparency and public participation in environmental decision making are so often found to be non-existent or ineffective, levels of public trust are low. As a result, as Tang Hao suggested, in the absence of effective channels for public participation, citizens' concerns frequently find their outlet in protest.

Conflicts such as the one over uranium processing in Guangdong point to the likelihood that China's attempts to meet its climate goals may clash increasingly with other ecological and social concerns. The 12th Five-Year Plan, while incorporating concerns about the environment and climate change, also promises a kickstart for China's nuclear industry—a move that is dubbed "Great Leap Forward" thinking by critics, such as prominent physicist He Zuoxiu, who fears the proposed boost is rash and unsafe. China plans a fourfold increase in its nuclear capacity, to at least 58 gigawatts, by 2020. The country currently has 17 nuclear power reactors in operation but another 30 being built, and more are about to start construction.[20]

Perhaps equally significantly, the 12th Five-Year Plan seems set to restart the push for energy from large hydropower on the country's southwestern rivers. Opposition to such projects gave birth to many of the current generation of China's environmental nongovernmental organizations (NGOs), which achieved an early significant victory in 2004 when they halted a cascade of dams on the Nu (Salween) River, Asia's longest undammed river.[21]

Now, the target to boost renewable energy to a 15 percent share of China's primary energy consumption by 2015 appears to depend on giving the green light to such stalled projects. The plan promises an extra 120 gigawatts of new hydropower, equivalent to more than one new Three Gorges Dam every year over the five years and, according to the advocacy group International Rivers, more than any other country has built in its entire history. This is no small worry, not only for those in China concerned about large-scale resettlement, possible damage to fisheries and biodiversity, and increased seismicity, but also for neighboring countries such as Myanmar, Thailand, and India, which are concerned about the possible downstream impacts on communities and ecosystems.[22]

China's proliferating social and environmental conflicts are unlikely to reach consensus any time soon. Instead, the challenge is for government to institutionalize greater transparency and forms of public participation in environmental decision making that can not only benefit green development, but also help to address a deepening social conflict that is being exacerbated by repressive policy responses.

Grassroots Hopes for a Beautiful China

Still, there are some hopeful glimmers in the smog, including a flourishing of new initiatives by Chinese green NGOs, journalists, and networked citizens, sometimes in coordination with enterprising officials who have recognized the need for more open and responsive government approaches to sustainability.

An unusually smoggy few weeks in Beijing in late 2011 saw aircraft grounded and roads closed as thick haze obscured all but the lowest buildings. Networked citizens in northern Chinese cities became concerned about not only the polluted air, but also the secrecy around official reporting of air quality data. Every year since 1998, when public reporting of air quality began, the Beijing government had increased the number of annual "blue sky days." This measure, based on the city's air pollution in-

Fredrik Rubensson

Smog in Harbin, December 2012.

dex, did not match people's visual perceptions of deteriorating air quality, or the accounts of online visual diarists, such as bloggers Lu Weiwei and Fan Tao, whose photographs attested to the worsening conditions.[23]

Nor did the measurement take into account airborne concentrations of PM2.5, fine particulate matter with a diameter of 2.5 micrometers or less that penetrates deep into the lungs. This data was being collected and shared hourly, not by the authorities, but, in a political plot twist, by the U.S. Embassy in Beijing, on its Twitter account @BeijingAir. Journalists and researchers compared the datasets and began to challenge the government data, exposing a yawning gulf between official and unofficial narratives on the severity of the pollution. An online "storm" of citizen complaints on microblogs called for the release of real-time information about airborne concentrations of PM2.5. One online poll, started by property developer Pan Shiyi, saw tens of thousands of signatories call for the government to release more accurate measurements.[24]

Innovative, citizen-science efforts sprang up as well. A project called

FLOAT Beijing attached tiny Bluetooth-enabled pollution sensors onto kites, traditionally flown by hobbyists in the capital, making an arresting art piece that also created a dynamic, air pollution dataset, available for free online. The environmental NGO Green Beagle helped organize residents to use home testing kits and post their own air quality readings online. Encouragingly, the Beijing government heard these calls for greater transparency, and in January 2012, it began releasing PM2.5 data. This led to some 73 more cities releasing real-time air quality information. Even the state news agency Xinhua praised the "stirring campaign" from citizens and the "satisfying response" from policy makers.[25]

These campaigns build on the efforts of pioneers like Ma Jun, a former investigative journalist for the *South China Morning Post* who founded the Institute of Public and Environmental Affairs (IPE) in Beijing. IPE collects publicly available information to build maps of environmental data, including on air and water pollution as well as the levels of data transparency in different cities. These data are being used by citizens to locate the sources of pollution near them, by residents' groups to challenge the transparency of their local authorities, by businesses to better understand the environmental impacts of their supply chains, and by journalists to conduct investigations.[26]

Many campaigns help to challenge the collusion between officials and polluters at the local level. In 2013, with characteristic humor, Chinese microbloggers began asking government officials to swim in their local polluted rivers. One businessman in eastern China offered his city's environment chief more than $30,000 to swim for 20 minutes in a local waterway—an offer that he illustrated with pictures of the foul river teeming with rubbish. The official declined.[27]

One of China's best hopes is that it might harness these emerging forms of public participation and open information, particularly in the new media context, to help address its environmental woes. For several years, China's Ministry of Environmental Protection has operated a hotline for citizens to phone in tip-offs about pollution incidents and environmental infractions; however, awareness and uptake has been low. More recently, China's environment authorities have begun using microblog accounts at different levels, in many cases to engage in two-way communication and to listen to public opinion.

The Environmental Protection Bureau in Chongqing, a large municipality of some 29 million people in southwestern China, has a microblog account for each of its 40 districts. The blogs are not only used to speedily disseminate environmental information (such as on air quality), but also intended to create greater transparency and improved responsiveness to public opinion and citizen complaints.

Conclusion

Whether the issue is water pollution or climate change, China has ambitious environmental targets, laws, and regulations—and there is political will at the center. But in the absence of strong citizen oversight and public participation, supported by greater government transparency, implementation will continue to be thwarted by structural problems, including collusion between local officials and polluters.

China will need to navigate new forms of grassroots public engagement if it is to address such structural issues and to improve environmental governance during its complex and ambitious transition to a cleaner, low-carbon economy. It will also need to contend with multiple, proliferating uncertainties—not least social ones—which will require citizen perspectives to be taken into account if frequent conflict is to be avoided.

Navigating these waters will require a commitment to full and early public participation in environmental decision making, which has been hampered by inadequate implementation of existing government laws and regulations. In the coming years, China will need to embrace open channels, unfettered by censorship, for concerned citizens to protect themselves against the consequences of poor decisions—and to express their visions of an Ecological Civilization.

CHAPTER 13

Assessing the Outcomes of Rio+20

Maria Ivanova

Ecosystems and economies are intertwined, and international cooperation is critical to addressing cross-border threats to the integrity of habitats and biomes. Economic and political effects of national policy decisions can reverberate around the world within days. Simply put, sustainability cannot be achieved without integrating environment and development at the international level. This was recognized as early as the 1970s, when governments convened at the landmark Stockholm Conference on the Human Environment to create the architecture for global environmental governance, defining sustainability as an economy "in equilibrium with basic ecological support systems" and recognizing the confluence of environmental, economic, and social concerns.[1]

But 40 years later, global environmental, economic, and social problems have become more prominent, acute, and urgent. Consequently, governments and citizens are putting growing pressure on international institutions to deliver results as effectively, efficiently, equitably, and quickly as possible. In 2012, nearly 50,000 people representing governments, nongovernmental organizations (NGOs), businesses, and citizenry from all over the world assembled in Rio de Janeiro, Brazil, for the largest-ever global environmental summit—the United Nations (UN) Conference on Sustainable Development, also known as Rio+20—to review accomplishments and reinforce commitments.

Weary of empty political promises, analysts predicted the breakdown of Rio+20 months before it started and deemed it a "colossal failure of leadership and vision" immediately upon its conclusion. Greenpeace dubbed the event's 50-page outcome document, *The Future We Want*, "the longest suicide note in history." Upon careful examination, however, it is clear that while the conference did not create a collective global vision for a radically different world, its outcomes are nonetheless significant and will likely shape global governance in the immediate decades to come.[2]

Maria Ivanova is an assistant professor and codirector of the Center for Governance and Sustainability at the University of Massachusetts Boston. In 2013, she was appointed to the new UN Secretary-General's Scientific Advisory Board. She thanks Natalia Escobar-Pemberthy and Gabriela Bueno for their valuable research assistance.

Rio+20 resulted in important conceptual, institutional, and operational outcomes that will have a direct impact in the context of the post-2015 development agenda. Conceptually, the conference created a new narrative of sustainable development, overcoming some limitations while reinforcing others. It also rekindled countries' political commitment to sustainable development, at least in rhetoric. Institutionally, the event created a new platform for integrating economic prosperity, social inclusion, and environmental stewardship through the reorganization of relevant UN structures. Operationally, it stimulated a slew of voluntary commitments from governments and other actors, with pledges exceeding $513 billion. Countries also agreed to create a set of Sustainable Development Goals that will guide action in the coming decades.

Conceptual Outcomes:
The Evolving Sustainable Development Narrative

Global narratives about the environment and sustainable development play an important role in shaping country priorities at the national level. Before the groundbreaking Stockholm Conference in 1972, for example, environmental ministries existed in only a handful of countries. The creation that year of the UN Environment Programme (UNEP) as the anchor institution for the global environment provided the conceptual vision and the support mechanism that enabled countries around the world to establish and equip such ministries.

Importantly, the prevailing view at the time saw protection of the environment as a precondition for development. Even though development was a clear priority for many countries, especially those that had recently gained independence, governments agreed that "the protection and improvement of the human environment is a major issue which affects the well-being of peoples and economic development throughout the world; it is the urgent desire of the peoples of the whole world and the duty of all governments." Over the ensuing decades, however, the focus shifted from environment as a precondition for development to development as a precondition for environmental protection.[3]

The Rio Earth Summit in 1992 confirmed sustainable development as the new aspiration, moving the needle of political priorities to the development dimension. Subsequent international summits—the Millennium Summit in 2000 and the World Summit on Sustainable Development in 2002—shifted the focus further in the direction of development as a precursor to environmental protection. At Rio+20, in 2012, governments stated that "eradicating poverty [is] the greatest global challenge facing the world today and an indispensable requirement for sustainable development," rather than poverty alleviation being an outcome of sustainable development. They also committed, however, to a new set of Sustainable Development Goals, opening the door to a rethinking of priorities.[4]

The Sustainable Development Goals will likely enter into force at the end of the 15-year period of the Millennium Development Goals (MDGs), which governments adopted in 2000 and which have shaped the human development agenda in the UN system. The eight MDGs address multiple dimensions of human well-being—with the main goal to eradicate extreme poverty and hunger—and incorporate policy areas ranging from education and health to gender equality, environmental sustainability, and the creation of a global partnership. (See Table 13–1.) One of the eight goals (#7) is related to environment; however, because it is articulated separately from the rest and in very broad terms that are difficult to monitor and measure, it reinforces the false dichotomy of environment versus development rather than promoting an integrated, holistic approach to sustainable development.[5]

The MDGs illustrate the power of global goals to provide meaning, purpose, and guidance, which can then translate into political attention and action. By offering a structure to focus advocacy, spur motivation, and target investment, the MDGs have improved the ability of countries to meet many of the targets. For example, extreme poverty has been reduced across all regions, including sub-Saharan Africa; worldwide, the share of people living on less than $1.25 a day dropped from 47 percent in 1990 to 24 percent in 2008, reflecting improved economic conditions for some 800 million people. The share of people with access to improved sources of water increased from 76 percent in 1990 to 89 percent in 2010, achieving the MDG target of halving the proportion of people without sustainable access to safe drinking water.[6]

Yet the MDGs also highlight the challenges that global goals present. The narrow focus on a limited set of goals restricts attention to only a select few issues and might distort risk and investment preferences. For example, the main focus of the MDGs is on traditional socioeconomic development, and the goals do not explicitly recognize the interconnections among the three dimensions of sustainable development (economic, social, and environmental). The environmental goal, #7—to "ensure environmental sustainability"—is not only distinctly separate from the other goals, but it includes only three environmental issues as targets—biodiversity, water, and urbanization.

The MDGs also have become the overarching development strategy steering investment (through official development assistance or other funds) into sectors identified as important in these eight global goals. Other country priorities therefore might be neglected. Moreover, because the MDGs apply only to developing countries, they do not recognize the monetary and

Table 13–1. UN Millennium Development Goals and Targets

Goal	Target(s)
1. Eradicate extreme poverty and hunger	• Halve, between 1990 and 2015, the proportion of people whose income is less than $1 a day. • Achieve full and productive employment and decent work for all, including women and young people. • Halve, between 1990 and 2015, the proportion of people who suffer from hunger.
2. Achieve universal primary education	• Ensure that, by 2015, children everywhere, boys and girls alike, will be able to complete a full course of primary schooling.
3. Promote gender equality and empower women	• Eliminate gender disparity in primary and secondary education, preferably by 2005, and in all levels of education no later than 2015.
4. Reduce child mortality	• Reduce by two-thirds, between 1990 and 2015, the under-five mortality rate.
5. Improve maternal health	• Reduce by three-quarters the maternal mortality ratio. • Achieve universal access to reproductive health.
6. Combat HIV/AIDS, malaria, and other diseases	• Have halted by 2015 and begun to reverse the spread of HIV/AIDS. • Achieve, by 2010, universal access to treatment for HIV/AIDS for all those who need it. • Have halted by 2015 and begun to reverse the incidence of malaria and other major diseases.
7. Ensure environmental sustainability	• Integrate the principles of sustainable development into country policies and programmes and reverse the loss of environmental resources. • Reduce biodiversity loss, achieving, by 2010, a significant reduction in the rate of loss. • Halve, by 2015, the proportion of the population without sustainable access to safe drinking water and basic sanitation. • By 2020, to have achieved a significant improvement in the lives of at least 100 million slum dwellers.
8. Develop a global partnership for development	• Develop further an open, rule-based, predictable, non-discriminatory trading and financial system. • Address the special needs of least-developed countries. • Address the special needs of landlocked developing countries and small island developing states. • Deal comprehensively with the debt problems of developing countries. • In cooperation with pharmaceutical companies, provide access to affordable essential drugs in developing countries. • In cooperation with the private sector, make available benefits of new technologies, especially information and communications.

Source: See endnote 5.

moral responsibility of industrialized countries and offer a weak approach to addressing issues of social justice, equality, vulnerability, and exclusion.[7]

With the MDGs set to expire in 2015, the Rio+20 conference engaged governments in debates about the post-2015 development agenda. Governments reaffirmed their commitment to sustainable development as the overarching goal but, in a positive step, shifted to a more integrated vision of what this entails. The Rio+20 outcome document, *The Future We Want*, substituted the traditional definition of sustainable development as having three distinct "pillars"—environmental, economic, and social—with a new narrative of the three "dimensions" of sustainable development. This change recognizes the fluidity and interconnectedness among these aspects, and opens up opportunities for more integrative forms of governance.

Still, there are problems. Close analysis of the text of *The Future We Want* reveals that the environment has almost disappeared as an independent concept. The term "environment" (and its multiple variants) is mentioned 70 times in the 50-page document, and 21 of those mentions occur in the catch-all descriptor "social, economic, environmental." "Development," on the other hand, appears 635 times, 239 of those in the phrase "sustainable development." The environmental discourse is therefore absorbed by, rather than integrated into, the development narrative.

The "green economy," one of the framework themes of Rio+20, fueled expectations for a radical restructuring of the global political economy that would reconcile economic growth with planetary boundaries, account for natural capital, and ensure planetary stewardship. The concept, however, elicited criticism both from countries striving toward capitalism (which regarded the "green economy" mandate as a threat to their national development strategies) and from countries rejecting capitalism (which saw it as the commodification of nature). While *The Future We Want* mentions the green economy as one tool among many in the quest for sustainability, it also acknowledges the need to move beyond gross domestic product (GDP) as a measure of human well-being. Thus, ideas about new indicators of progress and prosperity gained ground and legitimacy.

Institutional Outcomes: Reforming the Institutions for Environment and Sustainable Development

The United Nations was created in 1945 without an environmental body. Almost 30 years later, in Stockholm, governments established UNEP as the anchor institution for the global environment, and, another 20 years later, they created the UN Commission on Sustainable Development. Ultimately, however, the need to reform the institutional architecture for environment and sustainable development became a political priority, as "ever-growing concern over sustainable development and the proliferation and fragmentation of environmental initiatives eroded the embracing mandate of UNEP for environmental governance." Institutional reform was one of two top agenda items at Rio+20 and one of the conference's most significant outcomes.[8]

Rio+20 concluded a 15-year reform effort that had contemplated the need to change UNEP's institutional status from being a subsidiary organ of the UN General Assembly to being a specialized agency. UN specialized agencies—such as the World Health Organization, the International Labour Organization, and the Food and Agriculture Organization—are autonomous bodies set up independently and linked to the United Nations through special agreements in accordance with Articles 57 and 63 of the UN Charter. They are established through the adoption and ratification of intergovernmental treaties, and membership is universal, meaning that any country can join. Specialized agencies do not receive any funding from the UN regular budget, and their budgets instead include mandatory financial contributions assessed according to a particular scale.

In contrast, subsidiary organs are created under Article 22 of the UN Charter to address emerging problems and issues in international economic, social, and humanitarian fields. They have various formal designations—programmes, funds, boards, committees, commissions—and governance structures. They are created through a UN General Assembly resolution, and membership is limited and geographically representative. Funding comes exclusively from voluntary contributions, although some subsidiary organs may receive a small portion of funding from the UN regular budget. Subsidiary organs work directly through the United Nations, which gives them access to UN administrative and security services as well as a direct relationship with other UN offices and subsidiary organs.[9]

Although governments decided at Rio+20 to retain UNEP's formal status as a subsidiary organ, they did create a new institutional structure that combines some key attributes of a specialized agency while preserving the flexibility and advantages of a subsidiary organ. This approach offers several key advantages with regard to membership, mandate, financing, and delivery of services to stakeholders:

First, changing UNEP's governance structure gave it greater formal authority. With the creation of a new governing body comprising all UN member states—the United Nations Environment Assembly—UNEP became the only subsidiary organ in the United Nations with universal membership. Although, legally, UNEP always had the authority to engage with the UN system, governments, and civil society on environmental issues, in practice it had not always marshaled the clout necessary to command political attention and financial support, due in part to a legal mismatch in membership. With a limited membership of 58 states, UNEP faced challenges in claiming authority over global environmental conventions related to climate, biodiversity, etc. whose legal bodies, the Conferences of the Parties, comprised nearly all UN member states. Expanding UNEP's membership was a logical, feasible, and potentially effective legal measure to upgrade UNEP's institutional structure and authority.

Second, preserving UNEP's status as a subsidiary body allowed it to access greater, and more predictable, resources from the UN regular budget. One of the main arguments for transforming UNEP into a specialized agency was that this would help bring greater stability and predictability of financial resources. Rio+20, however, resulted in an innovative use of an existing financing source to serve the same function. Affirming the need for "secure, stable, adequate and predictable financial resources for UNEP," the Rio+20 outcome document and subsequent General Assembly resolutions committed contributions from the UN regular budget to UNEP's core operational needs, in a manner that adequately reflects UNEP's administrative and management costs. Governments also acknowledged that the budgetary resources that UNEP receives should correspond to the scope of its program of work and pledged to increase their voluntary contributions.[10]

Third, the review of UNEP's functions and mandate led to a recognition of the need to expand UNEP's role. Governments recognized that UNEP's engagement on the ground needed to be expanded so that it can play a greater role in helping countries build capacity and implement environmental commitments. Through these reforms, UNEP's role in global environmental governance evolved from a primarily normative role to an implementation role, as countries requested more comprehensive on-the-ground programs and greater regional and sub-regional presence from UNEP.[11]

Fourth, UNEP was mandated to improve its delivery of a range of measures beneficial to diverse stakeholders. These measures, as outlined in the outcome document, included: promotion of a "strong science-policy interface" in order to allow for scientific input and assistance during global decision-making processes; dissemination of environmental information and raising of public awareness; delivery of capacity-building and technology access to developing countries; and engagement with nongovernmental actors (called "major groups and stakeholders" in the UN context) in a more effective, meaningful way.[12]

All of this was accomplished without a lengthy treaty negotiation process that would have been required for changing UNEP's status to a specialized agency. Although a fair assessment of the effectiveness of these reforms can be undertaken only in the future, the reinforcement of UNEP's role as the leading global authority for the environment, and the political legitimacy conferred by all member states, are indicators of an improved, revitalized institution.

In the sustainable development field, institutional reform resulted in the abolishment of the UN Commission on Sustainable Development (CSD). The CSD fell short in fulfilling its mandate to review national plans for sustainable development and to set an integrated agenda for the UN system, and it was unable to engage all UN agencies and bodies in considering environment and economic issues, as was envisioned at the 1992 Rio Summit. Analysts criticized the CSD as a "talk shop" that delivered few sustainable development outcomes. UN agencies and civil society observers noted that "the Commission progressively lost its lustre and its effectiveness" and that the CSD itself was unable to follow up and implement its own decisions.[13]

Ultimately, the CSD failed in its core mission of integrating the three dimensions of sustainable development and did not produce the effective and timely global responses that were necessary. Through the CSD, however, multi-stakeholder dialogues became accepted UN practice, as the annual two-week sessions in New York brought together government officials and numerous other stakeholders to deliberate on issues such as forests, energy, water, and oceans. Although these sessions attracted mostly environmental officials rather than the envisioned cross-section of development, trade, environment, agriculture, energy, and foreign affairs ministers, they created a culture of engagement with civil society. As some observers point out, "without the Commission, sustainable development would not be at the stage of maturity where it is today" and the CSD was "instrumental in launching initiatives and introducing new topics into the intergovernmental debates."[14]

At Rio+20, governments decided to replace the CSD with a High-Level Political Forum on Sustainable Development. The purpose of this new entity is to build on the work of the CSD and to follow up on the implementation of sustainable development. Starting in September 2013, the Forum aims to convene heads of state and government every four years at the UN General Assembly, as well as to convene ministers annually under the aegis of the UN Economic and Social Council (ECOSOC). The Forum's main goal is to provide political leadership for the integration of the three dimensions of sustainable development. To this end, it is intended to work with UN agencies to support implementation of the Sustainable Development Goals and to productively engage major groups and stakeholders.[15]

Three main innovations characterize the new High-Level Political Forum: universal membership, greater visibility, and improved accountability. The Forum involves heads of state and government of all countries in the design and approval of sustainable development policies across governance levels. Starting in 2016, the Forum will conduct regular reviews of implementation of sustainable development commitments taken by states and UN agencies. Given that it is a newly established institution, the Forum's effectiveness and relevance will become apparent only in the next few years as it demonstrates its ability to engage member states to take action nationally and sister UN entities to take more coherent action internationally. Ultimately, the Forum will be judged by its successes in reducing the UN system's current fragmentation of environmental governance and in avoiding duplication of effort. More importantly, the Forum will have the important task of turning the principle of sustainable development into an actionable, concrete, and specific policy agenda.

The close relationship between the High-Level Political Forum and ECOSOC is not an accident: ECOSOC is one of the main bodies in the UN system tasked with shaping the economic and social development agenda and coordinating the activities of numerous agencies and funds. Although ECOSOC's involvement in the environmental field has not been very clear, at Rio+20, governments committed to strengthening its role in coordinating social, economic, and environmental policies across different institutions, thereby making it an important environmental player. The end result of the strengthening process of ECOSOC and the role of the Forum in helping it to advance the sustainable development agenda is not yet clear, and cooperation with environmental institutions such as UNEP will be key to providing a more coherent set of objectives and policies in the future.[16]

Operational Outcomes: Voluntary Commitments and Sustainable Development Goals

At the operational level, Rio+20 had two main outcomes. First, countries, companies, and citizens articulated a series of voluntary commitments to promote action around sustainable development. UN Secretary-General Ban Ki-Moon described the negotiated intergovernmental agreements as the "foundation" and the voluntary commitments as the "bricks and cement" in the global governance architecture, emphasizing the importance of both. Second, governments committed to create a set of Sustainable Development Goals to "address and incorporate in a balanced way all three dimensions of sustainable development and their interlinkages." Governments did not articulate the specific set of goals during the conference, but they stated in the outcome document the broad principles for global goal development. The challenge is to connect the articulation of the SDGs with the vision for followup to the MDGs after they expire in 2015.

Inspired by the system of partnerships that emerged from the 2002 World Summit on Sustainable Development, Rio+20 introduced a new mechanism—voluntary commitments by governments, corporations, NGOs, and citizens—to encourage the implementation of sustainable development policies. The UN has begun to provide greater structure around this new set of unilateral pledges to action by launching the Sustainable Development Knowledge Platform and the Sustainable Development in Action Registry. Close to 700 commitments totaling more than $513 billion were made at Rio, and, as of September 1, 2013, the Action Registry included 1,412 voluntary commitments, partnerships, initiatives, and networks for sustainable development created since the 2002 World Summit.[17]

In a comprehensive report assessing progress on these commitments, the Natural Resources Defense Council notes that about 58 percent were made by the private sector and civil society, 30 percent by governments, and 12 percent by UN organizations—with a total value estimated at $637 billion, nearly 1 percent of annual global GDP. This sum encompasses investments pledged in projects that are both internal and external to the actors making the commitments. For example, Microsoft committed to becoming carbon neutral by the end of 2013, an outcome it claims to have achieved. Bank of America pledged $50 billion in investments in renewable energy, energy access, and energy efficiency projects over a decade. The timelines for the commitments vary, but the majority—51 percent of the 1,412 commitments—aim to deliver results by 2015 or earlier, reflecting the goal of meeting the MDGs. The target deadline for another 16 percent of the commitments is 2022, a decade after Rio+20.[18]

The highest number of voluntary commitments, close to 300, is in the area of education. This reflects the launch at Rio+20 of the UN's Higher Education Sustainability Initiative, which aims "to get institutions of higher education to commit to teach sustainable development concepts, encourage research on sustainable development issues, green their campuses, and support sustainability efforts in their communities." The second most popular area for voluntary commitments is the green economy, where primarily governments have made pledges to action. In terms of resources, the most significant Rio+20 commitment was for sustainable transportation, where eight multilateral development banks pledged $175 billion in loans and grants to developing countries over 10 years to enhance sustainable transport in urban areas. A year after this commitment, some $17 billion, or 10 percent of the pledge amount, had been approved for projects.[19]

Among the major challenges for voluntary commitments, both by governments and by other actors, are accountability and the assessment of results. The UN Department of Economic and Social Affairs has proposed a voluntary accountability framework based on three pillars: 1) annual reporting, 2) updated registry, and 3) third-party independent review. Functional institutional arrangements will be critical. Moreover, an engaging public discussion about the voluntary commitments will make them more visible, likely leading to greater pressure for regular reporting and reviews and, ultimately, to their fulfillment at various levels of governance.

In another key outcome, Rio+20 resolved to establish an intergovernmental process to define a new set of Sustainable Development Goals, taking into account basic human needs, environmental sustainability, social equity, and governance tools. In doing so, governments recognized the relevance of development goals as useful instruments to frame action toward sustainable development. Although they did not elaborate on the specific goals, they agreed on a set of general characteristics—specifically, that the SDGs be comprehensive, universal, limited in number, ambitious, and easy to communicate.[20]

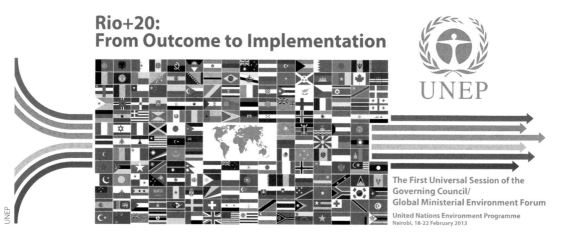

A major point of contention, however, was the mechanism for goal formulation, with governments debating the need for an expert-driven or a political, intergovernmental process. To implement this mandate, the UN General Assembly established a 30-member Open Working Group to articulate a proposal for the SDGs and guarantee the opportunity for international organizations, stakeholders, civil society, and UN agencies to provide input during the process. Governments participating in the Open Working Group and analysts have suggested multiple topics around which the goals could be constructed. The concurrence with the UN's Post-2015 Development Process, however, raises the challenge of integrating the two agendas. Ultimately, the global goal is "to end extreme poverty in all its forms in the context of sustainable development and to have in place the building blocks of sustained prosperity for all."[21]

In this context, the SDGs emphasize explicitly that such prosperity cannot be achieved without safeguarding the ability of the planet to maintain the conditions critical to human well-being. An innovative proposal from Colombia suggests that governments should focus on defining adequate targets and indicators across issues and subsequently cluster them to arrive at common goals. Some targets could appear in various goals, even if the overall number of goals is low. As governments negotiate the nature and number of goals, it is critical to set up systems for measuring progress as well as support for implementation. Concrete measurement strategies and mechanisms provide governments and international organizations with the necessary data and science-based information to evaluate progress and take corrective measures as required. (See Box 13–1.) Implementation support systems would ensure that countries develop adequate baselines for measurement and adequate methodologies for data gathering, create the necessary policies to integrate the goals into national planning and strategies, and provide the necessary financial and personnel backing.[22]

Importantly, obligations under the SDGs will be universal—they will extend to all countries, regardless of their level of development, unlike the MDGs, which apply only to developing countries. Contextualized, specific national and regional targets can be used to measure progress at the different levels, to complement the general approach of global goals. Ultimately, the definition and application of the new framework will require transparency, participation, and engagement from all groups.[23]

Conclusion

Implementing the outcome document of the Rio+20 conference, *The Future We Want*, presents different challenges for governments, international organizations, and other stakeholders. At the conceptual level, the acceptance of sustainable development as a core organizing principle in the UN system is the result of a long political process that started in the 1980s. At the institutional level, Rio+20 completed the years-long process of reforming the system of global environmental governance. Reform measures for UNEP and the new institutional architecture for sustainable development are now being implemented and will require regular and systematic monitoring, reporting, and assessment.

At the operational level, the United Nations system will face coordination issues if the SDGs and the post-2015 processes remain disconnected. Mechanisms to connect the two agendas are necessary in order to ensure a comprehensive approach to a global development agenda. Ultimately, the goal of these two processes is the same: the attainment of long-term human prosperity. In this context, the SDGs recognize explicitly that such prosperity cannot be achieved without safeguarding the ability of the planet to maintain the conditions critical to the well-being of humans and the other species with which we share the planet.

Box 13–1. A Policy Mechanism for Ensuring Sustainable Development: National Resource Sufficiency Evaluation

Recognition is growing globally that strong action is needed immediately to move toward environmental sustainability. A global high-level panel recently warned that, "We must act *now* to halt the alarming pace of climate change and environmental degradation, which pose unprecedented threats to humanity." But that sense of global urgency needs to be translated into global action. The gap between what political leaders want in terms of development (the "political mandate") and the resources that are realistically available to accommodate that development (the "reality mandate" often put forth by the scientific community) appears to be wide. Although most concrete efforts to promote sustainability have focused on technological evolution and resilience in the face of a changing environment, a strong case exists that bridging the gap will require absolute reductions in consumption and a reversal of population growth, or else measurable progress toward global sustainability may never occur.

The current debate on the United Nations' post-2015 development agenda, including the work of the Open Working Group (OWG) on the Sustainable Development Goals (SDGs), offers an opportunity to begin to close the gap between politics and the reality of the human predicament. The SDGs are expected to set goals and targets needed to facilitate and ensure progress in human development, while at the same time fostering a global transition to an "inclusive green economy" and "a sustainable century." The "S" (or sustainability) factor implicitly acknowledges the need to conduct humanity's future global development programs in balance with planetary limits. The OWG process, which is expected to be completed by September 2014, will play a key role in determining if the next human development agenda will represent economic development "as usual," or whether the SDGs will form

a new point of departure that leads to a more sustainable world.

Much depends on whether the proposed SDGs recognize the biophysical limitations to economic growth and the need for governance at all levels to consider the implications of such limits for efforts to eradicate poverty and reduce income inequality. Although it is implicitly understood (and sometimes even explicitly stated) that long-term human prosperity can be attained only by safeguarding environmental assets, the operational mechanism needed to accomplish this task—especially at the scale of the global socioecologic system—has not been proposed nor agreed upon.

One idea being offered to the OWG is Resource Sufficiency Evaluation (RSE), the use of established metrics to determine whether the current and projected demand for natural resources is sustainable. Scientifically based accounting methodologies such as life-cycle assessment (LCA) or input-output (I-O) modeling are already available to conduct resource sufficiency evaluations in a universally applicable manner. These methodologies, and the biophysical "balance sheets" that are generated, offer policy makers and the public a clearer understanding of ecological sustainability and what is needed to achieve it.

In contrast to the end of the twentieth century, natural resource scarcities and costs are now becoming an increasingly significant economic factor for most countries, and this significance will only grow as resource demands increase. By adopting RSE, countries can proactively address resource constraints and better plan for their economic future. Countries that understand their natural resource assets and limitations, and reduce their reliance on scarce resources, acquire a competitive edge in a now globalized world.

RSE provides an appropriate analytical

Box 13–1. continued

framework and policy response to the growing global imperative to better manage the balance between human activity and the natural resources required for long-term well-being. The Sustainable World Initiative, a nonprofit project associated with the Population Institute in Washington, D.C., is working with United Nations, governmental, and civil-society leaders to stimulate a discussion of RSE in the context of environmental governance. If that governance is ever to succeed in achieving true sustainability, it must begin with recognition of planetary limits—and efforts to reconcile them into economic development plans.

Countries will never know if they have enough resources to maintain human development—or can realistically expect these resources to be available externally—unless they first evaluate their resource demands and compare them to what is available. No one would think of driving a car or flying a plane without a fuel gauge. By the same token, policy makers at all levels of governance cannot adequately plan for the future without knowing whether they have the natural resources needed to realize their development agendas.

—Ed Barry
Director, Sustainable World Initiative,
www.swinitiative.org
Source: See endnote 22.

CHAPTER 14

How Local Governments Have Become a Factor in Global Sustainability

Monika Zimmermann

Growing population and urbanization are increasing the relevance of cities and local governments to the problems of sustainability. Half of the world's population now lives in cities, a share that is projected to increase to 75 percent within the next 30 years or so. A city such as Mumbai, India, governs more people than any of the 150 smallest United Nations (UN) member states. This intensifying urbanization will require the construction, within the next 40 years, of urban capacity, buildings, and infrastructure equal to all that has been built in the last 4,000 years.[1]

Countless cities will be affected strongly by climate change, while remaining obligated to provide basic human services and secure the feeding of their populations. Yet at the same time, their formal powers, portfolios, and resources are relatively narrow. Even countries with explicit decentralization processes shift far more duties than opportunities to their local governments.

The emerging importance of local governance raises some critical questions: Can cities, towns, counties, metropolitan areas, and other local units govern themselves and their social and economic development in ways that maintain, save, and improve the natural resources and ecosystems that enable all development? Can local governments influence national and global governance toward sustainability? And even more relevant: Do their actions result in global improvements?

Understanding how local governments have become a factor—maybe the key factor—in global sustainability efforts in recent years can help clarify the current discussions around climate governance, the UN's Sustainable Development Goals and the role of cities in achieving them, and urban sustainability in general.

Locally Global

"Local government" refers to public administrative units—the lowest tiers of government—and includes provinces, regions, departments, counties,

Monika Zimmermann is Deputy Secretary General at ICLEI–Local Governments for Sustainability.

prefectures, districts, cities, townships, towns, boroughs, parishes, municipalities, shires, and villages. Their leadership is locally elected or appointed by higher administrative authorities.

In recent years, common concerns about environmental protection and sustainable development have driven local governments to cooperate more closely across countries. This increase in international cooperation is related in large part to local governments' involvement in the global sustainability debate. The global role of local governments dates back only two decades or so—a measure of how the world has changed.

Several globally relevant organizations, such as ICLEI–Local Governments for Sustainability and United Cities and Local Governments (UCLG), are open to all interested local governments and are involved in global advocacy processes and improvements in global and local governance. Other groups, such as Metropolis, offer participation or membership to selected cities according to size (e.g., number of inhabitants). Networks focused on thematic and regional city cooperation include CITYNET in Asia, Mercociudades in Latin America, Eurocities in Europe, and Climate Alliance and the C40 Cities Climate Leadership Group in the field of climate protection.[2]

The growing movement of local governments is paralleled by a similar phenomenon among regional governments, some of which have also formed global organizations based on similar visions and concerns. Among the best known are nrg4SD (Network of Regional Governments for Sustainable Development) and R20 (Regions of Climate Action).[3]

Local and regional governments are often referred to as "subnational governments," but in many cases their character is mixed. Examples are the German city states, like Berlin, or highly urbanized states like São Paulo in Brazil. Within global geopolitical processes such as transnational negotiations and agreements, local and regional governments often cooperate closely and perceive themselves as counterparts to national governments and to the UN system. In part this is a necessity, as the multilateral system of cooperation among sovereign nations, the UN system, and related mechanisms do not define a role for local governments; they are instead perceived as part of, and represented through, their respective countries.

Defining a role for local governments within the debate on global governance for sustainable development is a challenge that many countries hesitate to take up. Meanwhile, many local governments are concerned about the increasingly discussed failure of the current mechanisms of global governance, especially (but not solely) the UN structure. (One 2011 proposal from ICLEI suggested convening a group of "United Actors" in parallel to the United Nations for an upcoming climate change conference, with local government taking a lead convening role and anchoring the United Actors in a future participatory framework for global environmental governance.) In general, limited progress at the level of national governments suggests both the need and opportunity for a much stronger role for cities and towns.[4]

A Growing International Role

Local and regional governments form a strong coalition of the concerned and are by no means simply subordinate arms of national governments. Local governments from different countries "act locally and argue globally" despite their varied political and economic systems and their often limited range of responsibilities. Their global cooperation is largely free of the usual patterns of national politics, interests, and approaches; almost all local governments that engage in international cooperative processes do so in a relatively open-minded fashion and by prioritizing joint goals, such as climate protection, biodiversity preservation, and sustainable resource management. The divide between industrialized and developing countries plays a much lesser role among local governments than among their respective national governments. When local leaders address the UN, they consciously do it on behalf of local governments in general rather than on behalf of a distinct group of developing countries such as the G77.

The reasons are straightforward. Sustainability is a common priority, and many representatives of local governments show strong commitment and leadership. The voluntary cooperation of the more-informed and interested, and the common commitment to providing good living conditions for people, are more relevant than defending abstract national interests. These motivations help explain why local governments often have been faster than national governments to take action on environmental initiatives. After the adoption of the UN Framework Convention on Climate Change (UNFCCC) in 1992, for example, it took local governments just eight months to convene the first Municipal Leaders Summit on Climate Change and to launch the ICLEI Cities for Climate Protection Campaign. It took national governments 13 years to put in place the global implementation mechanism, the Kyoto Protocol, and even then the United States, the largest emitter of carbon dioxide (CO_2) at the time, failed to ratify it.[5]

Similarly, local governments often show greater commitment and readiness to implement the goals and targets of international agreements. In particular, the advanced, forward-looking, and well-run local governments have proven that their sustainability commitments are not limited to isolated local actions but are taken within a global context and with the explicit goal of helping to reach globally set targets. If national governments would recognize and take active advantage of this tendency, they could reach their commitments more easily and more quickly.

The role of local governments in the global sustainability debate has widened over the last 20 years. Until the late 1980s, local governments did not factor significantly in global debates, nor were they seen as transnational actors. Although a relevant "twin-city" movement existed, it focused primarily on peace building and on cultural people-to-people interactions. Bilateral exchanges were prioritized, supported to some extent by national governments and a few existing global organizations for subnational governments. The International Union of Local Authorities, most strongly anchored in the Anglo-Saxon world and in central and northern Europe, took a more multilateral approach, while the United Towns Organisation, anchored in the French-speaking countries, focused mainly on partnerships between Russian and European cities (both organizations are now part of UCLG). International cooperation among multiple municipalities around specific themes was rare.

The founding of ICLEI by some 200 city leaders in September 1990 in New York marked a significant change: for the first time, elected city officials decided to build an international city organization for what we now call "sustainability." ICLEI's mandate from the start was to (1) network among environmentally concerned local governments globally, (2) motivate and support local governments to (jointly) act locally in areas of global concern, and (3) link local action to global UN processes. The creation of ICLEI was the key local government response to the emerging notion of sustainable development, as coined by the Brundtland Commission in 1987.[6]

ICLEI strongly influenced preparations for the 1992 UN Conference on Environment and Development, also known as the Rio Summit, by proposing wording for what became Chapter 28 of *Agenda 21*, the conference's key outcome document. The chapter called upon local governments worldwide to engage their communities in the development of a "Local Agenda 21," which gave birth to the global Local Agenda 21 movement. (See Box 14–1.)[7]

Growing Input to UN Processes

To the extent that national interests have allowed, UN organizations and the global processes under their influence have been supportive of this newly born movement of civil society, and especially of the growing voice of local governments. Following the Rio Summit in 1992, the UN Department of Economic and Social Affairs, the body designated to oversee implementation of the 1992 decisions, defined nine so-called Major Groups, one of which is local governments. A culture of greater openness, transparency, and dialogue has been started, in which the expertise of stakeholders is of increasing relevance. Local governments have now become significant and recognized players in UN processes.

Within the UN mechanisms to address climate change, for example, municipal observers are involved actively through the Local Government and

Box 14–1. Local Agenda 21: A Powerful Movement with Wide-ranging Impacts

Local Agenda 21 has been defined in many different ways, but ICLEI's definition is used commonly: Local Agenda 21 is a participatory, multisectoral process to achieve the goals of *Agenda 21* at the local level through the preparation and implementation of a long-term, strategic action plan that addresses priority local sustainable development concerns.

ICLEI has attempted periodically to analyze Local Agenda 21 progress on a global scale. In 1997, ICLEI helped inform the UN General Assembly Special Session tasked with a five-year review of *Agenda 21*, and in 2002 it worked with the Secretariat of the UN World Summit for Sustainable Development and the UN Development Programme's Capacity 21 Program to provide a second five-year assessment. In 2012, ICLEI carried out a comprehensive stock-taking aimed at identifying if, where, and how Local Agenda 21 had become mainstream. Whether it is called "Local Agenda 21," as in South Korea, Latin America, and several southern European countries, or "urban sustainable development," or just "local sustainability," Local Agenda 21 has unfolded vigorously in thousands of places around the world.

Groups of concerned citizens, religious groups, nongovernmental organizations, and others began interacting with additional stakeholders, such as business, science, or government agencies, to formulate how they wanted to pursue the (sustainable) development of their communities. Not only have many sustainability initiatives resulted from these processes, but in most cases no international process, development cooperation approach, or project can start without declaring stakeholder involvement as a key to its success.

As these processes were linked to and inspired each other across national borders, national civil societies were built at the same time as a global civil society movement grew. Based on the original call of Chapter 28 of Agenda 21, local governments have been seen as the units to manage these processes. In many cases, initiatives came from the public, and local leaders took them up more or less voluntarily. In many other cases, the credit goes to local governments, whether mayors, councils, or chief administrative bodies, for having brought the notion of Local Agenda 21 to their towns and cities.

More than 20 years after Chapter 28 was put in place, local consciousness about global and future impacts of today's actions—and inaction—has never been higher. The multilocal movement has prepared the ground for advancing national and international sustainability policies, and local sustainability processes have established themselves as hubs of social innovation. It is clear that sustainability needs a multilevel governance system with a multisectoral approach. It is time to move from national interests to global environmental justice.

Source: See endnote 7.

Municipal Authorities Constituency. When the first Conference of the Parties (COP) of the UNFCCC met in Berlin in 1995, ICLEI started a series of related local-government events and a Mayors Summit on Climate Change. Since then, each COP has been accompanied by a local government side event and a gathering of local leaders. The core message has always been the same: local governments are concerned about climate change and its impacts, are taking action to reduce greenhouse gas emissions themselves, and call on national governments to increase and accelerate their joint efforts to combat climate change. Over the years, this message has been widened and now includes required action on climate change adaptation as well as mitigation.

Each year, local leaders have based their urgings to national and international governments and organizations on reports about their own local activities and achievements. Even as national governments were still discussing the Kyoto Protocol, many local governments had already agreed on targets to reduce greenhouse gas emissions by 30 percent. And while national governments still cannot agree on joint efforts to reduce emissions, cities and towns have openly declared goals for themselves as "low carbon," "fossil free," or "climate neutral." From the very first local leaders' summit, the offer from local governments was clear: we *do* act locally and are ready to support national implementation of internationally agreed targets. No country can reach the urgently needed greenhouse gas reduction targets without strong support at the local level.

Overall, the last 20 years of global climate negotiations reflect well the growing role of local governments in international governance processes. (See Figure 14–1). Local actors, organized in large part by ICLEI, have mirrored global efforts at nearly every stage and have often stimulated the debate among nations with their own commitments.[8]

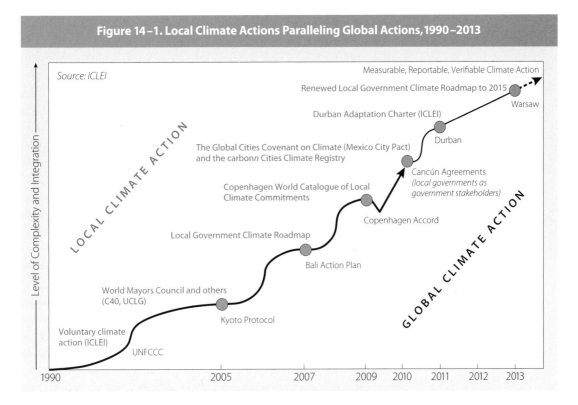

Figure 14–1. Local Climate Actions Paralleling Global Actions, 1990–2013

When delegates to the UN climate conference in Indonesia in 2007 agreed to the so-called Bali Roadmap, ICLEI gathered local government organizations to form the Local Government Climate Roadmap 2007–2009—the largest-ever coalition of local government networks—to call for a comprehensive post-2012 global climate agreement in which local governments are recognized, engaged, and empowered. The hope was that the Copenhagen climate talks in 2009 would result in a breakthrough for climate change. With more than 1,000 registered participants, ICLEI's local government delegation to Copenhagen was the second largest after the Danish NGOs. Local leaders from around the world came to encourage national governments in a series of organized dialogues within the Local Government Climate Lounge.[9]

When the Copenhagen conference ended in failure, disappointment with the lack of national government leadership reached the local level as well. Since then, local governments have changed (but not reduced) their advocacy strategies. In November 2010, just prior to the UN climate conference in Cancún, Mexico, many cities decided to show their leadership by adopting the Global Cities Covenant on Climate, also known as the Mexico City Pact. The agreement built upon the Copenhagen World Catalogue of Local Climate Commitments, a clearinghouse for more than 3,500 voluntary greenhouse gas emissions reduction commitments by local governments. Closely related was the launch of the carbon$n$ Cities Climate Registry (cCCR) as a tool to document these commitments and related actions. Both initiatives should have a lasting impact as crystallization points for committed cities and towns.[10]

Although no climate consensus was reached at the Cancún conference, the role of subnational governments was officially mentioned for the first time in a COP outcome document, and subnational governments were recognized as governmental stakeholders within the global climate regime. This has freed them from the paradoxical category of "nongovernmental." Labels are not really the issue, however. Local governments are fighting for the expectation that national governments will accept local governments as appropriate and efficient implementation partners and endow them with powers and access to resources—a role that is in their own interest of advancing the fulfillment of their global commitments.[11]

At the 2011 climate conference in Durban, South Africa, the Durban Adaptation Charter completed the mechanism of local government commitments. The Charter's content points to the close relationship between climate mitigation and adaptation needs, pulling countries in the same direction. Hundreds of local governments and their national associations have signed this local commitment to respond to climate change, linked to a call to reduce the sources of climate-altering greenhouse gases. The strong presence of local governments in Durban demonstrated again the cooperation of municipalities from both industrialized and developing countries, and encouraged local government networks to prepare a new advocacy phase.[12]

Most recently, at the 2013 climate conference in Warsaw, Poland, local government organizations joined forces to present the second, "renewed" phase of the Local Government Climate Roadmap, looking toward the Paris climate conference in 2015. The agreement highlighted synergies with processes focused on urbanization outside of the UNFCCC framework, and it made a stronger case for financial resources and direct access to global funds and market-based finance instruments. In general, it has become clear that many local activities are underfunded and that the local level needs investment for reducing emissions along with everyone else. Increasingly, transnational mechanisms such as the Global Environment Facility, the Green Fund, and even the World Bank and its related institutions are moving toward supporting local activities, and private investors are more willing to spend money on local climate action.[13]

As compared with the UNFCCC, municipal observers play a lesser-defined role in other UN agreements, such as the Convention on Biological Diversity, where they are grouped under "Sub-National Governments, Cities and Other Local Authorities." With support from ICLEI, however, these UN mechanisms and their related meetings and negotiations have increasingly recognized the implementation role of local governments. (See Box 14–2). Local governments also are involved actively in the UN's post-2015 development agenda, the main outcome of the Rio+20 conference in 2012. (See Box 14–3.)[14]

Pioneering Local Governments for Sustainability

Of the million or so local governments worldwide, only a few thousand are engaged actively in international sustainability projects and networking. Yet many of these governments serve as models that guide or inspire others, offering examples that innovative actors in not-yet-active cities can point to and follow. Case studies, internationally known "best cases," and leading mayors with high public profiles represent a potent way to accelerate progress. Local governments have demonstrated to national governments that action is possible at the local level, even when progress may be slow at the national level. Local-level action is successful because it is closest to people—at a level where conditions are best known. The stakeholders know each other, trust can be built, and potential failures have limited impacts.

The most successful way of encouraging action—whether on energy conservation, climate change adaptation and resilience, low-carbon development, water management, non-motorized mobility, or other issues—is for pioneering cities to lead by example, thus inviting similar action in other cities both at home and abroad. To promote this development, ICLEI has as-

---

Box 14–2. Local Government Involvement in the UN Biodiversity Convention

Local governments have played a growing role in the annual meetings, or Conferences of the Parties, of the UN Convention on Biological Diversity. In 2008, in parallel to the CBD negotiations in Bonn, Germany, ICLEI started a biodiversity initiative similar to the involvement of local governments in UN climate negotiations. Partnering with the CBD Secretariat and local host governments, the organization has since helped coordinate a series of biodiversity summits of local and subnational government leaders.

In 2008, the Mayors Conference on Local Action for Biodiversity produced one of the first documents to outline the important role of local governments in protecting biodiversity worldwide, titled *Cities and Biodiversity: Bonn Call for Action*. Two years later, the 2010 City Biodiversity Summit took place in parallel with the CBD conference in Nagoya, Japan. A key outcome was the *Aichi/Nagoya Declaration on Local Authorities and Biodiversity*, which provided support for the *Plan of Action on Subnational Governments, Cities and Other Local Authorities for Biodiversity (2011–2020)*, which had been adopted by all 193 parties to the CBD. For the first time, local and other subnational actors were mentioned in such a high-level UN document for the CBD and recommended as partners for national action plans.

In October 2012, the Cities for Life: City and Subnational Biodiversity Summit took place in parallel to the annual CBD conference in Hyderabad, India. This event built on the previous successes by examining and assessing the implementation status of the *Plan of Action*. A significant outcome was the *Hyderabad Declaration on Subnational Governments, Cities and Other Local Authorities for Biodiversity*, a pledge by the mayors and governors of local and subnational authorities to develop and implement local strategies toward the *Plan of Action*, and to achieve the 20 Aichi Biodiversity Targets. Similar activities are planned for 2014 CBD conference in Pyeongchang, South Korea, which will be hosted, for the first time, by a province (Gangwon).

These summits are only the visible portion of local government activities to support biodiversity protection. Municipalities depend on functioning ecosystems, and maintaining biodiversity is not only a goal for urban areas but a necessary reality. City development strategies are key to sustaining biodiversity, whether within their borders or in remote areas affected by development.

Source: See endnote 14.

---

sembled a network of model and satellite cities in model countries, especially for urban low-carbon development strategies. Supported with measurement instruments and tools, access to a pool of experts, and information exchanges both nationally and internationally, these cities are implementing multiyear plans with clear goals and assessment systems at hand.[15]

In a second key approach, ICLEI propagates the courage, pioneering spirit, and innovative local activities of these leading cities as a way to influence many others. This "mainstreaming" can be supported by developing methods, mechanisms, tools, and guides, and by demonstration of successes and impacts. Deepening and cementing this mainstreaming will require the creation of facilitative national laws and economic incentives and the handing off of increased responsibilities to local governments.[16]

Rio+20, the largest UN conference in history in June 2012, was expected to bring a global transformation of human civilization by ensuring the sustainability of human societies and global ecosystems. In practice, Rio+20 revised the setting of the global architecture of sustainability efforts by kicking off key processes, including creating a High-Level Political Forum to safeguard the event's outcomes more effectively, initiating development of a universal set of Sustainable Development Goals (SDGs) with a wider scope applicable to all countries, and addressing the UN Environment Programme as the global environmental authority. (See Chapter 13.)

In 2012, the UN Secretary-General established two new bodies in which local governments hope to play an important role. On the technical level, the Sustainable Development Solutions Network (SDSN) mobilizes expertise from academia, civil society, and the private sector and is structured around 12 Thematic Groups, including one on "Sustainable Cities." In its report to the Secretary-General's second new body—the High-Level Panel on the Post-2015 Development Agenda—this group gave a clear recommendation to develop a standalone goal on sustainable urbanization in the set of SDGs to be developed. One local government leader, Istanbul Mayor Kadir Topbaş, was invited to join the High-Level Panel, supported by the Global Task Force of Local and Regional Governments for Post 2015 and Habitat III, facilitated by UCLG and with ICLEI acting as one of the core partners.

In September 2013, the Local Authorities Major Group, with active participation from ICLEI and UCLG, was invited to organize a Special Event on Sustainable Cities as a side event to the UN General Assembly meeting, indicating the UN's interest in including or at least listening to the local level. U.S. mayor Frank Cownie of Des Moines, Iowa, and Lilia Rodriguez, director of international relations at Quito Metropolitan Municipality in Ecuador, participated as speakers. This was one of the first times that the General Assembly had specifically invited an event on cities and local governments.

The debate over whether to include a standalone SDG on sustainable urbanization (i.e., UrbanSDG) is an important element of the post-2015 discussion. Supporters of the idea, including UN-Habitat, the SDSN, as well as UCLG and ICLEI as core leaders of the Global Task Force of Local and Regional Governments, contend that the process should go beyond localization of SDGs, be based on positive experience in the past (in particular the implementation of Chapter 28 of *Agenda 21*), and anchor the role of local governments in a key UN document and implementation process. The UrbanSDG would strengthen the provision of supportive international and national framework conditions for local action and help to propel model urban sustainability projects and policies into the mainstream. The discussion of how an UrbanSDG could be formulated and how best to advocate for it is the main goal of the Communitas Coalition for Sustainable Cities and Human Settlements in the New UN Development Agenda, started jointly in 2013 by UN-Habitat, nrg4sd, the Tellus Institute, and ICLEI, and supported by the Ford Foundation.

Source: See endnote 14.

In the end, however, as much as local action can change the world and serve as a motivating and driving factor, it is also true that national governments cannot shirk their responsibilities. Despite the powerful effects of local action, it has limits and ineffectiveness. Even as many local governments invest in voluntary and sometimes symbolic action, in many cases the goals could be reached much more quickly and efficiently through national framework conditions, such as through national laws or building standards and nationally directed adjustments to economic conditions, such as energy prices.

The key to increased recognition and support of local action is better evidence of achievements and impacts. Documenting success and efficiency is the indispensable condition for a transparent verification of relevance. The carbon*n* Cities Climate Registry is a major step forward in this regard. This online system, developed by ICLEI and available to all local and (soon) regional governments globally, records local commitments, targets, policies, activities, and achievements. cCCR not only is a means to document the relevance and impacts of local action, but also can open doors for local actors taking part in the global carbon market and can increase their access to climate funds and finances.[17]

Even so, in order to support the relevance of local government and local civil society's contribution to the global sustainability targets, better measurements of impacts, achievements, and progress are needed that go beyond mainly describing activities. It is important to identify key indicators for climate change (beyond simply measuring CO_2 emissions), as well as for biodiversity, water, and other resources.

The strategy of global targets and local implementation hardly means ignoring the national level. On the contrary, it means mobilizing the energy and creativity of countless subnational entities with their own governance systems—their own leadership, sources of inspiration, resources, understanding of citizens' needs, and local solutions. Cumulative local actions can achieve tangible improvements in global sustainability. The challenge for the global governance system is to understand this huge potential and to form framework conditions within its exchange and decision-making mechanisms that encourage and unleash this potential to achieve direct improvements to our environment, ecological systems, and social well-being.

Economic Governance

CHAPTER 15

Scrutinizing the Corporate Role in the Post-2015 Development Agenda

Lou Pingeot

As multinational corporations grow larger and increasingly powerful, they have become actors to be reckoned with in international policy debates on poverty eradication, development, the environment, and human rights. At a time when governments seem unable and unwilling to resolve pressing challenges in multilateral settings, business is positioning itself as an alternative solution that is more flexible and efficient, and less bureaucratic, than states. Corporations, governments, and various civil society organizations are promoting multi-stakeholder initiatives and public-private partnerships as innovative models to tackle global issues.[1]

The World Economic Forum's report on the future of global governance, *Global Redesign*, posits that a globalized world is best managed by a coalition of multinational corporations, governments, and select civil society organizations. The report argues that countries are no longer "the overwhelmingly dominant actors on the world stage" and that "the time has come for a new stakeholder paradigm of international governance." The World Economic Forum's vision includes a "public-private" United Nations (UN), in which certain specialized agencies would operate under joint state and non-state governance systems, such as the Food and Agriculture Organization through a "Global Food, Agriculture and Nutrition Redesign Initiative." This model also assumes that some issues would be taken off the agenda of the UN system to be addressed by "plurilateral, often multi-stakeholder, coalitions of the willing and able."[2]

Similarly, the Oxford Martin Commission for Future Generations, an initiative designed to "identify ways to overcome today's impasse in key economic, climate, trade, security, and other negotiations" and chaired by former World Trade Organization Director-General Pascal Lamy, proposes to establish a "C20–C30–C40 Coalition" made up of G20 countries, 30 companies, and 40 cities that would work together to "counteract climate change." Although this "coalition of the working," based on "inclusive

Lou Pingeot is a policy advisor at the Global Policy Forum.

minilateralism," would report to the UN Framework Convention on Climate Change, it would not rely on binding commitments.[3]

The trend toward an increased role of corporate actors in global governance through various models of multi-stakeholder initiatives is also reflected at the UN level. Already in 2002, the World Summit on Sustainable Development endorsed "the concept of voluntary, multi-stakeholder initiatives to facilitate and expedite the realization of sustainable development goals and commitments." Several such high-profile initiatives are currently under way, addressing issues ranging from women and children's health ("Every Woman, Every Child") to sustainable energy ("Sustainable Energy for All"). This trend is supported by member states, as demonstrated by the resolutions of the UN General Assembly under the item "Towards global partnerships," which invite governments to continue to support UN efforts to engage with the private sector.[4]

Still, there are diverging views among governments, UN institutions, and civil society organizations about the legitimacy and effectiveness of the growing interaction between the UN and business actors. While some maintain that there is no alternative to this new model, others have raised concerns about the limits and risks associated with public-private partnerships and multi-stakeholder initiatives. Some civil society groups argue that corporate influence at the UN diverts the organization from tackling the root causes of environmental, social, and economic problems and puts its credibility and legitimacy at risk.[5]

Against this background, large multinational corporations are expected to play a growing role in and have expanding influence over the UN's post-2015 development agenda, as indicated in a series of reports written by business organizations as well as UN documents.

The outcome document of the 2012 Rio+20 conference, titled *The Future We Want*, called for the creation of a new set of Sustainable Development Goals (SDGs), which are meant to build upon the Millennium Development Goals (MDGs, set to be reached by 2015) and to converge with the UN's post-2015 development agenda. *The Future We Want* also mandated the creation of an intergovernmental Open Working Group (OWG) to develop a proposal for the new goals, as well as the creation of an intergovernmental High-Level Political Forum (HLPF) to provide political leadership, guidance, and recommendations on sustainable development. Both bodies were established during 2013. (See Chapter 13.)[6]

The UN is aiming to integrate the various "work streams" stemming from the post-MDG and post-Rio processes into a universal sustainable development agenda. In addition to the OWG and the HLPF, these include two initiatives by the UN Secretary-General. One is a High-Level Panel (HLP) established in July 2012 to advise on the global development framework beyond 2015, and the other is the Sustainable Development Solutions Network (SDSN) launched in August 2012, which aims to help overcome the gap between technical research and policy making and is to work with UN agencies and other organizations. Business groups, and in particular large multinational corporations, have been particularly active in the HLP and the SDSN.*[7]

Business also has a strong presence through the Global Compact, a voluntary corporate responsibility initiative at the UN launched by former Secretary-General Kofi Annan. In early 2011, the Compact created a new initiative with a select number of companies, the Global Compact LEAD, to implement the "Blueprint for Corporate Sustainability Leadership." Of the current 55 members of LEAD, 11 are from the mining sector and the oil and gas industry, 4 are electric or other utility providers, while just 1 hails from the alternative energy sector. Further, a number of business associations are involved in the post-2015 consultations, including the World Business Council for Sustainable Development (WBCSD), the World Economic Forum, the International Chamber of Commerce, and the International Organization of Employers.[8]

Corporate Perspectives and Governance Models

The business sector has an important role to play in the future of sustainable development, which will require large-scale changes in business practices. Some pioneering companies are already on the path toward sustainable development solutions (for instance in the area of renewable energy). (See Chapter 19.) But current business participation in the post-2015 process raises concerns that corporations may have undue (and unchecked) influence on policy making. The risks and side effects of corporate influence relate, on the one hand, to the messages, problem analyses, and proposed solutions, and on the other hand to the promoted governance models.

The corporate sector is feeding into the post-2015 agenda through a variety of reports emerging from business-led initiatives (such as the Global Compact) and from processes that have given an important space to business (such as the SDSN). The corporate sector reports, and to a large extent even the reports of the HLP and the SDSN, are striking for their lack of historical perspective on what caused the problems that the post-2015 agenda is meant to tackle. The SDSN report, for example, notes that "the business-as-usual (BAU) trajectory is marked by a failure of international coordination and

---

* The HLP is composed of 27 individuals, leaders from government, civil society, and the private sector. The SDSN is governed by a Leadership Council composed of individuals representing education and research institutions, corporations, foundations, civil society, and the UN. The SDSN also has 12 Thematic Groups of Experts, one of which focuses on redefining the role of business for sustainable development.

cooperation." It does not acknowledge, however, that powerful economic actors have benefited substantially from BAU, and thus have a strong interest in resisting far-reaching structural transformation toward sustainability—as is illustrated by the rising number of cases in which corporations are suing governments to weaken environmental, health, and social policies. A 2013 Global Compact report to the Secretary-General claims that "business is at the heart of virtually any widespread improvements in living standards," ignoring the pivotal role of governments in providing public goods and the role of unions and social movements in pushing for adequate standards and regulations. (See Chapter 21.)[9]

The various business reports present growth as the main solution for poverty eradication and as an indispensable condition to the realization of sustainable development. But growth alone has never had such unambiguously positive impacts. As the civil society initiative Participate notes, "inequality and distorted power relations prevent the dividends of economic growth from reaching the very poorest."[10]

The business discourse promotes a market-based approach to sustainable development, which assumes that incentive-driven voluntary commitments are preferable to binding commitments or "command-and-control" regulatory approaches. The HLP report, for example, promotes a limited form of corporate accountability based on the assumption that market forces will favor companies committed to sustainability over those that are not. But it is not clear that such an approach will yield the fundamental changes in consumption and production patterns that are needed and that many in civil society are calling for.[11]

Making the business case for sustainable development may be seen as a pragmatic approach. The 2013 report of the HLP strongly suggests that progress must be monetarily quantifiable and provide a good return on investment. This begs the question, however, of what to do when necessary efforts for the public good do not constitute a good investment for the private sector. This type of analysis conveys a vision of the world in which everything is seen through an economic lens, and in which people are foremost regarded as consumers or entrepreneurs, but less as multifaceted citizens. The Participate initiative, in response to the HLP report, argues that "economic democracy is as important for poverty eradication as political democracy."[12]

From a governance perspective, it is important to note—and also worrisome—that some of the key channels for corporate influence on the post-2015 agenda were not established through regular intergovernmental processes, and thus do not respond to usual intergovernmental mechanisms for accountability. Both the Global Compact and the more recent SDSN were initiatives of the UN Secretary-General. The Compact was launched without a mandate from the UN General Assembly, and was granted recognition only after the fact. It also falls outside of regular UN processes because of its extra-budgetary funding from the corporate sector and a small group of member states. This was flagged as problematic by the UN's internal watchdog, the Joint Inspection Unit, in 2010. A lack of transparency also surrounds the creation of the SDSN. Its sources of funding are not made public, and no clear criteria were established by which to select the corporate participants (who, with business associations, represent 21 of 73 SDSN Leadership Council members).[13]

Walmart

White roof and skylights on a retail store in Las Vegas help reduce energy use and have a lower heat-island effect than a darker roof.

There is concern that corporate influence in the post-2015 process is shifting the balance of power to the detriment of civil society. Part of the problem stems from a lack of clarity in UN processes around the concepts of "civil society" and "stakeholders," which come to encompass both nonprofit and for-profit entities. Yet direct participation in policy processes is only one of the many ways in which corporate influence can manifest itself. Access to policy makers, officially or behind-the-scenes, is also a key element of political influence. Through contributions to political campaigns and lobbying, some corporations have built tight connections with local and national policy makers, which can translate into influence in global policy processes.[14]

Although UN processes tend to refer broadly to the participation of "business" or "the private sector," in practice, large multinational corporations rather than small and medium-sized enterprises (SMEs) are the primary representatives of "business" in the post-2015 process. The Global Compact does offer some channels for SME participation. For example, SMEs are well represented in the Global Compact Local Networks, which are an important part of the organization's activities. However, the Compact gives big business special access to the post-2015 process through its LEAD initiative. When the Global Compact LEAD organizes a luncheon with the Secretary-General and other high-profile events, it provides a privileged access to political processes where small enterprises have no place at the table.[15]

Corporate-sector involvement in the post-2015 process also reflects an imbalance between different types of industries. The mining industry is particularly over-represented in both the Global Compact LEAD and the SDSN. Out of more than 30 corporate representatives involved in the SDSN

Leadership Council or thematic groups, 6 have ties to the mining industry, accounting for about 1 in 5 business representatives in this process. It could be argued that these companies are precisely the ones that should be involved because of their important impact on development, human rights, and the environment. However, the mining and oil and gas sectors also have the most incentive to delay or limit the transition to sustainable development, so as to protect their profit sources and ultimately their existence. (See Chapter 20.)[16]

The multi-stakeholder model of global governance rests on the assumption that the interests of governments, business, and civil society ultimately align, and that all stakeholders work together to achieve common goals. A report by the co-chairs of the SDSN (affiliated with Novartis and the WBCSD), for instance, states that "the international community, multilateral institutions, national governments, academia, civil society and business have got to work together towards a common agenda." Similarly, a joint report by the Global Compact and the WBCSD notes that "healthy societies and healthy markets go hand-in-hand."[17]

There is merit in—and considerable need for—fruitful cooperation. Yet such a model, with its emphasis on partnerships and consensus, can negate the existing conflicts among stakeholders, in particular between large multinational corporations on the one hand and social movements on the other. Labeling all participants "stakeholders," as if all were equal and had the same interests, can obscure the power imbalances between various sectors and the vast differences between their agendas. This creates the illusion that "win-win" solutions can be found if only all stakeholders sit at the table for a rational debate, and promotes a depoliticized model of governance that does not address the power structures inherent in the global economic system. The numerous cases in which corporations are suing governments on the basis of bilateral investment treaties, alleging that social, health, and environmental regulations harm profits, also challenge the notion that "we are all in this together."[18]

Making Business Participation More Transparent and Accountable

Avoiding "corporate capture"—the undue influence of business actors, and in particular large multinational corporations, on the post-2015 agenda—will require governance reforms and norm setting to make business participation as transparent and accountable as possible. It will also entail careful monitoring and evaluation of partnership activities and greater transparency of associated funding. UN member states need to adopt much more stringent criteria and rules for those who participate in multi-stakeholder initiatives, and for how these actors will be held accountable.

At present, international business associations can participate in UN processes as "nongovernmental organizations (NGOs)" on the grounds that they are nonprofit, even though they represent the interests of for-profit corporations. There needs to be a clearer distinction between public-interest NGOs and business-interest NGOs.

Some governments have supported the UN's outreach to the corporate sector even while seeking to keep civil society groups at bay on the grounds that the intergovernmental nature of the organization should be preserved. It is time for member states to speak out on the role they envision for the business sector in the post-2015 agenda and in the UN system at large. The recent initiative spearheaded by Ecuador (and supported by several member states as well as more than 100 civil-society organizations) in the Human Rights Council to advance a binding instrument to regulate multinational corporations may be signaling that the debate is shifting toward a much stronger recognition of business responsibilities.[19]

The UN should adopt a standardized, systemwide set of guidelines for its interaction with the private sector and all other stakeholders. This could take the form of a General Assembly resolution, comparable to the UN Economic and Social Council's resolution on the regulation of the consultative relationship with NGOs. This resolution should define partner selection and exclusion criteria. It should prevent actors who violate internationally agreed-upon environmental, social, and human rights conventions or otherwise violate UN principles (for example, through corruption, breaking of UN sanctions, proven lobbying against international agreements, evading taxes, etc.) from entering into collaborative relationships with the UN.

The UN should also adopt a systemwide conflict-of-interest policy. Corporate partners should disclose to the UN any situation that may appear as a conflict of interest. They should also disclose if a UN official or a professional under contract with the UN has any kind of economic ties with a corporate partner. Specific requirements in the code of ethics for UN employees could help address the potential conflicts of interest raised by the circulation of staff between UN entities and national governments, private foundations, corporations, lobby groups, and civil society organizations. A "cooling off" period, during which former UN officials cannot start working for lobby groups or lobbying advisory firms, could be considered.

Before the UN enters into new multi-stakeholder initiatives or partnerships with business actors, the possible impacts of these activities must be assessed systematically. This should include evaluating the added value of the initiative for the realization of the UN's goals; the relation between the risks, costs, and side effects and the potential benefits; human rights impacts; and the possible alternatives to the planned activities. Impact assessments and evaluations should be carried out by neutral bodies and the results of the investigations be publicly accessible.

A UN regulatory framework for partnerships, in particular with the business sector, will require capacity in the secretariats and at the intergovernmental level. Staff is needed for the additional duties of screening companies, legal advice, and monitoring and evaluation of partnerships. This task could be fulfilled, for instance, by the existing Joint Inspection Unit of the UN, if its financial resources and mandate were extended accordingly. For the monitoring and oversight of partnerships in the post-2015 development context, the High-Level Political Forum (HLPF) could become the hub.

The UN seeks extra-budgetary funding in a context where some member states have failed to pay their full dues and, in several instances, have cut their voluntary contributions. Since the 1980s, donor contributions, while generally increasing in amount, have shifted away from "core funding" toward voluntary earmarked funds, thus eroding the multilateral character of the organization. An increasing amount of funding also comes from nongovernmental sources, such as NGOs, philanthropic foundations, and the corporate sector. In 2012, $13.7 billion of a total $41.5 billion in UN systemwide funding, or just 33 percent, came through assessed (mandatory) contributions from member states. Half of all funding was in the form of voluntary contributions provided for specific purposes, and another 13 percent was voluntary funds for nonspecified use. (See Figure 15–1.)[20]

Member states have a key role to play in reversing this trend by providing adequate core funding to UN programs, and civil society groups need to advocate for adequate and reliable financial resources. At a minimum, the UN should disclose the funding it receives from the private sector more transparently. According to UN data, extra-budgetary resources from "Major Other Organizations, NGOs, Foundations, Private Sector" increased from $883 million in 2002–03 to $2.3 billion in 2008–09. But there is currently no systematic reporting of the funds that the UN receives in the form of extra-budgetary resources, and there is no disaggregated reporting to track the evolution of private sector funding.[21]

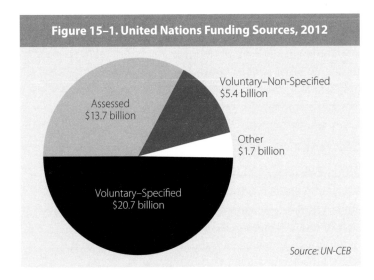

Figure 15–1. United Nations Funding Sources, 2012

Assessed
$13.7 billion

Voluntary–Non-Specified
$5.4 billion

Other
$1.7 billion

Voluntary–Specified
$20.7 billion

Source: UN-CEB

Better reporting is also needed for funds committed to multi-stakeholder initiatives, such as Every Woman, Every Child or Sustainable Energy for All. While these initiatives claim billions of dollars in pledges and investments, it is usually difficult to assess where the money has gone, whether it has been really new and additional to existing commitments, and what impact it has had. If these initiatives are going to be part of the post-2015 agenda, they require much more stringent reporting.

Civil society groups have an important role to play in this context. It will likely fall to them to highlight the context within which corporate influence on UN processes becomes problematic. They will need to operate from an understanding of the broader problems of increasing fragmentation of global governance, the weakening of representative democracy on the national level, the unpredictable and insufficient financing of public goods, and the lack of adequate monitoring and accountability mechanisms.

Civil society organizations engaged in partnerships with the business sector in particular need to carefully evaluate the impacts and side effects of these initiatives and to potentially reconsider their involvement. In a context where reporting requirements and accountability standards for public-private partnerships are low, it is difficult to assess their success or failure. Are they achieving their stated goals and do they contribute to sustainable development? Do they empower local communities and meet their needs? Civil society groups advocating for effective corporate accountability rules at the UN need to be able to answer these questions.

CHAPTER 16

Making Finance Serve the Real Economy

Thomas I. Palley

In the wake of the financial crisis of 2008, financial sector reform has been a major policy focus around the world. However, that focus has been almost exclusively on the issue of "stability" and preventing a repeat of the crisis. There has been little debate about the broader role of finance in shaping economic developments over the past 30 years, and inadequate attention is given to the challenges of how to remedy the massive economic inequities and problems relating to unemployment and growing indebtedness of many households.

This silence on the broader role of finance has economic and political consequences. The framing of the reform debate in terms of the narrow issue of stability shuts down the case for deeper systemic reform. Financial markets have a broader social purpose than just the efficient allocation of capital on behalf of shareholders. That broader purpose is to contribute to the delivery of "shared prosperity," which can be defined as full employment with rising incomes and contained income inequality. Today, we clearly do not have shared prosperity, and a big reason for that is the economic and political power of finance.

The structure of the economy affects whether the economy meets the needs of people, and in a broader sense, it influences the way that societies are governed. The rising influence of finance has distorted the public discourse and narrowed the range of those who are able to make themselves heard. This not only affects decision making in the economic and political realms, it also means that the growing urgency of reconciling the economy with environmental limits—creating the conditions for a shared and sustainable prosperity—has been largely neglected. For more than three decades, financialization has been an engine of an economy that gobbles up growing amounts of scarce resources even as it distributes the product in ever more unequal ways. In the future, the finance sector will need to be governed in ways that facilitate the transition to a more equal and sustainable economy.

Thomas I. Palley is senior economic policy advisor at the AFL-CIO.

In the United States, as around the world, the process of "financialization"—by which the financial sector has become the new master of the broader economy—needs to be tamed so that finance once again serves the economy and people's needs. Subjecting runaway financial institutions to rules and regulations driven by the public interest forms a critical part of overhauling governance processes.

Finance and the Destruction of Shared Prosperity

To understand how finance has undermined a shared and sustainable prosperity requires some historical context. Prior to 1980, the U.S. economy could be described as a Keynesian wage-led growth model. Under the logic of this model, growth in economic productivity drove growth in wages, which fueled demand. That drove full employment, which provided the incentive to invest, which drove productivity growth, and so on. (See Figure 16–1.)

Within this economic model, finance was essentially a form of public utility governed by New Deal regulation. The role of finance was to (1) provide business and entrepreneurs with finance for investment, (2) provide households with mortgage finance for home acquisition, (3) provide business and households with insurance services, (4) provide households with saving instruments to meet future needs, and (5) provide business and households with transactions services.

After 1980, however, the Keynesian wage-led growth model and the public utility model of finance were gradually pulled apart and dismantled. A first critical change was the implementation of economic policies that helped sever the link between productivity growth and wages. A second critical change was the dismantling of the New Deal system of regulation—through deregulation—combined with a refusal to regulate new financial developments and innovations. As a result of severing the once-strong link between productivity growth and wages, average hourly wages and compensation stagnated after 1980 despite continuing productivity growth. (See Figure 16–2.)[1]

The new model can be described as a "market fundamentalist" policy box that fences workers in and pressures them from all sides. (See Figure 16–3.) On one side, the corporate model of globalization has put workers in international competition via global production networks that are supported

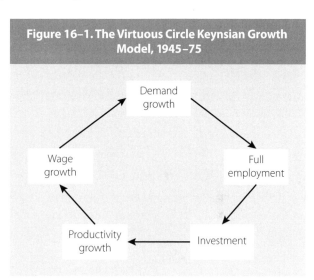

Figure 16–1. The Virtuous Circle Keynsian Growth Model, 1945–75

Demand growth

Wage growth

Full employment

Productivity growth

Investment

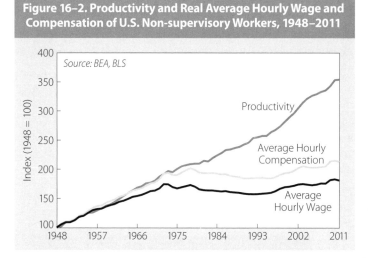

Figure 16–2. Productivity and Real Average Hourly Wage and Compensation of U.S. Non-supervisory Workers, 1948–2011

Source: BEA, BLS

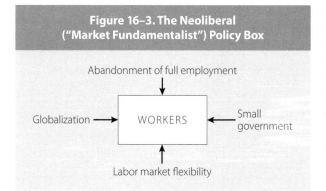

Figure 16–3. The Neoliberal ("Market Fundamentalist") Policy Box

by free trade agreements and capital mobility. On the other side, the "small" government agenda has attacked the legitimacy of government and pushed persistently for deregulation regardless of dangers. From below, the agenda of labor market flexibility has attacked unions and labor market supports such as the minimum wage, unemployment benefits, employment protections, and employee rights. And from above, policy makers have abandoned the commitment to full employment, a development reflected in the rise of inflation targeting and the move toward independent central banks controlled by financial interests. The result is a new system characterized by wage stagnation and income inequality in which the problem of demand shortage has been papered over by debt-financed consumption and asset price inflation.

Finance has played a critical role in both creating and maintaining the new economic model, whose main characteristics are worsened income distribution, the increased importance of the financial sector relative to the real economy, and the transfer of income from the real economy to the financial sector. During the past 40 years, the financial sector has increased both its share of gross domestic product (GDP)—reaching more than 20 percent in 2007—as well as its share of profits relative to the non-financial sector. (See Table 16–1.)[2]

The process whereby financial sector interests have come to dominate the economy is widely referred to as "financialization." This process had three main conduits, related to the financial market structure, corporate behavior, and economic policy. (See Figure 16–4.)

First, finance (commercial banks, investment banks, hedge funds, insurance companies, mutual funds, etc.) used its political power to promote the policies on which the new model rests. Thus, finance lobbied for financial deregulation; supported the shift of macroeconomic policy away from focusing on full employment to focusing on inflation; supported corporate globalization and expanding international capital mobility; supported privatization, the regressive tax agenda, and the shrinking of the state; and supported the attack on unions and workers.

More specifically, globalization policy created a global economy through trade agreements like the North American Free Trade Agreement (NAFTA) that lacked effective labor and environmental standards. The regressive tax agenda was evident in the decline in corporate income taxes, the shifting of the tax burden onto lower income households via increased payroll and sales taxes, and the lowering of top personal tax rates. The attack on workers was exemplified by the decline in the minimum wage and by labor laws that favored corporations against workers trying to form unions. The shift to a focus on inflation was evident in the U.S. Federal Reserve's prioritization of inflation concerns over unemployment concerns.

Second, finance took control of American business and forced it to adopt financial-sector behaviors and perspectives. This change was accomplished via increased actual and threatened use of hostile takeovers, hedge fund activism, and increased use of massive stock option awards for top management that aligned management's interest with that of Wall Street. The resulting change in business behavior was justified using the rationale of shareholder value maximization. The result was a widespread use of leveraged buyouts that burdened firms with unprecedented levels of debt; the adoption of a short-term business perspective and impossibly high required rates of return that undercut long-term real investment; growing reliance on off-shoring and the abandonment of a business commitment to community and country; and the adoption of exceedingly generous Wall Street pay packages for top management and boards of directors.

Third, the deregulated financial system provided the credit that financed borrowing and created asset price bubbles. Examples of these bubbles include the stock market and Internet bubbles of the late 1990s, and the commercial real estate and housing price bubble of the 2000s. These

Table 16–1. Growth of the U.S. Financial Sector, Selected Years, 1973–2007

Year	Financial Sector Output as Share of GDP	Financial Sector Profits as Share of Non-financial Sector Profits
	percent	
1973	13.6	20.1
1979	14.4	19.7
1989	17.9	26.2
2000	20.1	39.3
2007	20.4	44.6

Source: See endnote 2.

Figure 16–4. Main Conduits of Financialization

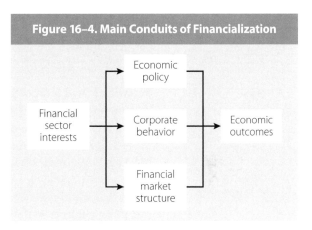

Table 16–2. Growth of U.S. Household Debt, Selected Years, 1973–2007

Year	Household Debt as a Share of GDP
	percent
1973	45.3
1979	49.9
1989	60.5
2000	70.3
2007	98.2

Source: See endnote 3.

bubbles effectively filled the "demand shortage" that had been created by years of wage stagnation and increased inequality. Instead of being able to rely on purchasing power, many households ended up financing their purchases by going ever more deeply into debt. Household debt rose as a share of GDP from 45.3 percent in 1973 to 98.2 percent in 2007, just prior to the most recent financial crisis. (See Table 16–2.)[3]

Viewed in this light, financialization is at the very core of current economic difficulties. Finance drove the policies that undermined shared prosperity, and then fueled a 30-year credit bubble that papered over the demand shortage caused by worsening income distribution. That created an unstable financial system that collapsed when the credit bubble burst. And now, as it emerges from the depths of the financial crisis, the U.S. economy is stuck in stagnation because of deteriorated income distribution and the massive structural trade deficit that, together, undercut the domestic demand needed for full employment.

Putting Finance Back in the Box

Restoring shared prosperity will require re-establishing a link between productivity growth and wages and having economic policy commit to working toward full employment. Moreover, this will have to be accomplished within the additional constraint of environmental sustainability. This is a massive task requiring a range of policies related to labor markets, the international economy, the public sector, the environment, and macroeconomic policy. Given the critical role of finance, such a transition also requires regaining control over finance so that it again serves the real economy, rather than the real economy serving finance.

One part of the challenge is political and concerns campaign finance reform. The political power of finance rests on money, which is why it is so critical to reduce the role of money in politics; absent political reform, finance will be able to distort the democratic process and block necessary economic policy reform. A second part of the challenge is changing corporate behavior. This requires a reform of corporate governance that makes business more accountable, changes incentives that promote current business practice, and recognizes the interests of stakeholders other than shareholders. (See Chapters 15 and 19.)

A third challenge is to regain control over financial markets. Figure 16–5 illustrates a four-part program for putting financial markets back in the box so that they promote shared and more-sustainable forms of prosperity rather than destructive speculation.

The top edge of the box indicates the need to restore a commitment to full employment, abandon a rigid ultra-low inflation target, and recognize that monetary policy can permanently influence the level of economic activity. The left edge of the box concerns the need for tough regulations that impose appropriate capital and liquidity requirements on financial institutions, and also bar banks from engaging in speculative activity using government insured deposits—the so-called Volcker rule.

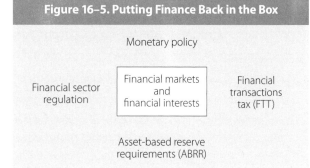

Figure 16–5. Putting Finance Back in the Box

Monetary policy

Financial sector regulation

Financial markets and financial interests

Financial transactions tax (FTT)

Asset-based reserve requirements (ABRR)

Regulation also must be enforced, which speaks to the importance of a good government agenda that ensures the integrity and operational efficiency of regulatory agencies. The right edge of the box concerns the need for a financial transactions tax (FTT), which can raise revenue, help shrink the financial sector to more appropriate and healthy proportions, and discourage damaging speculative transactions.[4]

Lastly, the bottom edge of the box advocates that the Federal Reserve institute a system of asset-based reserve requirements (ABRR) that covers the entire financial sector. ABRR require financial firms to hold reserves against different classes of assets, and the regulatory authority sets adjustable reserve requirements on the basis of its concerns with each asset class. By adjusting the reserve requirement on each asset class, the central bank can change the return on that asset class, thereby affecting incentives to invest in the asset class.[5]

The U.S. housing price bubble showed that central banks cannot manage the economy with just interest rate policy targeted on inflation and unemployment. Doing that leaves the economy exposed to financial excess. Interest rate policy therefore must be supplemented by balance sheet controls, which is the role of ABRR.

ABRR provide a new set of policy instruments that can address specific financial market excess by targeting specific asset classes, leaving interest rate policy free to manage the overall macroeconomic situation. ABRR are especially useful for preventing asset price bubbles, as reserve requirements can be increased on overheated asset categories. For instance, a housing price bubble can be targeted surgically by increasing reserve requirements on new mortgages. That makes new mortgages more expensive without raising interest rates and damaging the rest of the economy.

Finally, ABRR can be used to promote socially desirable investments and "green" investments that are needed to address climate change. Loans for such investment projects can be given a negative reserve requirement that can be credited against other reserve requirements, thereby encouraging banks to finance those projects in order to earn the credit. In sum, ABRR provide a comprehensive framework for collaring the financial sector and ensuring that it promotes shared prosperity.

Conclusion: Beyond Orthodox Economics

We live in an age of market worship. Orthodox economics fuels that worship, and it also gives special standing to financial markets, which are represented as the most perfect form of market. Although there is some critique of the functional efficiency and casino aspects of financial markets, this stops far short of a deeper critique of financialization. Consequently, orthodox diagnoses of the financial crisis and policy recommendations stop far short of what is needed to put finance back in the box.

The economic evidence shows the need to make finance serve the real economy, rather than having the real economy serve finance, as is now the case. It can be done. The challenge is to get a hearing for policies that will do so. Meeting that challenge requires getting new economic ideas on the table, which is why the debate about economics and the economy is so important. However, the road to policy change runs through politics. Putting finance back in the box also requires breaking the political power of finance, which is why campaign finance reform, electoral reform, and popular political engagement are equally important.

CHAPTER 17

Climate Governance and the Resource Curse

Evan Musolino and Katie Auth

The energy sector is the world's single largest driver of climate change, accounting for roughly 70 percent of global greenhouse gas emissions. Limiting emissions by reducing our dependence on fossil fuels will require the active participation of diverse, and often conflicting, stakeholders, including policy makers, scientists, industry leaders, and consumers. The difficulties inherent in rallying such groups to combat a complex, long-term problem like climate change make it a "super wicked" public policy challenge, testing not only our capacity to innovate technological solutions but also—perhaps most importantly—our capacity to govern.[1]

So far, global leaders and delegates to the United Nations Framework Convention on Climate Change (UNFCCC)—the international governance structure designed to prompt and enforce the global response—have failed to enact significant change, despite near-universal scientific consensus on the existence and causes of human-induced climate change, widespread public support for climate mitigation in countries around the world, and growing momentum behind grassroots climate activism in the United States and elsewhere. Given this lack of action, are there alternative means of leveraging social pressure and pushing for meaningful action that would have greater success?

Given the strength of the fossil fuel industry, its political influence, and the extent to which our economies and infrastructure have become dependent on its products, the concept of the "resource curse" offers one potential way to understand the immense challenges facing democratic governance and international climate cooperation. Traditionally, the curse has been understood as a socioeconomic phenomenon that negatively affects poor countries dependent on resource extraction, including by impeding democratic governance or enabling political repression. Now, however, and in the context of climate governance, similar effects can be observed in some of the world's most stable industrialized democracies, including Australia, Canada, and the United States.

Evan Musolino is a research associate and project manager with the Worldwatch Institute's Climate and Energy Program. **Katie Auth** is a research associate with the Climate and Energy Program and the Institute's regional lead for sub-Saharan Africa.

The Traditional Resource Curse and Its Impacts on Governance

The resource curse theory has been a major component of international relations since it was put forward in the mid-1990s to explain the paradoxical observation that countries with abundant natural resources—particularly non-renewable resources like oil and minerals—often fail to achieve the economic growth and development that might be expected. Instead, strong dependence on such resources often results in economic stagnation, increasing social stratification, and a failure to invest in long-term development needs. Although these impacts have been linked with dependence on the exploitation of natural resources of various kinds, oil has particularly acute effects because of its central role in the global economy and the opportunities it affords for a high return on investment.[2]

Much of the established literature focuses on the economic impacts of the resource curse, yet the theory also posits that economic dependence on oil resources can have detrimental impacts on national governance, discouraging investment in public priorities and creating incentives for or enabling government corruption and authoritarian rule. A regime supported by massive oil revenues has little need to cultivate popular support or respond to its citizens' demands, and can therefore use resource earnings to enrich a small elite, neglecting broader development priorities such as education and public health. Elements of this paradox have been observed in troubled countries like Angola, the Democratic Republic of the Congo, Nigeria, and the oil states of the Persian Gulf. Angola has been cited as a prime example of how developing countries with significant oil resources are among those "most prone to poor governance, armed conflict, and poor performance in economic and social development."[3]

Certainly, the style of governance in any given country reflects a wide range of economic, social, and historical factors. Many of the world's oil-producing developing countries gained independence only recently, or had weak governing institutions to begin with. In such cases, although the discovery and exploitation of oil resources may not *cause* repressive governance, it may enable or aggravate it, providing a guaranteed financial cushion that allows governments to ignore or suppress popular demands for increased accountability. In some cases, significant oil earnings have allowed repressive regimes to remain in power longer than would have been the case otherwise.[4]

Nevertheless, the correlation between economic dependence on oil and gas resources and poor governance can be illustrated by numerous international indicators. World Bank data indicate that of the 30 national economies most dependent on oil and gas resources, 27 rank below both global and regional averages in the annual "Voice and Accountability" index, which assesses citizens' ability to select government representatives, exercise freedom of expression and association, and access a free media. Heavily oil- and gas-dependent countries also perform poorly in other international benchmarks for government accountability, including the Reporters Without Borders Press Freedom Index, an annual ranking of the "freedom to produce and circulate accurate news and information" in 179 countries worldwide. Of the world's 30 most oil- and gas-dependent economies, only 3—Kuwait, Norway, and Trinidad and Tobago—fall within the Index's top 100. (See Figure 17–1.)[5]

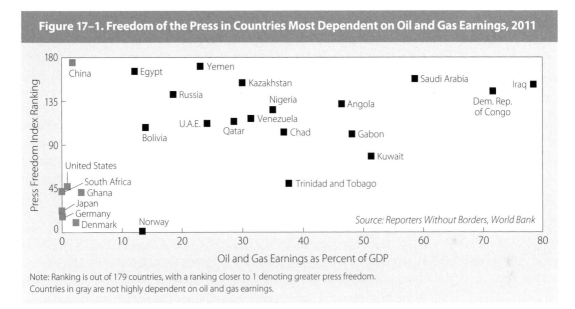

Figure 17–1. Freedom of the Press in Countries Most Dependent on Oil and Gas Earnings, 2011

Source: Reporters Without Borders, World Bank

Note: Ranking is out of 179 countries, with a ranking closer to 1 denoting greater press freedom. Countries in gray are not highly dependent on oil and gas earnings.

Many oil- and gas-dependent countries with limited accountability to their citizens also fail to mitigate the negative health and environmental impacts associated with the extraction and use of fossil fuels. In a study of the oil industry's impact in Nigeria, Amnesty International concluded that oil exploration has resulted in violations of the right to an adequate standard of living (including food and water), of the right to gain a living through work, and of the right to health. With hundreds of oil spills occurring each year, the study concluded that the Nigerian government's lack of accountability played a chief role in perpetuating these damages, and the absence of government transparency remains a major stressor in the Niger Delta. According to Amnesty, the Nigerian government continually fails to enforce its own laws and regulations and, in designating a partner of the oil industry as its regulator, has developed a regulatory scheme that "fundamentally conflicts with the concept of an independent regulatory body."[6]

Correlations between economic dependence on fossil fuel extraction and undemocratic governance can be observed even in strong economies where governments do choose to spend resource earnings on public priorities. In some cases, this spending reflects an effort to stifle popular criticism rather than to improve democratic rule. Political scientist Michael Ross, using statistics gathered from 113 countries between 1971 and 1997, concluded that oil wealth can inhibit democratization in part through its "taxation" and "spending" effects: oil wealth allows governments to relieve social pressures and increase patronage spending, both of which can dampen demands for reform.[7]

Following the start of the Arab Spring in December 2010, which eventually toppled repressive governments in several countries in the Middle East and North Africa, the government of Saudi Arabia nearly doubled national spending in order to placate its population and solidify central power. This was enabled by increased oil revenue tied to a jump in the global oil price from less than $70 per barrel to more than $100 per barrel. Some speculated that this market shift was not purely coincidental but rather was driven by Saudi Arabia working with its fellow Organization of Petroleum Exporting Countries (OPEC) members to assert their influence over global oil prices in an effort to increase revenue.[8]

In addition to its impacts on domestic governance, evidence suggests that oil dependence can be associated with a decreased likelihood that countries will engage actively in global governance, and that such countries have developed their own norms for international cooperation. Although petro-states are deeply integrated into the global economy and rely heavily on foreign markets, oil-rich nations generally can access foreign markets with greater ease and without concession, granting them the freedom to operate autonomously with little fear of losing potential buyers. This is in contrast to countries that lack such abundant resource endowments and that must seek markets for goods with a more elastic demand.[9]

Petroleum products already benefit more widely from import duty exemptions than any other product group, further reducing the incentive for oil-producing nations to seek new agreements guaranteeing market access and allowing them to protect their domestic industries. The world's dependence on a steady supply of fuels also can work to shield such states from censure by the international community. Despite engaging in actions such as "expropriating foreign investors, flouting human rights, and financing terrorism and armed rebellions in foreign countries," petro-states in which these situations have occurred manage to remain actively engaged in international markets, belying the predicted increase in political and legal cooperation as countries increase their economic integration.[10]

Despite the multitude of examples linking economic dependence on resource extraction with a decrease in democratic governance, the correlation is neither causal nor inevitable. Many established states with strong democratic traditions have managed to exploit their extractive resources without sacrificing responsive governance, choosing to rely on extensive public input to ensure that oil revenues support public priorities and benefit the country's population in both the short and long term. Norway, which has chosen to apply its oil revenues to ethical and sustainable long-term foreign investments, is a case in point. (See Box 17–1.)[11]

The New Face of the Resource Curse

Historically, industrialized, democratic nations with strong governance institutions and diversified economies have been considered immune from both the domestic and international impacts of the resource curse. Yet the economic importance of fossil fuel extraction in resource-rich countries such as Canada, Australia, and the United States presents similar—although perhaps less extreme—challenges to governance, with negative effects on both climate mitigation and diplomacy.

Canada, once considered one of the most environmentally progressive industrialized nations, has recently scaled back its political ambitions to address climate change, despite evidence pointing to strong public support for climate action. National polling indicates that in 2007, 46 percent of the Canadian public viewed the implementation of government standards and regulations as the most essential way to fight climate change; by November 2012, 59 percent of Canadians favored such measures.[12]

Yet in December 2011, Canada became the first country to withdraw from the Kyoto Protocol,* shelving the previous administration's plans to meet its Kyoto targets by implementing mandatory emission cuts for large factories and power plants, improving fuel efficiency, facilitating the purchase of emission reduction credits, and assisting municipalities and provinces in local efforts. In addition to adopting a much weaker greenhouse gas reduction target, the government reduced funding for Canada's climate change plan and cut several key programs, including the Wind Power Production Incentive. On issues of weighing environmental protection and economic development, a 2013 public opinion poll reported that 60 percent of Canadians support protecting the environment, even if that poses a risk to economic growth.[13]

Although the factors behind Canada's policy shift are varied and complex, the country's energy industry likely plays an important role. According to a 2010 report from the MIT Joint Program on the Science and Policy of

---

* In 2012, Japan, New Zealand, and Russia joined Canada in withdrawing; they are currently the only nations to have made commitments under the first Kyoto commitment period that did not make new commitments under the second period from 2013–20.

Box 17–1. The Norwegian Oil Fund

Unlike many of its fellow petro-states, Norway has managed to avoid many of the challenges commonly associated with a high dependence on oil revenue. Although the factors behind this success are numerous and diverse (and include a long tradition of strong democratic governance and institutions), Norway's efforts to ensure that its resource revenues benefit the public over the long term—and to develop a system allowing for public ownership and management of the nation's oil revenue—provide a model for other countries looking to avoid the resource curse.

In 1990, Norway created the Government Petroleum Fund (now named the Government Pension Fund) to manage the country's growing oil wealth. Commonly referred to as "The Oil Fund," it is now the world's largest pension fund—valued at some $800 billion at the end of 2013—and the single largest European stockholder. The fund, designed to "facilitate government savings to finance rising public pension expenditures and support long term considerations in the spending of government petroleum revenues," channels oil revenue to long-term, exclusively foreign investments.

The Fund directs a small share of oil earnings back into domestic programs such as infrastructure development and education, but great care is taken to ensure that the money is saved rather than spent in the near term. The reasons are twofold: first, long-term investments are necessary in order to protect against a projected shrinking of future oil revenue, and second, limited domestic spending is necessary to shield Norway from the ill effects of a resource-based economy—such as the declining competitiveness of the manufacturing sector, often referred to as "Dutch disease"—that have plagued other resource-dependent states.

The investment strategy seeks opportunities to achieve a high rate of return at moderate risk through firms promoting sustainable economic, environmental, and social development. Likewise, the Fund includes strict ethical guidelines to ensure that investments target only institutions that adhere to certain standards of operation, and cannot be directed to companies that contribute to killing, torture, deprivation of freedom, or human rights violations. Beyond these overarching principles, the Fund has published specific guidelines on how companies are expected to operate in respect to the country's high-priority areas of children's rights, climate, and water.

One of the most important aspects of the Fund's design is the inclusion of a strict set of transparency requirements, including public disclosure of the Fund's goals and regulations, management, and holdings. Norway has made nearly all aspects of the Fund's operations public, and each of the agencies sharing responsibility for its operation is held accountable to the public as well as to its cost managers. Overall, this has led to the operation of one the world's best-functioning sovereign wealth funds with respect to both its ethical standards as well as its 13 percent return on investment.

Source: See endnote 11.

Global Change, progress in global climate policy would threaten Canada's lucrative oil sands industry. The study found that in scenarios where developing countries participate actively in climate policy, "there appears to be little role for Canadian oil sands at least through the 2050 time horizon." If worldwide demand for petroleum fell as a result of progressive international climate policy, oil sands would not be competitive with conventional petroleum, and demand would decline. "The niche for the oil sands industry," the report concludes, "seems fairly narrow and mostly involves hoping that climate policy will fail."[14]

Australia's government, too, has sought to reverse key progressive climate policies in recent months, including the country's carbon reduction target and carbon tax, as well as its Climate Commission, which could leave Australia with virtually no national climate strategy moving forward. Although the Australian public is far from unified on the causes of climate change, approximately 90 percent of the population believes that there is an urgent need to combat the problem. Polls indicate that domestic public opinion increasingly favors Australia taking a leading role in the search for climate solutions, and popular support for such action actually increased in 2013 for the first time in half a decade.[15]

Despite widespread global opposition to Australia's proposed rollback of its carbon tax, at least one foreign government voiced support. In November 2013, Canada praised the actions taken in Australia, declaring that the "Australian Prime Minister's decision will be noticed around the world and sends an important message." Australia and Canada recently joined together to block the creation of a Commonwealth Climate Change Fund, which, if implemented, would have provided financial assistance to help small-island states and African countries in the Commonwealth deal with the impacts of climate change.[16]

The emergence of new technologies that enable major emitters such as China and the United States to rapidly expand their domestic energy production adds to this disturbing trend. In many ways, the United States is considered the global face of climate change inaction, given its rampant fossil fuel consumption and contentious politics. Many in the climate movement argue that funding from private interests—the fossil fuel industry chief among them—has a destructive impact on national governance in the country, particularly in the energy sector. (See Chapters 11 and 20.)

The fossil fuel industry pours significant sums into the U.S. political system. Although the direct influence of this support on legislative decisions is subject to debate, the industry's heavy spending is well documented. Over the two-year term of the 111th Congress, the fossil fuel industry spent an estimated $347 million on lobbying and campaign contributions. Over that same period, the government awarded approximately $20.5 billion in subsidies to the industry. Close personal ties between the industry and those tasked with its regulation have been cited as a cause for concern. Despite the Obama administration's strong rhetoric in favor of climate governance and efforts to regulate emissions, U.S. oil and gas exploration is at an all-time high, with the International Energy Agency predicting that the United States will become the world's top oil producer by 2015.[17]

Escaping the Curse

In the realm of environmental management, it has long been considered the job of national governments to enact regulations and promote policies to safeguard their citizens. Across most of the industrialized world, destructive behaviors once considered common practice—such as polluting rivers with industrial waste—are now strictly regulated, often as a result of public pressure. One might argue that the response to climate change should work the same way: that is, once countries are aware of the serious negative impacts associated with fossil fuel combustion—and are pressured by growing public demand for reform—they should enact regulations to reduce emissions and ensure a habitable planet.

This supposition, which assumes both the responsibility and responsiveness of individual countries, currently shapes the international community's approach to climate change. As the science on climate change improves, governments should act on this new information to protect their citizens. This approach is codified in the UNFCCC, which states that countries "should take the lead in combating climate change and the adverse effects thereof." The annual UNFCCC Conferences of the Parties (COPs) provide a forum for countries to engage in national climate diplomacy and are designed to ultimately produce an international agreement to address global climate change when the parties gather for the twenty-first time in Paris in 2015.[18]

This belief in the ability of countries to facilitate change is not unfounded. Under the UNFCCC process, several countries—such as Denmark—have emerged as climate leaders. Many others have taken action unilaterally or in collaboration with other countries. At the regional level, the European Union made significant progress by introducing the EU Emissions Trading System, notwithstanding the system's current struggles. Yet despite these successes, a global pact remains elusive and existing national pledges to reduce emissions remain markedly insufficient. In the long build-up to COP 21, it appears that most countries have not acted in the way many had hoped, or with the urgency required to avoid significant climate change.[19]

Despite the predominance of national governments in the international climate governance regime, recent analyses point to the significant role of nongovernmental entities—particularly in the fossil fuel and cement industries—in emitting greenhouse gases. Of the top 90 major carbon producers, 50 are privately owned corporations, 31 are state-owned corporations, and only 9 are countries. Overall, investor-owned corporations have been responsible for 21.7 percent of carbon dioxide and methane emissions since 1750, with state-owned corporations responsible for an additional 19.8 percent. Although national regulations can play a role in overseeing corporate behavior within domestic borders, they fall short in influencing the overall operations of multinational corporations. Finding ways to engage directly with energy industry emitters represents an opportunity to transcend national borders and influence climate action on a broader scale.[20]

Private industry needs to play a significant role in achieving necessary emissions reductions. Some have suggested that this is not just an opportunity to broaden the discussion, but that these organizations have "an ethical obligation to help address climate destabilization." If the current international approach to addressing climate change is expanded to include actors beyond national governments, corporate energy interests could become partners in the search for a solution, rather than observers or vilified symbols of obstruction.[21]

Conclusion

For climate change activists, an expanded understanding of the resource curse's negative impacts on governance can help illuminate some of the significant obstacles preventing national governments from taking action. Domestically, citizens in oil-dependent states often have a severely limited capacity to influence government decisions. Internationally, despite their critical role in the world economy, energy exporters often hesitate to engage in issues of global governance. Although these challenges were once considered only in respect to governments in the developing world, expanding energy interests in industrialized, democratic countries have contributed to reversals in positions on international climate governance and to the rolling back of progressive domestic measures, even in the face of increased public pressure for governmental climate action.

Unfortunately, the international community's reliance on national governments to solve the climate crisis runs head on into these problems, contributing to the challenge of negotiations already fraught with contention. The examples discussed here suggest that both developing and industrialized countries with strong resource extraction industries—even those that have managed largely to avoid the economic consequences of the resource curse—will continue to face severe governance challenges. They will likely remain either unwilling or unable to challenge powerful fossil fuel interests to the degree required to address climate change, even if public support for climate mitigation increases.

This should prompt climate activists and all those in search of effective levers to influence climate policy to search for new tactics. Although public movements to pressure political leaders must play an essential role, they likely will not be sufficient. To drive real change, governments and the public must find ways to actively engage private industry. This can and should take multiple forms, including efforts to both confront these actors and constructively engage them. The recent grassroots movement advocating for divestment from fossil fuel interests has spread to cities, religious institutions, and hundreds of colleges and universities across the United States, hinting at new and potentially effective strategies to shift pressure directly onto industrial actors. Other strategies, including those aimed at shifting consumer behavior, may also be powerful tools for climate activists.[22]

In addition to these more confrontational approaches, there are more collaborative ways of engaging with corporate actors, and some of this is already taking place in various UN processes. (See Chapter 15.) In order to be effective, the climate movement must continue to search for innovative ways to engage directly with fossil fuel energy interests, using a combination of both consumer-driven pressure and a willingness to expand traditional governing structures beyond the realm of national governments. The failure of governments to solve the climate crisis alone—combined with the important role that energy corporations play in the evolving global economy—necessitates that these private actors have a seat at the table. While this certainly does not guarantee success, it could be a significant step forward in building a governance framework that is best equipped to address the fundamental nature of climate change.

CHAPTER 18

The Political-Economic Foundations of a Sustainable System

Gar Alperovitz

It is increasingly obvious that the current political and economic system, globally but especially in the United States, is largely incapable of addressing the big sustainability challenges that the world faces today. Many of our economic and ecological problems are global in nature, but the United States faces some unique challenges that make the crisis even more acute. Unlike in Western Europe and Japan, where population is projected to be relatively constant, the U.S. population is set to grow by at least 100 million—and likely 150 million—by 2050. Where and under what conditions these people will live present serious challenges to sustainability planning. U.S. cities today are so spatially and economically unstable that anything beyond superficial sustainability planning is impossible.[1]

Adding to this challenge is America's "throwaway city" habit: as jobs move in and out of cities in uncontrolled ways, the country literally throws away housing, roads, schools, hospitals, and public facilities—only to have to build the same facilities elsewhere at great financial, energy, and carbon costs. All the while, the instability makes it impossible to carry out coherent regional planning. Detroit and Cleveland are dramatic examples; the population of Detroit, a city once home to nearly 2 million, has slumped to barely 700,000, while Cleveland's population has declined from a peak of 915,000 in 1950 to about 390,000 today. Of the 112 largest U.S. cities in 1950 (those with populations over 100,000), 56 had experienced population decline by 2008. The people moved elsewhere, where facilities had to be built anew to serve them and were built under conditions that were inherently prone to future instability and disruption.[2]

Beyond the challenge of throwaway cities is the sprawling of U.S. metropolitan areas, which has direct impacts on carbon emissions per person, a key measure of sustainability. A 2009 report published by the International Institute for Environment and Development showed that cities have significant ecological advantages over suburban areas. New York

Gar Alperovitz is Lionel R. Bauman Professor of Political Economy at the University of Maryland and cofounder of The Democracy Collaborative. The author would like to thank Steve Dubb for important research and other contributions to this essay.

Paul Sableman

Suburban sprawl in Houston, Texas.

City, for example, emitted 7.1 tons of carbon per person per year on average, compared to 23.9 tons per person nationwide in the United States. Likewise, Londoners emitted 6.2 tons of carbon per person annually, compared to a British national average of 11.2 tons. So it ought to be of major concern that a 2010 study of residential construction in the 50 largest U.S. metropolitan areas found that between 2003 and 2008, suburban areas accounted for the majority of new construction in nearly every metro area (and more than 85 percent of new construction in nearly half of the areas).[3]

Public policy could, in theory, alter these and related trends. But getting serious about sustainability requires focused attention on why public policy support has, at best, been able to slow but not stop ecological deterioration. The roots of this challenge lie in the growing concentration of wealth and income and the consequent self-reinforcing capture of the machinery of politics to serve private ends. The best-laid plans to foster sustainability will fall apart if social and economic pressures cause residents to leave for the sprawling suburbs or perhaps to a fast-growing metropolitan area in another region.

Equally important, there are strong reasons to believe that the politics of sustained "green" mobilization at the local and metropolitan level cannot work unless there is a baseline of economic health, and indeed unless bolstering economic security is a central part of such mobilization; when the economy goes sour, all other priorities fall by the wayside. Even the real urgency attached to the global climate crisis seems less immediate to elected officials and their constituents than the here-and-now pain of unemployment and insecurity when times are tough. But the urgency of the climate crisis itself means that we can no longer afford for environmentalism to go out of fashion when the economy is struggling.

To illustrate the extent of the problem: in the United States, membership in private-sector organized labor has fallen from 35 percent of the labor force in the 1950s to 6.9 percent today (counting the public sector, the share is at 11.9 percent and falling). This fact is central to issues of sustainability, because at the center of the traditional progressive political success has been the ability to contain the corporation, economically and politically, through political mobilization—especially aided, abetted, and bolstered by the organizational and financial power of labor unions. The "main finding" of international research on the relationship of union membership to political outcomes in industrialized countries, the late Seymour Martin Lipset and Noah Meltz observe, is straightforward: "Support for unions is associated with social democratic strength."[4]

Studies of European social democracy by Emory University sociologist Alexander Hicks also reveal a "[near]-perfect relationship between mid-century [social] program consolidation and working class strength in five major areas of social insurance policy—retirement, work-injury, illness, unemployment, and child-rearing compensation." Even when labor unions and environmentalists have bickered over specific issues, the power of labor has been critical to the election of progressive political leaders; Senator Gaylord Nelson, the founder of Earth Day, was a labor lawyer.[5]

Confronting the realities of labor's decline is not easy. But these realities mean that unless some new institutional source of political capacity is developed, advocates of economic justice and sustainability can expect to face continued hard times. Real wages for 80 percent of American workers, for instance, have risen only trivially for at least three decades. At the same time, income for the top 1 percent has jumped from roughly 10 percent of all income to roughly 20 percent. Virtually all the gains of the entire economic system have gone to a tiny, tiny group at the top for at least three decades.[6]

Fifty million Americans live in officially defined poverty, a rate higher than in the late 1960s—another disturbing trend marker. Moreover, if we used the standard common throughout the industrialized world (which considers the poverty level to be half the median income level), the number would be just under 70 million and the rate almost 23 percent. This is to say nothing of an unemployment rate that, if properly measured to include involuntary part-time work and discouraged workers, is stuck in the range of roughly 15 percent.[7]

What Does Justice Require?

A politics capable of altering ecological and social justice outcomes must be a politics that addresses these foundational issues. Wealth and income inequality have ill effects for rich and poor alike. This is most notably documented in *The Spirit Level: Why More Equal Societies Almost Always Do Better* by British epidemiologists Richard Wilkinson and Kate Pickett, which demonstrates that regions with greater income equality have better health outcomes (life expectancy, etc.) for people of all social strata.[8]

Moreover, because knowledge and wealth tend to be produced collectively and incrementally, the enormous inequities of today are largely undeserved, since most of the technology, on which wealth creation depends, was created by others before we were born. Who deserves the benefits of the steam engine? Even the invention of the computer pre-dates many of us now alive. We all benefit from the economic value of this inheritance, regardless of whether we are industrious or slovenly. Simple justice requires that much of society's welfare should benefit the vast majority who are the logical inheritors of the technologies created in previous eras, often with significant public support.[9]

If this idea is surprising, perhaps it is because of the widespread misperception that there is a strict dichotomy between private and public, market and state, free individual activity and coercive government power. Most people believe that, on one side, there is the private marketplace where individuals work and make productive contributions, and then receive rewards—wages, benefits, wealth, etc.—that are roughly equal to the value of what they contribute to the economy. On the other side, there is the "dictatorial" realm of government, which confiscates individual earnings for the greater good.

But this view is profoundly unrealistic about the sources of value and growth in an advanced society. "Private" market activity—and pre-tax income—is already *highly socialized* in many ways before governments start "spreading the wealth around" by tax policy. Modern research has demonstrated that the overwhelming share of each individual's gains is actually *unearned surplus* derived largely from technological gains made in the past, "an increase in output that is not commensurate with the increase in effort and cost" contributed by today's current market actors, as economic historian Joel Mokyr observes.[10]

There are many obvious examples of such collective subsidy in the marketplace. Government-funded research and development (responsible for, among other things, the Internet), as well as government-created markets (through procurement) provide a huge collective subsidy for private gains, a key public foundation of private wealth. Public education is another example: some experts judge that 15 percent of total productivity gains during the twentieth century resulted from advancing education levels in the workforce, as free, universal K-12 schooling became the norm.[11]

Lacking a better empirical understanding of the economic impact of common assets—most importantly our expanding inheritance of scientific and other forms of productive knowledge and know-how—public debate will continue to be controlled by moral arguments pitting strong assumptions of individual "deservingness" in the private economy against equally strong assumptions of "undeservingness" in the development of social policy.

Yet if we are serious in holding that contribution matters, then society "deserves" far, far more. Herbert Simon, a Nobel-laureate economist, employed this concept in a forceful attack against growing inequality: "If we are very generous with ourselves, I suppose that we might claim that we 'earned' as much as one-fifth of [our income]. The rest is the patrimony associated with being a member of an enormously productive social system, which has accumulated a vast store of physical capital, and an even larger store of intellectual capital—including knowledge, skills, and organizational know-how held by all of us."[12]

One of the most influential and penetrating advocates of these ideas, Leonard Trelawny Hobhouse, understood the moral task in a way that resonates strongly today. As he wrote in his 1911 book *Liberalism*, "The true function of taxation is to secure to society the element in wealth that is of social origin, or, more broadly, all that does not owe its origin to the efforts of living individuals." An "individualism which ignores the social factor in wealth" is no individualism at all, but rather a type of "private socialism" that "deprive[s] the community of its just share in the fruits of industry and so result[s] in a one-sided and inequitable distribution of wealth."[13]

Private socialism occurs when the wealth generated by common assets is not shared generally but is captured by a small minority, and when, at the same time, the public absorbs the losses when the wealthy fail. At a time when the top 400 individuals in the United States have a combined net worth of $2 trillion—a good third more than the poorest three-fifths of the U.S. population combined!—we can safely conclude that private socialism has run amok.[14]

Building an Alternative

As resistance has grown to the widening gulf between the top 1 percent and the rest of the population, more Americans have looked to community wealth building as the place to begin developing an alternative. The central idea is simple: people join together through some form of public, community, or employee-owned business to meet local needs and thereby regain a measure of local economic democracy and control. Community wealth building institutions include community development corporations, community development financial institutions, social enterprises, community land trusts, employee-owned enterprises, and cooperatives.

All of these institutions pool capital in ways that build wealth, create living-wage jobs, and anchor those jobs in communities. The efforts also provide a new approach to challenging corporate power—a strategy that changes who owns, controls, and benefits from the underlying economic wealth of the system. It displaces private capital by developing community ownership of business. Profits flow to workers, consumers, or the community, rather than to outside investors.

Nonprofit social enterprise is a community wealth building strategy through which nonprofits secure resources to pursue their missions when government support is inadequate. In San Francisco, for example, a group known as REDF (formerly the Roberts Enterprise Development Fund) has helped boost the business activity of 50 social enterprises that have employed 6,500 people and earned revenues of more than $115 million. Three-fourths (77 percent) of social enterprise employees interviewed two years after first being hired were still working. Average wages increased by 31 percent and monthly incomes by 90 percent.[15]

In Grayland, Washington, Coastal Community Action—a nonprofit agency that operates a range of housing, food, healthcare, and employment programs—has built a six-megawatt wind farm that sells energy to the electrical grid, generating enough power to satisfy the energy needs of more than 1,500 households. The nonprofit estimates that its ownership of the $14 million project generates $720,000 in unrestricted income each year, enabling it to increase service delivery options, lessen its local dependence on outside funding, and meet more of its community's needs.[16]

In Seattle, Pioneer Human Services offers drug- and alcohol-free housing, employment, job training, counseling, and education to recovering alcoholics and drug addicts. Founded in 1963, it employs 1,000 people and finances 99 percent of its $70 million budget through fees for services and earnings generated in the manufacture, distribution, and sale of products. Businesses include retail cafés, sheet metal fabrication, aerospace precision machining (the group is a contractor for Boeing), wholesale food distribution, and contract packaging. Not only do these enterprises build community wealth and finance social services, but the businesses themselves are central to Pioneer's mission of helping people on the margins of society stay out of prison and off the streets, enabling Pioneer to employ more than 700 men and women drawn from the ex-offender, homeless, and drug-recovery populations that it serves.[17]

Community development corporations (CDCs), formed initially in the 1960s in a crucible of urban riots and rural neglect, now are community wealth builders across the United States. CDCs can be found in virtually every major city. A Massachusetts study found that between 2003 and 2012, Massachusetts-based CDCs created or preserved more than 13,000 homes and 22,000 jobs, and generated $2.7 billion in economic investment. A 2010 national study found that, over the previous two decades, CDCs produced more than 1.6 million units of affordable housing nationwide.[18]

Community development financial institutions (CDFIs), first given federal recognition in the 1990s, aim to build wealth in low-income communities by providing financing where conventional lenders fear to tread. Even in the face of a weak economy, assets in U.S. community investing institutions have soared from $25.8 billion in 2007 to $61.4 billion in 2012.[19]

Community land trusts provide still another powerful illustration of community wealth building. Beginning in the 1960s and 1970s, pioneers like Bob Swann in western Massachusetts and Charles Sherrod in Georgia struggled against huge odds to develop modest land trust efforts, often also involving other concerns, such as respect for environmentally sound land use practices and rural community development. Today, hundreds exist; in Irvine, California, the city's strategic plan calls for 5,000 units of housing to be developed using land trust strategies.[20]

Artist's rendering of a net-zero, mixed-income development of 11 homes, 2 rental units, and an office/resource center built by Lopez Community Land Trust, Lopez Island, WA.

Image courtesy of MITHUN

Trusts of this kind keep the ownership of land underlying housing in nonprofit or public ownership. Appreciation in land values is split between the homeowner and the trust, thereby avoiding gentrification. A study of a community land trust in Burlington, Vermont—the nation's largest—found that during its first two decades, 61.9 percent of residents who sold their land trust home after an average residency of six years were able to take on traditional homeownership. Meanwhile, the increased equity that the trust retains enables it to continue providing affordable housing to future generations. In a down market, community land trusts are even more important, as they can keep people in their homes. A 2011 study found that at the end of 2010, land trust homeowners were 10 times less likely to be in foreclosure proceedings (0.46 percent of all units) than conventional homeowners (4.6 percent).[21]

Employee ownership is yet another powerful community wealth building strategy. The National Center on Employee Ownership estimates that in 2009, there were 9,800 companies owned in whole or part by workers through their pension contributions, a form of ownership known as an employee stock ownership plan (ESOP). As of 2009, 10.3 million Americans were employee-owners of companies owned in whole or part by ESOPs, with net assets of $869 billion. The average employee-owner had an ownership stake of over $84,000.[22]

Employee ownership has powerful economic stabilizing effects: between 2000 and 2008, while the number of manufacturing jobs fell 29 percent in the state of Ohio, employee-owned manufacturing jobs declined only 1 percent. Across the United States in 2010, 12.1 percent of all workers were laid off in the previous 12 months; by contrast, only 2.6 percent of workers who were employee-owners lost their jobs. In addition, employees at ESOP companies have, on average, 2.5 times more retirement benefits than employees at comparable companies that are not employee owned. Depending on the industry, wages are 5–12 percent higher than those at jobs in comparable non-employee-owned companies. Productivity at employee-owned companies is also higher (which is why ESOP companies can provide higher wages and better benefits). On average, productivity increases 4–5 percent in the year after an ESOP is adopted; over a given 10-year period, ESOPs have 25 percent faster job growth than comparable non-ESOP companies.[23]

Perhaps the most visible form of community wealth building is the cooperative. More than 130 million Americans are currently members of a co-op or credit union. Because many Americans own shares in more than one co-op or credit union, the total number of co-op memberships in the United States exceeds 350 million. A 2009 University of Wisconsin study found that nearly 30,000 cooperatives in the United States account for more than $3 trillion in assets, $514 billion in total annual revenue, and 856,000 jobs.[24]

In Oberlin, Ohio, what David Orr calls an "integrated or full-spectrum sustainability" approach drawing on such efforts aims to build a sustainable economy, become climate-positive, restore a robust local farm economy supplying up to 70 percent of the city's food, educate at all levels for sustainability, and help catalyze similar efforts across the United States at larger scales.[25]

International Developments

Many examples of community wealth building can also be found in other countries, including the well-known worker cooperative movement in Argentina, which has seen hundreds of worker co-ops open their doors over the past two decades. Many of the co-ops, known as *empresas recuperadas de trabajo* (worker-recovered businesses), emerged out of the economic collapse in 2001, in which workers occupied abandoned factories and put them back into operation under worker self-management. (See Box 18–1.)[26]

Another prominent non-U.S. example is provided by the Mondragón cooperatives. In 1943, a Spanish priest, Father José María de Arizmendiarrieta (1915–1976), founded a technical school in the small city of Mondragón, in the Basque region of Spain. In 1956, five graduates of that school helped found a worker cooperative called Ulgor, which initially employed 24 people and produced kerosene stoves. In 1959, a cooperative bank, Caja Laboral, was founded, which proved to be a critical vehicle for financing ongoing expansion. United today by a worker-controlled holding company, the Mondragón Cooperative Corporation has grown to consist of 257 businesses, including the Eroski retail chain with over 200 hypermarkets, super-

Box 18–1. Ten Years On: Argentina's "Recuperated" Worker-Owned Factories

In the late 1990s, Argentina entered an economic depression that led to shattering unemployment, inflation, default on foreign debt, and ultimately the fall of the government. One response to the economic upheaval was the *empresas recuperadas*, or "worker-recovered businesses." This movement gained enormous momentum after the collapse, when foreign investors saw their businesses in Argentina's industrial sector crumble, and consequently closed up shop. Workers at some of these factories, who already knew how to run the businesses and operate the machines, refused to let their former workplaces lie cold and vacant while they were out of work. One by one, they began to occupy their factories and demand the right to work (protected under Argentina's constitution) and to resume production as worker-owned cooperatives.

Their logic was simple: since their labor produced all the added value for the products and because their employers had walked away from their businesses, it was their only option and also their right to run the factories themselves, under horizontal direct democracy. Once a group of workers decided to take over a factory, a long and often complicated judicial process awaited them. They camped out, sometimes for months, in or near their workplaces to ensure that the former bosses didn't gut the factory and sell the machines in the middle of the night. Early in the process, many occupations were repressed and police turned violent as they tried repeatedly to evict the entrenched co-op members. But the process has now become more streamlined and normalized.

Some 300 businesses now operate under worker control as cooperatives in Argentina. Not all are "recuperated," but each has been motivated and inspired by the promise of direct democracy in the workplace and the struggle of their fellow workers. Times remain difficult for everyone, and prosperity is still an intangible goal for many small business owners. Many of the co-ops keep daily operations running smoothly, but they have few savings and no access to bank loans or the pension programs that are available to traditional businesses. As a result, in addition to the difficult economic conditions, these businesses cannot use many standard tools of the financial system because banks do not recognize their management structure. Worker-owners are often fiercely proud of how far they have come and are both undaunted and realistic about the challenges in the future.

The economy remains shaky in Argentina, with an annual inflation rate of around 30 percent. Investing with Argentinean pesos is unwise. The frustrating lack of transparency surrounding most monetary transactions complicates efforts to conduct business at all—making it all the more remarkable that a small cadre of worker-owned businesses has begun to flourish in the economic rubble. Worker ownership means banding together with a few other workers and operating a business without a formal boss, making decisions about production and hiring with a one-worker, one-vote direct democracy. In Argentina, worker ownership demands trust against all odds, requiring faith in neighbors and

continued on next page

markets, and convenience stores, and Mondragón University, which offers training in management and business skills. With a workforce of more than 80,000, Mondragón, according its website, had revenues in 2012 exceeding €14 billion ($19 billion).[27]

Consumer cooperatives are used widely internationally as well. The

Box 18–1. continued

communities when the general economy and the job market cannot be relied upon to provide a living.

This movement has inspired immense hope for many around the world who saw factory occupation and recuperation as the beginning of a paradigm shift—a chance to build a new system within the broken shell of globalized capitalism. In the United States, this flood of energy and idealism was captured in *The Take*, a film by Naomi Klein and Avi Lewis, which outlines the struggle of one cooperative to gain

control of production in their former workplace. For many, the Argentine cooperative movement offers lessons and inspiration for building a production system that embodies some fairly radical principles while remaining connected to mainstream markets. Whether those lessons will survive translation to other venues remains to be seen.

—Nora Leccese
Senior economics major,
University of Colorado Boulder
Source: See endnote 26.

United Kingdom, the birthplace of the modern cooperative movement, is home to the world's largest single co-op, The Co-operative Group, which runs a broad range of member-owned enterprises, employs more than 90,000 people, and is co-owned by 7.6 million members. The Co-operative Group operates more than 2,800 food stores, 750 pharmacies, and 300 bank branches, as well as the U.K.'s largest funeral home company.[28]

Some countries have both strong consumer and worker co-op movements. In the Emilia Romagna region of Italy, roughly 60 percent of the region's 4.4 million inhabitants belong to at least one co-op. In the region, around 80,000 people work for a worker-owned cooperative. This is around 6 percent of the total workforce and 1.8 percent of the total population. In Bologna, 10 percent of the population works for a co-op.[29]

In Japan, the Seikatsu Club network includes both consumer co-ops and worker collectives, with a total membership of 307,000. The Seikatsu Club began in 1965, when several young activists organized 200 women (mostly homemakers) in Tokyo to form a buying club for the daily purchase of 300 bottles of milk. Over time, the cooperative extended the collective buying system to a range of other products, such as rice, fruit, frozen fish, household appliances, clothing, toys, and travel tickets, and developed its own product line of over 60 items.[30]

The Seikatsu Club now owns several dairies, a beef ranch, and a soap factory. It also has helped its membership, mainly middle-aged, middle-class Japanese women, to re-enter the workforce through worker collectives, the first of which was launched in 1992. Today, 582 such collectives employing more than 17,000 people engage in such activities as food distribution, food preparation, catering, recycling, childcare, and education.[31]

Next Steps

Clearly, enormous challenges remain. Building institutions of community and worker ownership is critical to create a stable economic base that allows for a sustainable economy to emerge. But to be effective, this economic institution building must be linked to a political strategy—so that growing economic power translates into effective political power.

There are four key factors at work. First, unless new ways to generate public revenues are found, transition funding—both nationally and locally—will continue to be limited. Second, a serious strategy must put forward a positive, coherent, integrated community-building plan to develop the local economy, create jobs, and build a tax base that can finance existing services and generate new revenues without unduly burdening working-class taxpayers. On the other hand, as we have just seen, practical precedents for the elements needed are now available. And politically, a positive plan can help unite key constituencies, such as community activists and unions, around a common agenda.

The specifics, of course, will differ by community. But what is needed is a longer-range strategic understanding. One cornerstone would be to build upon existing precedents in the ownership of land and worker-owned enterprises to promote locally anchored jobs and investment. Many tools can be employed to support this strategy, such as leveraging local government procurement (and encouraging university and hospital participation) to buy more goods and services locally, especially by supporting the development of worker/community-owned and anchored businesses.

The fact that these initiatives are based in local, everyday experience offers possibilities for long-term changes in the foundations of political and democratic cultural development over time. The economic security of individuals is essential to building political support for a sustained green transition. If low-income and minority constituencies fail to embrace the green economy, politicians will continue to place other priorities higher.

Related to this is that local economic stability is a prerequisite for a sense of community and—critically—for durable democratic decision making. Without such stability, the local population is tossed hither and yon by uncontrolled economic forces that undermine any serious interest in the long-term health of the community. And to the extent that local budgets are thereby heavily stressed, local community decision making is so financially constrained as to make a mockery of democratic processes.[32]

The ultimate goal of these strategies is to undermine and eventually replace the destructive "grow or die" imperative inherent in the current market-driven system. To do so, it is necessary to confront the systemic dynamics that promote a continued focus on growth. Former presidential adviser James Gustave Speth has bluntly observed that "for the most part, we have worked within this current system of political economy, but working within the system will not succeed in the end when what is needed is transformative change in the system itself."[33]

The local, foundational development of new, democratized ownership forms is critical because, in many ways, these are the beginning points for larger ongoing strategies. Ultimately, however, a longer-term systemic approach would have to apply similar principles to municipal, state, regional, and national level institutions. In the United States, many of the most important innovations of President Franklin Roosevelt's New Deal were merely scaled-up applications of principles that had been developed in projects undertaken in the state and local "laboratories of democracy" during the prior decades. As the current economic and ecological crisis deepens, a similar process may project local learning into ever-larger spheres of impact.

The most likely next stage areas for larger-scale democratization in the United States are banking (where future crises are predicted by many experts, and interest in public banking such as that of North Dakota has been building at the state level) and health care (where the current private system consumes almost twice the share of GDP as in many advanced systems elsewhere—with far worse outcome results). There are also likely to be challenges in systemically critical industries. The United States nationalized the General Motors Company during the last crisis, only to sell it back to private investors once the taxpayer had accepted the burden of change. In future crises, different outcomes may well become possible if a prior build-up of ideas of democratization has occurred at the local level.[34]

Taken together, the various emerging strategies suggest a long, slow developmental arc that is gradually gaining momentum in the wake of the failure of conventional politics and economics. The path to building a truly democratic economy may be long, but the growing base of community wealth building institutions provides many building blocks that, over time, suggest the quiet, foundation-laying development of the basis for a sustainable, and community-sustaining, economy.

CHAPTER 19

The Rise of
Triple-Bottom-Line Businesses

Colleen Cordes

A remarkable new breed of business is volunteering to be held publicly or even legally accountable to a triple bottom line: prioritizing people and the planet, while also promoting profits. This emerging movement is still a small phenomenon relative to the total global economy, but it continues to expand, led by mostly small and medium-sized companies in the United States, and to a lesser degree in Canada and Chile. Almost all are privately held, although a few major corporations have recently become connected through subsidiaries they have acquired.[1]

Some of these companies have been lobbying—often successfully—for new "benefit corporation" statutes that specifically allow them to incorporate with an official requirement to strive for a positive public purpose beyond financial success. Others have been seeking out and publicizing sweeping assessments of their companywide social and environmental impacts by independent third parties. And a significant number have been doing both.

This entrepreneur-led movement for an ethics-infused capitalism challenges business as usual, and there are already signs of pressures to dilute it. Yet whether its vision and values can be scaled up broadly, rapidly, and rigorously is a vital question.

Given the supersized global impacts of for-profit enterprises, sustainable economies are likely to remain elusive without substantial shifts in corporate norms. The conventional model for publicly traded corporations—regardless of the personal morality of the people who manage and direct them—has been a myopic focus on maximizing short-term financial returns for a relative few, typically investors and top executives. Attention to the consequences for everyone else on the planet, and for the planet itself, has been frequently downgraded to a matter of doing the minimum necessary to stay within the bounds of the law.

In pursuing that single-minded goal, many large companies have spent their way to outsized influence over governments at every level, resulting

Colleen Cordes is a public policy consultant and director of outreach and development for The Nature Institute of Ghent, New York.

too often in lax regulations, low legal standards for corporate behavior, and even lower public expectations. In many cases, this has set up a race to the bottom, in terms of corporate ethics. Even community-minded companies have felt pressured to cut ethical corners to compete. These excesses, especially in the absence of an alternative corporate vision to fire the imaginations of governments, citizens, and corporate leaders themselves, have translated into incremental and frustratingly slow progress toward environmentally and socially just societies. Put simply, the conventional economic model—amoral capitalism—and the willingness of so many investors and consumers to tolerate it are two of the most challenging threats to preserving a livable human future.

In the last few years, however, public restlessness around the world has been growing with revelations about the environmentally reckless behavior and fundamental social inequities that the conventional model has bred. Among the groups that have turned up the heat are grassroots activists and organized labor, a growing number of concerned investors and concerned customers, and national and international nonprofit groups advocating for human rights and ecological protections. They have prodded an increasing number of large, multinational corporations to acknowledge their companies' social and ecological responsibilities and to track their impacts. Over the last 15 years, for example, the number of businesses of all sizes that choose to self-assess how sustainable their operations are, using widely accepted social and environmental standards, and to publicly disclose their results has been growing rapidly, especially in Europe and Asia.[2]

But simply tracking results, often with an eye to preventing serious missteps that could destroy a brand's reputation in a hurry, will not speed the world to sustainable economies with the haste that is so urgently required.

From Shareholders Only, to All Stakeholders

The more dramatic movement noted above, led by privately held small and medium-sized companies, is aimed at something much more substantial: a new way of doing business. It is based on an expanded sense of mission that includes positive social impacts, environmental sustainability, and, in its most visionary form, not just sustained financial profits but also contributions to a more broadly and fairly shared prosperity. These companies represent a growing community of sustainability-minded entrepreneurs who champion high new standards for corporate governance and operations. They insist that for-profit businesses, as centers of enterprise and resources in a needy, crowded world, can and should be as committed to "doing good" as to "doing well" financially.

Especially in the United States, these companies are taking their vision to legislative bodies, promoting changes in corporate law to allow businesses that are so inclined to choose a new corporate form that specifies that their managers and directors must pay attention to company impacts on people and the planet, as well as to the generation of profits. The recent rapid adoption of statutes creating a new business status known as "benefit corporation" in the United States, in state after state, underlines the pivotal role that governments can play in laying the groundwork for such a shift in corporate focus.

A benefit corporation is a legal form, established under corporate law, that requires a company to state clearly in its original or amended articles of incorporation that it has a general purpose of having a positive impact on society and the environment and that its board of directors, in making decisions, is required to take into account the interests of multiple stakeholders in addition to the financial interests of its shareholders. The stakeholders that it must consider, by law, include the company's own workforce and that of its suppliers, its customers, the local community and general society, and the local and global environment.[3]

Benefit corporations also are required to report annually and publicly on their overall social and environmental impact as assessed against a transparent, credible, and independent third-party standard. Proponents of this new corporate form say that it essentially bakes a triple bottom line into a company's DNA. That frees companies from the fear of shareholder lawsuits if their decisions fail to maximize shareholder value because of some competing interest of other stakeholders, such as workers. Under current corporate case law in the United States, for example, corporate directors often are assumed to be liable in such suits, although legal scholars vigorously debate that point. Corporate attorneys, however, are likely to counsel directors to handle this question conservatively.[4]

Incorporation as a benefit corporation is intended to provide directors and managers with the legal cover they need to establish that they actually have a fiduciary responsibility to consider the interests of all stakeholders—not just shareholders. Formalizing a company's social and ethical purposes under this legal framework also makes it more likely that its good intentions will survive the departure of its founders or major spurts of growth, and that its directors will have the legal backbone to fend off buyout offers from conventional corporations without the same commitment.

Origins and Rapid Growth of Benefit Corporations

The movement for such a new legal form began in the United States under the leadership of a U.S. nonprofit called B Lab, which developed model legislation with the pro-bono help of William H. Clark, Jr., a Pennsylvania lawyer who specializes in corporate and business law, and other lawyers. The state of Maryland was the first to enact such a law, in 2010, which was

Table 19–1. U.S. Movement for Benefit Corporation Laws

States with Statute	Date Effective
Arizona	December 2014
Arkansas	August 2013
California	January 2012
Colorado	April 2014
Delaware*	August 2013
Hawaii	July 2011
Illinois	January 2013
Louisiana	August 2012
Maryland	October 2010
Massachusetts	December 2012
Nevada	January 2014
New Jersey	March 2011
New York	February 2012
Oregon	January 2014
Pennsylvania	January 2013
Rhode Island	January 2014
South Carolina	June 2012
Vermont	July 2011
Virginia	July 2011
Washington, D.C.	May 2013

*See Box 19-1 for how Delaware's statute differs from the general model that others follow.
Source: See endnote 6.

approved by lawmakers with strong bipartisan support, as it generally has been in other states.[5]

The benefit corporation movement continues to advance most rapidly in the United States: by November 2013, 18 states and Washington, D.C. had signed such legislation into law. (See Table 19–1.) A similar statute was adopted in Delaware, the state of incorporation for more than a million businesses, including many of the largest U.S. companies. Delaware in 2013 enacted a revision to its corporate code that creates a new corporate structure called a "public benefit corporation," whose requirements for transparency and public accountability are less rigorous than the version that other states are passing. (See Box 19–1.) Additional benefit corporation proposals were being discussed or had been officially introduced in another 13 states.[6]

Because some states do not keep records on how many companies elect to become benefit corporations, it is impossible to report the current total accurately. B Lab, however, compiles whatever information it can find, including that which states do make available. Based on its data, there were at least 344 benefit corporations in the United States (not counting Delaware's PBCs) by mid-October 2013. Most benefit corporations are either small or medium-sized businesses. But they do include a few larger companies that are privately held, such as the outdoor apparel and accessory firm Patagonia, Inc., which reportedly had annual sales of $540 million for the year ending April 2012, and King Arthur Flour, an employee-owned, 223-year-old company with reported sales of $84 million in 2010.[7]

In Delaware, another 44 companies had filed as "public benefit corporations" (PBCs) by mid-October 2013, according to B Lab's records. One of the first to register in the state was Method, a green cleaning supplies company that had merged not long before that with one of its main competitors, the Belgian company Ecover. The merger produced a privately held company with more than $200 million in annual sales. So while Ecover is privately held, it represents new European engagement in the PBC and benefit corporation community.[8]

Outside the United States, B Lab has partnered with an organization in

Box 19–1. Public Benefit Corporations in Delaware

Delaware's requirements for public benefit corporations (PBCs) differ from other states' typical requirements for benefit corporations in several ways. For example, PBCs do not have to formally commit to a general purpose of having a positive social and environmental impact. Instead, a PBC must identify in its certificate of incorporation one or more specific public benefits that the company is intended to produce. Delaware's statute also states that PBCs are intended "to operate in a responsible and sustainable manner."

Boards of directors of PBCs, however, are required in the company's certificate of incorporation to "manage or direct the business and affairs of the public benefit corporation in a manner that balances the pecuniary interests of the stockholders, the best interests of those materially affected by the corporation's conduct, and the specific public benefit or public benefits." The requirement to balance earnings with the "best" interests of those affected by the company is arguably a higher standard than what is required of directors under other states' statutes.

Yet Delaware's PBCs, unlike benefit corporations in other states, are not required to make available to the general public reports of their overall impact on all stakeholders. Instead, they must report at least every two years to their stockholders with self-assessments of how well they are doing in promoting the specific public benefit they have identified and "the best interests of those materially affected by the corporation's conduct." Their self-assessment does not need to be based on a third-party assessment tool.

There is no mention of the environment as a stakeholder, nor of trying to have a positive environmental impact as a necessary goal. The only reference to "environment" is inclusion of an "environmental" benefit in a list of examples of the kind of specific public benefits a PBC can choose to pursue. Only stockholders who own at least $2 million in market value or at least 2 percent of shares are allowed to challenge directors if they feel that the company is failing to pursue its special obligations as a PBC. In other states, any shareholder concerned that directors are not fulfilling their obligations to pursue positive social and environmental impacts can bring action against them. (In neither case can non-shareholders sue directors on such grounds, nor can directors be held financially liable for damages on such grounds.)

B Lab officials consider the Delaware law a significant advance and include it in their count of benefit corporation statutes. But they and other proponents of the benefit corporation movement also say they hope the Delaware law over time will be strengthened, especially in terms of transparency and reporting requirements. Some note, however, that Delaware generally designs corporate statutes with publicly traded companies in mind, since so many companies incorporate there, and that for those corporations the requirement to report to stockholders is essentially one to report to the general public. Such information would have to be publicly available, for example, if the company's stock were traded on the New York Stock Exchange.

The words "responsible and sustainable" as an intended goal seem ambiguous, and over time this could be interpreted as meaning financially sustainable, not environmentally responsible, given the absence of an environmental mandate in the law. Overall, given the differences between Delaware's PBCs and Delaware's dominance as the state of incorporation for so many corporations, it will be important for public interest advocates to monitor how the movement develops in that state and to support efforts to strengthen the statute there.

Source: See endnote 6.

Chile, Sistema B, to expand the movement to South America. Sistema B has worked in Argentina, Brazil, Chile, and Colombia, exploring with local partners whether the legal infrastructure is in place to allow companies to write or amend their articles of incorporation and bylaws to stipulate that they require themselves to consider the interests of all their stakeholders in making decisions. Jay Coen Gilbert, one of the three co-founders of B Lab, reported in 2013 that a national legislative proposal to provide that option was moving forward in Chile.[9]

B Lab is exploring additional regional partnerships to help expand its scope. That includes assistance in researching and developing, where needed, proposals to provide the legal infrastructure to protect companies seeking to establish a fiduciary responsibility for directors and officers to consider the interests of a broad range of stakeholders.

Certified B Corps and Other Third-Party Certifications

B Lab encourages companies to seek benefit corporation status through its own third-party certification process. The nonprofit assesses interested businesses in terms of their companywide environmental, social, and governance practices, and designates those that meet its requirements as "Certified B Corporations," or more informally as "B Corps." The community of Certified B Corps represents a growing advocacy group for the benefit corporation movement.

B Lab requires a B Corp, within a certain period of time, to establish legally in its governing documents that the board and officers must consider the interests of all stakeholders—not just shareholders—when making decisions (provided the company is headquartered in a location where such a requirement is now legally available). This could be done by filing for benefit corporation status within four years after a benefit corporation law has been passed in the company's state of incorporation, or before the company's two-year certification needs to be renewed, whichever is later. Or it could be done by amending articles of incorporation under a constituency statute, if one exists, before the certification period ends.

In the United States, corporate codes in 30 of the 50 states include constituency statutes, which allow but do not require companies to consider the interests of other stakeholders besides shareholders in making decisions. But there is little case law to determine whether directors and officers who make decisions based on the interests of other stakeholders, such as workers or the local community, would be held liable if such decisions failed to maximize profits. In places where neither legal option exists (either filing as a benefit corporation or amending articles under a constituency statute), Certified B Corps comprise a natural constituency to advocate for a benefit corporation statute. That increases the potential for country-by-country advocacy for this new kind of company with a general purpose of positive social and environmental impact.[10]

By October 2013, B Lab had certified 855 B Corps worldwide, up from 78 in 2007, the first year such certification was available. Annual gross revenues for all Certified B Corps in October 2013 were about $6.3 billion, and together these businesses employed some 33,000 people, according to B Lab. About half that number worked for the largest 25 Certified B Corps. Nearly 16,700 companies, interested in either becoming B Corps or getting a sense of their own impacts, had used B Lab's assessment tool.[11]

Certified B Corps were present in 27 countries as of October 2013, with the majority of this activity in the United States, followed distantly by Canada and Chile. (See Table 19–2.) As the number of certified companies in a country begins to grow, B Lab plans to work with those companies that are interested in exploring the need and opportunities for revisions in their home countries' legal infrastructure. A few companies in Australia have expressed an interest in this.[12]

B Lab reported in late 2013 that Ecover, the Belgian parent company of Method, was in the process of becoming a Certified B Corp itself and that B Lab expected a "B Lab Europe" to launch in 2014. There is interest in other parts of the world as well in building up a global community of B Corps that could build momentum for legislation similar to benefit corporation statutes in the United States, particularly in places where such legal encouragement for triple-bottom-line businesses is not yet in place.

In addition to B Lab, other major organizations are also promoting benefit corporation legislation in the United States. The new American Sustainable Business Council, which represents more than 73 business associations (which in turn represent more than 165,000 businesses) has established as a major public policy goal the promotion of the benefit corporation movement. The U.S. nonprofit Green America, which created the first U.S. green business network in 1983, also has advocated in support of benefit corporations. Green America has approved some 3,550 businesses for its Gold Green Business Certification, based on its assessment of their companywide environmental, social, and governance practices.[13]

Both B Lab and Green America are dominant players in a notable trend related to the benefit corporation push. In the United States and elsewhere, businesses are increasingly aspiring to hold themselves to higher standards of overall social and

Table 19–2. Global Reach of Certified B Corporations

Country	Certified B Corporations
	number
United States	648
Canada	86
Chile	37
Argentina	17
Colombia	13
Australia	12
Mexico	7
Brazil	6
United Kingdom	4
Guatemala	3
India	3
Kenya	2
New Zealand	2
Tanzania	2
Afghanistan	1
Belgium	1
China (Hong Kong)	1
Costa Rica	1
Ireland	1
Italy	1
Mongolia	1
Netherlands	1
Nicaragua	1
Peru	1
South Korea	1
Turkey	1
Vietnam	1
Total B Corps	855
Number of Countries	27

Source: See endnote 12.

environmental impact by trying to earn companywide certification from third-party organizations on a wide range of relevant criteria. Evaluating and comparing the rigor of third-party certifications is an area ripe for engagement by the nonprofit community of public interest advocates and environmental activists.

As with benefit corporations, most for-profits seeking companywide, third-party certifications are small to medium in size, although some companies with revenues in the hundreds of millions, such as Patagonia and Seventh Generation, are also included. Initially, such certifications were most often sought and won by small enterprises (often very small) that could be less constrained by outside investors' interest in maximizing profits. Over time, however, as the sustainability movement has expanded and as a much broader range of companies understands the advantages of public recognition for corporate social responsibility, larger companies are expressing interest as well.[14]

Almost all companies certified by B Lab and Green America are privately held. But the movement for companywide certifications includes a few subsidiaries of major multinational corporations (see examples below). And at least one Certified B Corp—Rally Software Development—is now listed on the New York Stock Exchange. Rally had won B Lab's companywide, third-party certification before going public in 2013 with a successful initial public offering that sold 6.9 million shares.[15]

Acquisitions by Larger Corporations Pose Complications

A few large corporations have recently moved to acquire smaller companies that had gained attention for their economic success in the sustainable business community. One of the first companies to file in Delaware as a PBC, for example, was Plum Organics, which just a month earlier had become a fully owned subsidiary of Campbell Soup Company, a Fortune 500 company that had revenues of about $8.1 billion in fiscal year 2013. Plum Organics, a major producer of organic baby and toddler foods, chose "working on hunger and malnutrition" as its specific public purpose. This appears to be the first instance of a large, publicly traded company becoming directly connected to the benefit corporation movement.[16]

Group Danone, a Fortune Global 500 company, recently bought a 92 percent stake in Happy Family, a fast-growing Certified B Corp that also produces organic foods for babies and children. Because Happy Family is incorporated in Delaware, the fact that the state now has passed its PBC statute means that B Lab will expect the company to either file for that status or forfeit its eligibility to renew its B Corp certification.[17]

Ben & Jerry's, a popular ice cream producer, since 2000 has been a wholly owned subsidiary of Unilever, a Fortune Global 500 company with annual sales of more than $70 billion in 2012 and more than 1,000 brand names. In 2012, Ben & Jerry's was the first wholly owned subsidiary to become a Certified B Corp, with support from its parent company. B Lab, in awarding that certification, created new requirements for transparency that it will expect of other wholly owned subsidiaries as well. These include posting online the full results of the companies' B Lab assessments and relevant portions of their governing documents to demonstrate that their directors are legally obligated to take into account the interests of all stakeholders. (Divisions or individual brands of larger corporations are not eligible to be Certified B Corps.)[18]

One recent controversy suggests how complicated these relationships may prove to be. Campbell Soup Company in 2013 contributed more than $384,000 to help the Grocery Manufacturers Association mount a massive campaign to defeat a ballot initiative in Washington state to label genetically modified organisms

Poster used in the Washington State ballot initiative campaign.

(GMOs) in food products. Campbell's gave half of that amount after its subsidiary, Plum Organics, had registered as a Delaware PBC. Anti-GMO groups that supported the initiative but were vastly outspent had publicized Plum Organics as a line of baby food products that parents trying to avoid GMOs could safely buy. For the businesses involved, the fallout from the revelation of Campbell's political involvement in fighting GMO labeling, which many public interest groups support, was negative. For the benefit corporation movement itself, the episode raises serious questions about whether large corporations that acquire smaller benefit corporations without a shared commitment to their new subsidiaries' particular social and ecological values will end up diluting both the identity and potential of the whole movement.[19]

Other big questions challenge the movement's long-term potential. Could it really ever be possible for large, publicly traded companies to fully embrace such a model? And, if so, can they do so quickly enough and in large enough numbers to help communities and countries shift to truly sustainable economies as rapidly as the situation demands?

The legal sufficiency of benefit corporation status has not yet been tested in the courts, as the statutes have been enacted so recently. Also, some nonprofits, while not necessarily opposing this new form, have raised concerns that benefit corporations may seek special tax treatment or other exceptional treatment from governments, or may compete for limited funding from do-good investors and donors, at the expense of the nonprofit community.

They are also concerned that benefit corporations could amass enough capital to underbid nonprofits for government contracts just long enough to drive them out of business and take over the public service space themselves, with no historical understanding or commitment to that space.[20]

Skeptics worry that if benefit corporations vie for business in areas that are now in the public sector, this could expose common-property resources such as water, parks, and public transportation to the inequities of the marketplace. (See Chapter 9.) Some public interest advocates also oppose the idea of elevating the role of the market in solving social problems. That work, they say, rightfully belongs to citizens, making decisions together within democratic institutions. The potential for abuses of this new model for "greenwashing" also exists. This is especially the case given how attractive fast-growing companies that appeal strongly to socially conscious consumers—i.e., most prominent benefit corporations—are to larger corporations that seek profitable acquisitions but that are not necessarily committed to advancing a triple bottom line.

"I'm all for people pushing these models, and am happy some people are grappling with such experiments," commented Charlie Cray, director of the Center for Corporate Policy in Washington, D.C. "But I don't think the people who talk about it are facing up to the reality of the magnitude of the problem of corporate domination of the economy, how much of public life has been captured as a result, and the importance of building strong institutions outside of the market sphere to take on the corporate-friendly ideology that so pervades society today."[21]

On the other hand, nonprofits themselves rely heavily on donations from wealth generated under the conventional model of amoral capitalism. Many do not provide public assessments of their own overall social and environmental impacts, based on a third-party evaluation tool, as benefit corporations do. So the movement for benefit corporation statutes offers a chance to rethink the ethical operation of nonprofits as well. It is still early in the movement. But this experiment is spreading rapidly and its potential for good—especially if nonprofits advocating for human rights and environmental activism engage directly with this movement to help shape and promote it in the most responsible forms possible—seems too strong to be ignored.*

Conclusion

The benefit corporation movement is part of a broader movement for redesigning economic activity in more socially responsive and Earth-friendly

---

* Opportunities exist here to promote much higher ethical standards for business practices that may cause serious social and ecological harms but that are now generally tolerated, such as advertising, which is based increasingly on intrusive corporate surveillance and a relentless push for impulse buying.

ways. Other practical experiments are evolving, for example, around more community-oriented banking, the expansion of worker and consumer co-operatives, more equitable and locally based ways to capitalize enterprises, and an expansion of approaches and resources available for socially responsible investment. (See Chapter 17.)

The most progressive examples promote new institutional structures for systematically sharing a company's financial returns more equitably with a broad range of stakeholders responsible for generating that wealth—including local communities, or future generations who will not have access to the natural resources that were used. (See Chapter 8.) That is an evolution from the conventional goal for profits to be "owned" only by the company, and it goes further than the typical understanding of the "triple bottom line" in the benefit corporation movement, although that movement does value worker ownership. Also, benefit corporations, while required to consider the interests of a wide range of stakeholders, are not required to actually include representatives of those stakeholders in their decision-making processes—a fact that some critics point to in questioning the real difference that this movement can make.

What is striking about the benefit corporation movement, though, is its promotion of a higher legal standard for corporate behavior that businesses themselves are seeking. At this point, that standard is one that only a relatively few companies have chosen voluntarily to function under. But it is still a minor revolution, with considerable potential to entice other companies to join in moving corporate culture and practices toward the triple bottom line—and to inspire the public to expect and eventually demand that other companies join them.

The fundamental acknowledgment in state law after state law that (1) it is possible for for-profit companies to strive to adhere to a broad set of social and ecological ethics, (2) the public will benefit when they do, and (3) there are enough entrepreneurs wanting to be held to such a standard to warrant such laws—all add up to an essential, unprecedented step toward a more sustainable economy.

Noting the progress of the movement so far, B Lab's Jay Coen Gilbert predicts that, "[i]n a generation's time…most of the Fortune 500s will be benefit corporations."[22] Although a tidal wave is not likely anytime soon, Gilbert foresees several small waves of change that would make a larger shift possible over time. Over the next few decades, he suggests, more large multinationals are likely to become connected through mergers or acquisitions of smaller benefit corporations. Big corporations also will recognize the value of the "halo effect" that they can earn by giving preference to benefit corporations in their procurement supply chains. And rapidly growing, privately held companies could provide their own wave of disruptive innovation by choosing to become benefit corporations or public benefit corporations while still privately owned, and then, as they grow and need more capital, going public. Their initial public offerings would tap into public demand for investment opportunities that can have a positive social impact, even at the expense of not necessarily maximizing profits.

All of those developments would then set the stage for the next major milestone: one or more Fortune 500 companies that already have enthusiastic major investors on board will seize the public relations opportunity to be first to file as a benefit corporation or public benefit corporation. Even more likely, a few corporations will take the step together. Just after Delaware enacted its new PBC statute, some "key pillars" of the environmental community reached out to B Lab and indicated that they believed the movement was now "ready for prime time," and they wanted to engage to help move it forward. B Lab's hope is that major environmental organizations will bring to the table with them some of the large corporations that they have been building working relationships with in recent years and that are the most forward thinking.

Although it could take years for a Fortune 500 benefit corporation to emerge, such conversations—and broader advocacy by citizens and public interest groups—could begin now to firm up and speed up that possibility. To date, the movement has been driven by B Lab and by committed businesses and corporate lawyers, who have tried hard to present the idea to lawmakers as an uncontroversial, pro-business measure. But if the benefit corporation movement is to realize its potential for transformational change, there will need to be more active engagement by strong advocacy groups, sharing both their corporate-watchdog expertise and grassroots skills in mobilizing concerned citizens to help shape it, protect and increase its rigor, and scale it up quickly.

"In a political sense, the surging popularity of [benefit corporations] will change the way people think about business," observed Jamie Raskin, the progressive Maryland lawmaker who pushed through the first benefit corporation statute. "We can have a market economy without having a market society, and we can have prosperous corporations that act with conscience. Our besieged labor unions and nonprofits should bolster these businesses—green, local, progressive, entrepreneurial, community-focused—as an alternative to an economy controlled by massive state-subsidized corporations that are too big to fail and whose executives are too rich to jail."[23]

CHAPTER 20

Working Toward Energy Democracy

Sean Sweeney

We face an energy emergency of global proportions. Projected massive increases in fossil fuel use in the coming years will make efforts to control climate change virtually impossible from a practical standpoint. Fossil fuel corporations are using their growing wealth and power to assert an "extreme energy" agenda; this includes using far-riskier extraction methods to get to difficult-to-reach and highly polluting "unconventional" fossil fuels (such as oil from tar sands, natural gas through hydraulic fracturing, and coal through mountaintop-removal mining). The extreme energy agenda has serious implications for communities, workers, the climate, and the environment. Fossil fuel corporations are also using their wealth and power to oppose or delay efforts to address climate change and to create a more equitable, democratic, and sustainable energy system.[1]

Although proponents of the fossil fuel agenda argue that it will create or save jobs, the promised employment gains have not emerged: new technologies allow companies to produce the same amounts of fossil fuel with fewer workers. In the United States, more than 400,000 miners mined nearly 600 million tons of coal in 1943; in 2010, less than 90,000 miners produced nearly 1.1 billion tons, and union membership has fallen to barely 15,000 working miners. Moreover, many workers in the energy sector do not have union representation and lack basic workers' rights, a problem that has become more severe as both exploration and extraction have shifted toward developing countries and the former Eastern bloc. In general, neoliberal energy policies have caused working conditions in the sector to deteriorate, particularly in relation to wages, health and safety, and employment security.[2]

The energy emergency encompasses other serious social issues as well. Even though more energy is being generated and consumed with each passing year, more than 1.3 billion people worldwide are without electricity access and another 1 billion have unreliable access. At least 2.7 billion people lack access to modern, non-polluting fuels. In many countries,

Sean Sweeney is the codirector of the Global Labor Institute, a program of the Worker Institute at Cornell University in New York.

privatization of energy has caused price increases, declining quality and service, and underinvestment.[3]

A transition to a clean, renewables-based, low-carbon energy system that meets essential social and environmental priorities needs to occur as quickly as possible. Between 2004 and 2011, global investments in renewable energy (excluding spending on mergers and acquisitions) surged sevenfold, from $39.5 billion to $279 billion, and a growing number of countries are adopting policies to mandate, guide, and support the deployment of renewables. But 2012 saw a 12 percent fall in investments, to $244 billion, and the decline continued in 2013, with third-quarter investments that year 20 percent lower than in 2012. Michael Liebreich, chief executive officer of Bloomberg New Energy Finance, concludes that the "loss of momentum since 2011 is worrying."[4]

Both installed capacities and production of renewables have expanded substantially, although in some cases starting from a very small base. In the case of hydropower, most of the current capacity was built over the past half century, and the bulk of production occurs at huge facilities that hardly deserve the moniker "sustainable." Geothermal power, too, has been used in a small number of countries for some decades. But the wind, solar, and biofuels industries have all risen to prominence within the last one or two decades. (See Table 20–1.)[5]

Table 20–1. Global Capacity or Production of Selected Renewable Energy Technologies, 2000 and 2012

	Installed Capacity/Production		
Renewable Energy Technology	2000	2012	Percent Growth (2000–12)
Capacity			
Wind power (gigawatts, GW)	17	283	1,565
Solar PV (GW)	1.4	100	7,043
Concentrated solar power (CSP) (GW)	0.35	2.5	620
Solar hot water (gigawatts-thermal)	44	255	480
Geothermal power (GW)	8	11.2	41
Production			
Hydropower (terawatt-hours)	2,662	3,673	38
Ethanol (billion liters)	17	83.1	389
Biodiesel (billion liters)	0.8	22.5	2,713

Source: See endnote 5.

In their own right, these are impressive rates of growth; however, renewable energy use is not growing fast enough relative to the world's enormous, and expanding, appetite for energy. The growth in renewables merely supplements the use of fossil fuels, which itself continues to increase. Today, "modern renewables" such as wind and solar contribute just 9.7 percent of global energy consumption (while traditional biomass, used by the world's poor, accounts for 9.3 percent, and nuclear power for 2.8 percent). The U.S. Energy Information Administration projects that world energy consumption will surge 56 percent between 2010 and 2040, and that fossil fuels will still account for nearly 80 percent of total energy use by that year. The current regulatory and market-based approaches to promote renewable energy and energy conservation are totally inadequate given the challenge of climate change and the need to reduce emissions dramatically.[6]

So far, the kind of global political framework that is needed to drive a truly green transition has failed to emerge. Few observers expect international negotiations to produce a global climate agreement that is both equitable and capable of meeting science-based emissions reduction targets. The political paralysis in the face of environmental degradation and the climate emergency also extends to the incapacity of most governments to even begin to address the problems of unemployment, precarious work, and persistent poverty in many regions of the world. They are symptoms of the same problem: a clash between the priorities of political elites and corporations on one hand, and the needs of the masses of people for a truly socially and environmentally sustainable society on the other.

The Need for Energy Democracy

In recent years, a new discourse on sustainability and the green economy has begun to emerge among labor unions and other social movements. (See Chapter 21.) It opposes the idea that putting a price on natural resources is key to solving the profound ecological crisis that we face as a species. This new discourse informs the idea of "energy democracy" proposed here. It shares the view that the economic and environmental crises are two sides of the same coin, and that they must be addressed simultaneously.[7]

Current regulatory and market-based approaches—including carbon markets and taxes—have failed because they do not confront the power of the corporations and have not been able to impede the rush toward rising energy demand, rising fossil fuel use, and rising emissions. (See Chapter 11.) A timely and equitable energy transition can occur only with greater energy democracy, which requires that workers, communities, and the public at large have a real voice in decision making, and that the anarchy of liberalized energy markets is replaced with a comprehensive and planned approach. This does not rule out a targeted deployment of carbon taxes and other "polluter pays" options, but such an approach is at best secondary or supplementary.

The alternative path of energy democracy steers clear of the neoliberal framework and also pivots away from the centralized power generation model that was built around fossil fuels several decades ago. Energy democracy is a public sector approach: it allows space for community-owned/operated, decentralized, or on-site generation, but it also sees an important role for "reclaimed" and restructured public utilities. Renewable energy technologies—particularly solar photovoltaics (PV)—have the potential to completely transform the global energy system by 2030 and also change the political and class relations around energy production and consumption. But the transition must be planned and coordinated in a democratic manner.

Undoubtedly, the political obstacles to energy democracy are enormous. Part of the fight will consist of a struggle to change perceptions about what is real and what is possible, and to assert an internationalist vision that is based on cooperation and sharing. Energy democracy can be the vehicle for a new set of values and a new sense of purpose—values grounded in solidarity, sufficiency, and true sustainability.

The quest for energy democracy entails three broad and strategic objectives: (1) resisting the agenda of large energy corporations, (2) reclaiming to the public sphere parts of the energy economy that have been privatized or marketized, and (3) restructuring the global energy system in order to massively scale up renewable and low-carbon energy, aggressively implement energy conservation, ensure job creation and local wealth creation, and assert greater community and democratic control over the energy sector. By addressing some of these issues, a compelling agenda for energy democracy could emerge in the years ahead.

Resisting the Dominant Energy Agenda

An indispensable part of a new, democratic approach is to resist the agenda of large fossil fuel companies and their political allies. Since the dawn of the fossil fuel age, many of these companies have grown into huge entities with a global presence. Their revenue and profit streams, and the critical role that fossil fuels continue to play in almost all aspects of the world economy, lend them substantial political influence and staying power. As of 2012, fossil fuel-producing companies and utilities represented 19 of the world's 50 leading corporations, accounting for 48 percent of the revenues and nearly 46 percent of the profits of this top-50 group. (See Table 20–2.)[8]

Their agenda has several key characteristics. It is marked by the continued expansion of fossil fuel use; by the aggressive development of extreme forms of energy whose extraction puts communities, workers, and the environment at great risk; by the perpetuation of national-level and World Bank

Table 20–2. Revenues and Profits of the World's 50 Largest Corporations, by Industry, 2012				
	Revenues		Profits	
Industry (Number of Companies)	Billion dollars	Percent of Top 50	Billion dollars	Percent of Top 50
Fossil Fuels/ Utilities (19)	4,482	48.0	258	45.7
Finance and Insurance (11)	1,520	16.3	132	23.5
Motor Vehicles (7)	1,182	12.7	68	12.0
Retail (2)	592	6.3	21	3.7
Electronics (4)	588	6.3	53	9.4
Telecommunications (3)	372	4.0	15	2.6
Others (4)	603	6.5	18	3.2
Top 50 Corporations	9,339	100.0	564	100.0

Note: Numbers may not add up to 100 percent due to rounding.
Source: See endnote 8.

subsidies and support for privatization and marketization of the energy sector; and by either outright opposition or, at best, a weak commitment to effective climate protection policies. Resistance to this agenda can occur in numerous ways: at the level of policy making, in the workplace, by raising public consciousness about the energy emergency confronting humanity, and by building alliances among various groups and social movements.

Opposing individual projects that present serious risks to workers, communities, and the environment, and that do not meet basic energy needs, is crucial. This kind of resistance can educate the public and galvanize the movement. But this cannot be the only approach. A successful energy transition will require a policy shift of major proportions, and it will include bold measures to deal effectively with the wealth, assets, and political leverage of the large energy corporations.

Resisting the fossil fuel agenda does not mean uncritically embracing the agenda of large companies that are developing renewable energy and other low-carbon energy options. Already, the indiscriminate pursuit of biofuels has led to devastating "land grab" practices to secure land for large-scale renewable energy developments. In Oaxaca, Mexico, for example, communities are resisting the plans of large wind companies that seek to profit from the development of mega wind farms with little regard for the needs, land rights, or cultural heritage of local residents.[9]

Rising energy demand is opening up new areas of the world to fossil fuel

Peter Blanchard

A demonstrator objecting to the continued expansion of the tar sands, at the Canadian parliament, Ottawa.

extraction, including the Arctic, the deep oceans, the Alberta tar sands, and shale rock formations in numerous countries. China's insatiable demand for energy is, in the United States alone, leading to increased coal extraction in the Powder River Basin in southeastern Montana and northeastern Wyoming, as well as to planned coal export terminals in Washington state. If completed, these projects will lead to devastating "carbon lock-in" as well as to serious environmental and social impacts.[10]

The expansion of fossil fuels and related infrastructure comes with the lure of new jobs. But although these projects create jobs initially, the export of these resources in raw form often brings little lasting value to the communities concerned. In Canada, the petroleum industry directly employed 16,500 workers in the decade to 2011, mostly in the Alberta tar sands. But the export of unrefined tar sands oil (diluted bitumen) to the United States and beyond will lead to a loss of jobs in Canadian refineries. Moreover, the demand for tar sands has raised the value of the Canadian dollar, making Canadian manufacturing less competitive and leading to the loss of more than 500,000 jobs nationwide in the past decade, according to the Canadian Center for Policy Alternatives.[11]

Social opposition to the further development of extreme energy—including coal and tar sands exports in North America—is increasing. In Canada, First Nations, coastal communities, and some unions have thus far blocked the proposed Northern Gateway pipeline that is intended to bring diluted bitumen from tar sands to the Canadian west coast for export to Asian countries, especially China. Resistance to the west coast coal terminals is growing, led by indigenous peoples who have refused to accept monetary offers from coal companies that wish to use their ancestral lands to transport and store millions of tons of coal. Many Canadian unions and several U.S. unions (in transport, retail, health, and domestic care) also have opposed the Keystone XL pipeline, which would connect the Alberta tar sands to heavy crude refineries in Texas and to global energy markets.[12]

Many of these movements are reactive and defensive, however. It is important to address broader policy issues, including through more proactive measures. Social movements, unions, and other allies can play an important role in convincing local organizations to take up demands for energy democracy, including advocating more strongly for an energy system that protects workers' rights and builds community power.

Unions and their allies also need to join with other groups to fight for well-paying jobs through the development of low-carbon infrastructure, such as expanding public transport systems that reduce emissions, improve air quality, and promote public health and safety, or by pursuing serious energy conservation. In the United States, the United Auto Workers now supports national fuel efficiency standards (something it was reluctant to do for many years), and many unions support initiatives to reduce the use of heating oil and electricity in buildings. Unions in building services, such as the Service Employees International Union, are training building superintendents in energy efficiency. Large Canadian unions (such as Unifor and the Canadian Union of Public Workers, CUPE) now support a moratorium on fracking for shale gas.[13]

The fight for energy democracy needs to engage mainstream environmental groups that typically embrace a more technology-driven "ecological modernization" approach to environmental issues. U.S. environmental groups, for example, have tended to prioritize legislative efforts hammered out in backroom deals rather than bottom-up solutions involving broad alliances. (See Chapter 11.) Many of these groups have been overly confident in the power of private markets and the political process to drive the "green economy," and many environmental leaders have been reluctant to advocate for non-market approaches that might open the door to fundamental change. But the rising political power of fossil fuel companies and the deepening climate crisis is opening up possibilities for new and bolder approaches at the level of policy and organizing. Many smaller renewable energy companies would likely prosper under favorable government procurement agreements that strong public sector involvement would require.[14]

Reclaiming the Energy System for the Public Benefit

Reclaiming the energy system for the broader public interest entails a three-fold challenge. It involves: (1) returning to public control parts of the energy sector that were once public but have since been privatized and/or marketized, (2) restoring principles of public service and responsiveness to public needs to energy entities that are currently publicly owned but are today run like private companies, and (3) reasserting the right to develop a new socially owned and fully unionized, renewables-based energy system that can begin to seriously address social and environmental challenges.

The fight for energy democracy can draw both insights and strength from the recent successes of the broader movement to protect and reclaim public services. Resistance to privatization has been intense in many countries, such as Argentina, Ghana, India, and Indonesia. Protests have halted privatization proposals in Ecuador, Paraguay, Peru, and South Korea. In Iraq, the Federation of Oil Unions (formerly banned under Saddam Hussein's regime) led a successful fight to halt the transfer of Iraqi oil operations to foreign multinationals. Even in China, workers have protested the sale of a public power plant in Henan. (See Chapter 12.)[15]

Privatization has led almost invariably to underinvestment, loss of jobs, reductions in wages and union coverage, worsening working conditions, and falling quality of service. And where privatization has occurred, public control has normally been replaced by oligarchies. In the United Kingdom, six private corporations dominate the power generation sector, owning 71 percent of generating capacity and 96 percent of the residential electricity market. In the Philippines, the neoliberal-inspired Electric Power Industry Reform Act (EPIRA) "brought about a transition from government monopoly to an enhanced private monopoly—worse, a hundred percent increase in power rates." In India, World Bank policies have produced disastrous results, including major power cuts and high levels of electricity theft.[16]

The case for reversing privatization is stronger today than it was in the past. Unions and their allies can draw on the body of knowledge and experience that has accumulated over the past 30 years to build public support for reversing privatization. For many years, the London-based Public Services International Research Unit (PSIRU) has documented struggles against privatization. Moreover, public opinion is beginning to shift: in the United Kingdom, fully 69 percent of residents want energy to be renationalized, according to a September 2013 poll.[17]

Efforts to oppose privatization can learn from the experiences in the water sector. Privatized water services have been "remunicipalized," or taken back into public ownership, in a handful of U.S. cities and in several Latin American countries, including Argentina, Bolivia, Colombia, and Uruguay. Although the return to public ownership has not always been smooth, unions and local communities are developing new forms of public service delivery, such as "public–public partnerships" (PUPs), for which basic principles of operation were adopted in Paso Severino, Uruguay, in 2009. Groups like Public Services International, the Red Vida network, the Transnational Institute, and Food & Water Watch have been active in promoting PUPs as an alternative to privatization and to public-private partnerships.[18]

PUPs in renewable energy are also possible. In Germany and the United States, many energy utilities are community controlled (roughly 20 percent of power in the United States is generated by municipally owned energy utilities). Renewable energy technologies lend themselves to the growth of energy cooperatives that can then network in a similar manner as has been witnessed with water. But more needs to be done to explore these possibilities and to build alliances with worker-community organizations focused on water rights and service provision.[19]

Private-to-public reversals in the energy sector are rare in comparison to the water sector, but they have occurred in Argentina, Bolivia, and Germany. In Germany, the remunicipalization of energy has moved forward at a steady pace, and the country now boasts the highest share of renewable energy use in the European Union. Although German municipalities ceded control of power generation in the 1980s and 90s, many have since chosen to reclaim their local grids, resulting in a major expansion of direct municipal provision of energy services. PSIRU reports that between 2007 and mid-2012, more than 60 new local public utilities (*stadtwerke*) have been set up and more than 190 concessions for energy distribution networks—the vast majority being electricity distribution networks—have returned to public hands. In total, about two-thirds of all German municipalities are considering buying back both electricity generators and the distribution networks, including private shareholdings.[20]

The city of Munich, for example, has decided that all of its energy will come from renewables by 2025, and that all of it will be generated by the public sector, because the private sector cannot be relied on. This was articulated powerfully in 2011 by Dieter Reiter, a Munich city councilor, when addressing an international conference of economists:

> Energy supply was one of the key sectors affected by privatization of formerly public enterprises. Today, energy supply is characterized by oligopolies of private energy suppliers. There is practically no competition on price. The transition to renewable energies is made rather reluctantly and only as a consequence of massive state subsidies and regulatory requirements.… The example of Munich shows how the transition process can be sped up if a city owns a utility company. By 2025, our utility company aims to produce so much green energy, that the entire demand of the city can be met. That requires enormous investments—around 9 billion euros by 2025—and can only be successful if the long-term goal is sustainable economic success rather than short-term profit maximization.[21]

Those who refer to Germany's successes in advancing renewable energy often appear unaware of, or perhaps reluctant to acknowledge, the role of public authorities in challenging privatization and intervening on behalf of the broader public.

The extent of the marketization of publicly owned or controlled entities, however, means that the task of building a democratic and sustainable energy system cannot be reduced to the issue of public versus private ownership. In the case of Argentina, the government moved to reclaim Repsol's 51-percent stake in the partially publicly owned oil company YPF in 2012, but then proceeded to enter into a partnership with Chevron in 2013 to exploit the country's considerable shale gas reserves. Under marketization, publicly owned companies are induced to behave as if they were private businesses. This means that they are focused on maximizing sales and profits and, in many instances, on investing overseas. Serving the public interest or the common good is not necessarily their principal motivation.[22]

In South Africa, the state-owned energy company Eskom behaves like a private multinational; its operations are spread throughout southern Africa and other parts of the world. Eskom's assets totaled $33.1 billion at the end of March 2010, and it pays its CEO $1 million per year. The company's new power plants are being financed by a range of foreign banks based in Europe and South Africa, as well as by multilateral institutions such as the African Development Bank and the World Bank. Unions in South Africa are campaigning for Eskom and other state-owned companies to honor the commitments made in the Freedom Charter and to serve the public good.[23]

Similarly, the Chinese company Sinopec is a major overseas investor in the Canadian tar sands, shale gas, and other forms of extreme energy. In 2000, Sinopec emerged after the Chinese government invited Morgan Stanley to turn its most promising operations into a company that would be listed on world stock markets. Sinopec invests overseas as a means of ensuring that China has supplies of energy to meet its rising demand.[24]

Achieving energy democracy will entail a wholesale reorientation of existing public companies, a redefining of the political economy of energy around truly sustainable principles, and a new set of priorities. Some unions have talked in terms of reclaiming or resocializing entities that were once privatized or marketized, such as the National Union of Metalworkers (NUMSA) in South Africa and CUPE in Canada, but most unions remain locked in anti-privatization battles of a more defensive nature.[25]

Restructuring the Energy Sector

Renewable energy systems operate under two main models: centralized generation, which includes structures such as utility-scale wind farms and remote central-station solar power plants, and decentralized generation, which refers to renewable generation located on existing buildings or vacant land close to the point of electricity consumption. Efforts to build support for the scaling up of renewable power within a democratic framework will need to explore the social benefits and limitations of both systems. What is best for jobs, stable communities, and the environment? What is most suited to systems of democratic engagement?

Decentralized generation is likely to be more conducive to local control. Half of Germany's wind power and three-quarters of German solar installations are locally owned. Decentralized generation also can create more jobs than utility-sized projects per million dollars invested, and can redefine the role and purpose of energy in a way that puts social and environmental needs before profit and accumulation. Yet local control is no panacea. Left to their own devices, communities and municipalities could choose to stick with fossil fuels or make a "unilateral declaration of independence" and thus try to opt out of any broader transition plan. There is no guarantee that the transition will be plain sailing or politically painless—but in some cases, the moral appeal of national referenda on climate protection or energy transition could provide national or regional governments with a certain *imprimatur* for proactive measures (such as research and development support or similar forms of assistance) that can reinforce the process of transition.[26]

Newly constructed home in Germany with nearly total solar panel coverage.

Tim Fuller

A significant challenge facing unions is that millions of their members work in the current fossil-based energy system, and unions are perhaps better positioned to establish a presence in centralized renewable energy systems than in decentralized systems. Union members are presently more likely to perform the work of constructing new utility-scale, remote central-station power facilities, at least in the United States and probably in other industrialized countries. In contrast, most community-based, local energy projects involve contractors that are local and mostly non-union. This further ties unions to the present centralized system.[27]

Some unions also note that many whom today advocate for distributed generation wish to further liberalize the energy system and undermine the unionized and regulated public utilities. The idea of opening the door to countless numbers of small energy producers also has attracted the support of key environmental organizations that traditionally have been less concerned with worker issues. And while it is true that utility-scale renewable energy projects are attractive to large private energy companies, this does not automatically mean that projects of this size have no place in a sustainable energy system or that small, local-scale, decentralized energy projects will not be owned or serviced by large private corporations. In Greece, implementation of the country's feed-in tariff saw the proliferation of installation companies that imported cheap solar panels from China, but when the tariff was adjusted downward dramatically, larger energy companies from Spain and Germany moved in, their eyes on the longer term.[28]

Addressing Energy Poverty

Some unions in the developing world see the potential for distributed generation as a means to promote electricity access for all. India's New Trade Union Initiative, for example, is developing a campaign for sustainable, affordable, and renewable energy, and notes how centralized power has served the country's dominant producers but not served the people. India's energy consumption is rising dramatically, but more than 400 million residents continue to lack electrical power, and more than 668 million depend on traditional biomass for cooking. Unions in the Philippines have put forward similar arguments.[29]

Renewable energy technologies open up off-grid and mini-grid potential for energy access in poor rural areas. Small hydropower plants, small wind turbines, biogas and other forms of bio-energy, and a range of solar technologies are potentially important tools in the fight against energy poverty. Fulfilling the potential of these technologies will depend on the willingness and capacity of local and national governments to arrange the financing, develop the human skills, and oversee the development and deployment of these technologies. Thus, although distributed energy may provide the most likely means of ending energy poverty, this will happen only if it is developed by public authorities committed to providing affordable access to electrical power.[30]

Energy democracy will require taking greater control over global supply chains so that developing renewable energy in a country or region leads to job creation and social benefits close to home. Today, just a handful of countries and a few dozen companies dominate the global market for solar PV, wind turbines, and many other renewable technologies. Under these conditions, the scaling up of renewable energy will mean that only a limited number of companies and countries—such as Germany, Spain, and China—will enjoy the bulk of the jobs and other economic benefits associated with equipment manufacturing and infrastructure development. (Installation and operations and maintenance work is likely to be more localized.)[31]

In a bid to establish a domestic renewable energy industry—and the associated employment—some governments have adopted "domestic content requirements" that compel manufacturers or project developers to source a specified share of equipment (or a portion of overall project costs) from domestic suppliers. These suppliers can be domestic firms, local subsidiaries of foreign-owned companies, or joint ventures between domestic and foreign-owned firms, but the key is for suppliers to invest locally rather than import equipment. Countries that have either implemented or are considering such policies include Brazil, Canada, China, Croatia, India, Italy, France, Malaysia, Morocco, South Africa, Turkey, and Ukraine. To be successful, domestic content policies need to be linked to a learning-by-doing process and to be part of comprehensive industrial, research and development, and training and skill-building policies.[32]

Unions are already engaged in the fight to control and localize supply chains. In Ontario, unions like CUPE and Unifor have supported domestic content provisions. But several countries have brought complaints against such policies before the World Trade Organization (WTO)—including against Ontario's Green Energy and Green Economy Act. The WTO's ruling against Ontario in December 2012 (and the subsequent rejection of Canada's appeal in May 2013) could be a harbinger for policies in other countries.[33]

Conclusion

To say that human civilization stands at a crossroads is hardly an overstatement, given the threat of climate change and the likely breach of planetary limits. The struggle to control and radically change how we produce and consume energy will be a crucial political battle in the next decade or two. Presently, nearly all of the economic power and political arrangements are on the side of the fossil fuel companies committed to an "extreme energy" agenda that will expand the use of fossil fuels—including unconventional (read: dirtier) fuels like tar oil and shale gas—in the expectation of tremendous shareholder profit. For the sake of climate stability and social equity, ordinary citizens, unions, and social movements need to organize alternatives to this deeply destabilizing and polarizing agenda.

Meanwhile, renewable energy is poised to grow spectacularly in many countries, but even the most optimistic global assessments are insufficient from the perspective of mitigating climate change. The energy transition that the world needs desperately will happen only if energy policy is brought under greater democratic control, with social and community ownership, and if changes in the system are carefully planned and coordinated. It is technically possible; it needs to be made politically irresistible.

Liberalized energy markets and marketized public utilities and energy companies have led to competition, whereas greater cooperation is needed. As efforts to remunicipalize utilities from Boulder, Colorado, to Berlin, Germany, suggest, this must give way to new public and community-based entities producing renewable energy for social need and not simply for private gain. The struggle to reclaim and restructure the world's energy system is already under way, but it has barely begun. Another energy system is possible, but not inevitable. Energy democracy can and should be a call to arms for unions and other social movements. There is, it seems, no alternative.

CHAPTER 21

Take the Wheel and Steer! Trade Unions and the Just Transition

Judith Gouverneur and Nina Netzer

The transition toward sustainable societies has consequences for labor markets worldwide—and it places trade unions in a dilemma. On the one hand, unions have to fulfill their core functions of fighting for adequate wages, more employment, and better working conditions. On the other hand, they are confronted with the fact that, in light of changing planetary realities, traditional responses to the threat of job loss address only the symptoms of the problem, but do not offer a real cure.

Parts of the trade union movement, as well as some individual unions, have accepted the reality that they need to become active participants in the transition toward sustainability. Most prominently, this stance is reflected in the "just transition" model, which reflects the trade union movement's ambiguous role in the transformation process. (See Box 21–1.) The concept makes a strong point of advancing the social dimension of sustainability; however, it is not a real departure from growth-based policies, but is based instead on the assumption that a low-carbon economy can be brought about mainly through technological innovation.[1]

The just transition approach does acknowledge that technological change is not socially neutral. Nora Räthzel and David Uzzell, who have been investigating the role of trade unions in the context of globalization and global environmental degradation, observe that, "if workers are not to become the victims of technological change, technological and social transformations need to go hand in hand." Trade unions need to claim a central role in designing and targeting the transition process. At the same time, a successful transition that takes seriously the three dimensions of sustainability—economic, ecological, and social—cannot be achieved without a strong trade union movement.[2]

No Jobs on a Dead Planet

The transition toward sustainability requires fundamental changes to our current growth-fixated economy that is built on the exploitation of finite

Judith Gouverneur works in the global policy and development department of the German political foundation Friedrich-Ebert-Stiftung (FES) and is editor in chief of the FES Online Platform on Sustainability, fes-sustainability.org. **Nina Netzer** oversees international energy and climate policy at FES in Berlin.

Box 21–1. The Just Transition Framework

The concept of a "just transition" is a trade union approach to fighting climate change. It was first mentioned at the end of the 1990s in Canadian union articles as "an attempt to reconcile the union movement's efforts to provide workers with decent jobs and the need to protect the environment." Since then, the idea has become an established tool for the trade union movement. It aims to smooth the shift toward a more sustainable society and to provide hope for the capacity of a "green economy" to sustain decent jobs and livelihoods for all.

Adopted unanimously at the 2nd International Trade Union Confederation Congress in 2010, one of the concept's purposes is to strengthen the idea that environmental and social policies are not contradictory but, on the contrary, can reinforce each other. In contrast to other approaches aimed at reconciling economic growth and climate protection, the just transition approach puts a strong focus on the social dimension. Consequently, a just transition is to be inclusive of all stakeholders and guarantees that its unavoidable negative employment impacts and social costs have to be shared by all. A recent key achievement was the United Nations Framework Convention on Climate Change's recognition of the just transition concept.

Source: See endnote 1.

resources and fossil fuels. It will necessarily lead to job losses in some sectors (such as emission-intensive industries) and to benefits in other sectors (such as renewable energy industries). In general, a socioecological transformation can be expected to have four broad impacts on labor markets:

• Job substitution, where employment will shift within or between sectors, such as from fossil fuels to renewables;
• Job elimination, where there will be no direct replacement for certain jobs, such as those in the European coal sector and in the oil refining industry;
• Transformation and redefinition of existing jobs, such as in industrial sectors that are oriented toward energy or resource savings; and
• Job displacement as a consequence of carbon leakage, such as relocation of enterprises to other countries that have laxer constraints on greenhouse gas emissions.[3]

No one can predict precisely how these different impacts will be distributed across the economy, when they will appear, and how they will influence each other. According to the International Labour Organization (ILO), 38 percent of all workers worldwide are employed in carbon-intensive sectors, such as fossil fuel extraction and industrial manufacturing. On average, they are relatively low skilled. That means that badly managed transitions run a high risk of leading to unemployment or wage cuts in carbon-intensive sectors and to a general increase in income disparities.[4]

In contrast, a transition process that is organized according to principles of a just transition—one that is socially just and bottom up—can create very positive effects on labor markets. Economic sectors such as energy-efficient housing, public transport, recycling, sustainable forestry, and renewable energy offer huge employment opportunities. According to the ILO, green policies, if combined with job support, could help raise employment by 14.3 million worldwide. An estimated 11.7 million of these jobs would be created in developing countries and 2.6 million in industrialized countries, helping to reduce social inequalities. Beyond these positive employment effects, green jobs offer the potential to create more decent jobs because growth in green sectors generally leads to more jobs requiring higher qualification, as well as broad-based distributional effects that unlock the potential of deprived areas and groups. To realize this potential, however, decisive political action is needed.[5]

In recent years, the employment impacts of moving toward greater sustainability have been aggravated by the consequences of the global financial and economic crisis of 2007–08. According to the ILO, an estimated 50 million jobs have been lost worldwide since 2008, on top of the 200 million people already unemployed and 1.5 billion people in vulnerable jobs. The crisis has had the biggest impacts on developing countries and on vulnerable groups such as women, youth, small-scale farmers, and workers in the informal sector.[6]

When the repercussions of the global financial crisis necessitated financial stimulus packages around the world, the idea of a "green economy" came to prominence. The so-called Global Green New Deal was touted as a way to stimulate growth to a level that would be sufficient to restore national economies to pre-crisis levels, while also promoting climate protection. Although the varying approaches to the green economy share the aim of developing a new economic model, the suggested reforms differ widely in their quality and scope. (See Table 21–1.)[7]

These different approaches to the green economy reflect in part the positioning of relevant actors, such as various United Nations agencies and international institutions. The standpoints of these actors, however, often do not reflect direct alignment with a particular approach but instead combine features of the different concepts. (See Table 21–2.)[8]

Trade Unions as Reluctant Agents of Change?

The trade union movement sees the answer to the economic and sustainability crises in promoting the employment potential of a green economy. In comparison to green-economy or green-growth approaches pursued by other actors, trade unions (summarized under the term "just transition") have strengthened the social dimension of the socioecological transforma-

Table 21–1. Green Economy Approaches: An Overview

Concept	Assumption	Underlying Paradigm
Green Growth	Greening of the existing economy through efficiency and technological innovation will lead to mitigation of climate change and to economic growth.	Acknowledges the existence of ecological boundaries but does not question the current growth-based economic system. No call for substantial shrinking of economic growth or radical redistribution of growth.
Green Development	Greening of the existing economy has to be supplemented by a strong focus on the social pillar of sustainable development.	Advocates a new model of production and consumption that results in improved human well-being and social equity, while significantly reducing environmental risks and ecological scarcities. Considers changing the existing concept of welfare from a monetary understanding of increased welfare based on GDP growth to indicators of broader well-being.
Sustainable Development	The green economy has to be part of sustainable development and has to consider the social dimension, especially the needs for international equality and poverty reduction.	Acknowledges potential conflicts between development and environmental protection, especially for developing countries. Approaches to sustainable development have to be country-specific and based on a fair burden sharing between industrialized, emerging, and developing countries according to the principles of common, but differentiated, responsibilities and the right to development.
Green Jobs	Greening of existing economy has to be supplemented by a strong focus on the discussion of labor standards.	Green jobs and the promotion of the green economy are pivotal for achieving an economic and social development that is also environmentally sustainable. The discussion of labor standards and labor as a input factor of production is important to sustain growth in green economic sectors.
Post-Growth/Degrowth	Green growth fails to address the causes of today's economic and environmental crises, and instead seeks solution by a "greenwashing" of capitalist structures—that is, an expansive cultural model that follows the logic of capitalist accumulation, growth fixation, over-consumption, and exploitation of resources.	The current primacy of efficiency and eco-innovation has to be supplemented by a focus on sufficiency. Industrialized countries as well as global middle and upper classes have a moral duty to discuss options of degrowth. Besides challenging the centrality of GDP as an overarching policy objective, a shrinking of economic activities also is necessary.

Source: See endnote 7.

tion in connection with a strong focus on employment issues. Although trade unions acknowledge the important role that they play at the intersection of labor and sustainability issues, they remain reluctant to accept their potential role as a main driver of a green transformation process.[9]

This persistent indecision is reflected in the trade union movement's choice of measures to address the sustainability crisis. Of the three basic

Table 21–2. Selected Proponents of the Green Economy		
Actor	**Key Report**	**Definition**
United Nations Environment Programme (UNEP)	*Towards a Green Economy: Pathways to Sustainable Development and Poverty Eradication* (2011)	A green economy is one "that results in improved human well-being and social equity, while significantly reducing environmental risks and ecological scarcities."
United Nations Economic and Social Commission for Asia and the Pacific (ESCAP)	*Green Growth, Resources and Resilience* (2012)	Green growth is "a strategy that seeks to maximize economic output while minimizing the ecological burdens."
Organisation for Economic Co-operation and Development (OECD)	*Towards Green Growth* (2011)	Green growth means "fostering economic growth and development, while ensuring that natural assets continue to provide the resources and environmental services on which our well-being relies."
International Labour Organization (ILO)	*Green Jobs: Towards Decent Work in a Sustainable, Low-Carbon World* (2008)	"Jobs are green when they help reduce negative environmental impact ultimately leading to environmentally, economically, and socially sustainable enterprises and economies."
Research and Degrowth Network (R&D)	Various reports	Sustainable degrowth is a downscaling of production and consumption that increases human well-being and enhances ecological conditions and equity on the planet.

Source: See endnote 8.

strategies toward sustainability—consistency (eco-innovation), efficiency, and sufficiency—trade unions focus primarily on consistency, that is, the restructuring of economies through ecological innovations and technologies, as well as elements of efficiency, that is, measures aimed at decoupling economic growth and environmental damage through enhanced efficiency and resource productivity. By contrast, the more system-challenging question of sufficiency—how lifestyles and business need to change to end the overuse of goods, resources, and energy—has been largely neglected.

This is understandable insofar as the trade union movement, with its traditional goals of advancing worker interests, is deeply anchored within an economic system that bases wealth generation on continuous growth of production and consumption. In view of the worldwide sustainability crisis, however, trade unions have to face the fact that especially industrialized countries have a moral duty to discuss options of degrowth, or how the current primacy of consistency and efficiency can be supplemented by a focus on sufficiency. As a consequence, in addition to challenging the centrality of gross domestic product (GDP) as an overarching policy objective, trade union strategy development must also include the downscaling of production and consumption.

Such downsizing is a challenge for trade unions not only because they need to redefine their understanding of labor in shrinking economies, but also because a socioecological transformation will completely disrupt union organizational structures. The traditional bastions of trade union activism and membership, such as coal mining, steel, and automotive industries, are mainly polluting and energy-intensive industries, and trade unions are reluctant to forsake jobs in these sectors in the course of the socioecological transformation. This is compounded by the fact that trade union structures have to be developed from scratch in the emerging green sectors. Small and medium-sized enterprises, such as those specializing in energy-efficient building retrofits, predominate in this sector, but they typically lack established links to trade unions, and many do not even have a workers' council.

Overall, this is not an easy step for the labor movement to take. It will likely cause friction within the movement, posing more serious difficulties to certain unions than to others. In preparation for the United Nations climate change conference in Bali, Indonesia, in 2007, the International Trade Union Confederation stated that: "Trade Unions are aware that certain sectors will suffer from efforts aimed at mitigating climate change. Sectors linked to fossil fuel energy and other energy-intensive sectors will be profoundly transformed by emissions reduction policies." International unions tend to take a more progressive stance and to develop broader visions than local unions in carbon-intensive sectors that are confronted with the daily hardships of restructuring processes or job losses—for which advocating for sustainability may at times mean advocating against their own sector and labor force.[10]

Reorganizing Work

Trade unions face the challenge of integrating just transition measures (whether at an individual company or across an entire industry) and formulating a broader concept and guiding principles for sustainable work. Unions can reshape the discussion about the socioecological transformation by bringing to bear their organizing capacities and their expertise in the fields of social, labor, and industrial policy.

A main starting point is related to the reorganization of work. Work continues to be a central part of life, and the way we organize and distribute it has important impacts on social inclusion and identity-forming processes. Today, however, work fulfills its social functions less and less. From both an ecological and social perspective, the way we currently organize work is failing—and certainly not working sustainably. During the last several decades, work has become increasingly precarious, flexible, and informal. This has led to a steady weakening, and in some areas even a complete breach, of the basic promises upon which the acceptance of the current work organization is based: first, the reasonable expectation to be able to contribute meaningfully to society through one's work, and, second, to receive proper recognition for making this contribution, both in a material and non-material sense. Yet although these expectations are not being met, people have not abandoned them.[11]

Rather, the multiple crises have opened the space for a critical reexamination of the way in which work is organized, reviving public debates about the (social) value of work and labor market structures. This is important—and it is a mandate for action, as Begoña María-Tomé Gil of Spain's Union Institute of Work, Environment and Health (ISTAS) argues: "Environmental unionism will have to redefine what labor should be in order for it to meet true human needs. Labor should not be reduced to a mere process of earning a living, in the same way as modern unionism should not be limited to bargaining for better salaries for the labor force in the capitalist market."[12]

Current debates about sustainable work showcase core areas in which trade unions could become important drivers of a socioecological transformation. But they also indicate how unions have failed to firmly steer the debate in a direction that is favorable to the core values of the union movement, and that offers a chance to consolidate and strengthen union influence. The pressure caused by the economic and financial crisis has given some visibility and even political traction to what had been a mostly academic conversation on sustainable work. Yet the issue has been framed first and foremost in economic terms. The multiple green-jobs approaches, which have quickly gained prominence among an array of institutions and actors, including trade unions, barely address the social aspects of gainful employment.

By contrast, a second strategy discussed in the context of labor and sustainability offers broader leverage to achieve social sustainability. It takes into account the meaning that the organization of work has both for individuals and for social distribution patterns of income, as well as for income-related factors such as health or education. The concepts discussed in this vein take a critical view of the ability of growth and efficiency to provide viable long-term solutions. They include debates about sustainable models of prosperity and alternative meanings of work beyond employment-centered perspectives.

Although the models vary in the details, they have in common a broader concept of work combined with working-hour reductions and appropriate social protection schemes. Suggestions for a fair (re)distribution of gainful employment are paired with the acceptance and recognition of all forms of work, including caregiving or community work. These approaches acknowledge the feminist critique of labor concepts focused on the male-dominated standard employment relationship. They integrate other critical perspectives about fundamental changes in the world of work, such as the blurring of meaningful boundaries between work and life, and rising demands for individual time management and flexibility with regard to job location and working hours. In this way, the discussion about sustainable work has been removed from a strictly academic circle and now is advanced by different social actors, including globalization-critical movements like Attac, parts of the labor movement, feminist groups, and churches.[13]

Demonstrators in Vancouver, Canada, on Defend Our Climate Day, November 2013.

These new actors have given momentum to the debate but have failed to establish extended concepts of work as a serious political alternative. This is where trade unions could make an important contribution by openly communicating and discussing such models. On a more concrete level, trade unions are indispensable when it comes to securing equitable (re)distribution of paid work, which requires continuing education and training, adapting social protection systems to the changed concept of work, limiting the intensification of work, and regulating staffing levels in company and wage agreements.[14]

It is obvious that a reorganization of work as a key concept of social sustainability cannot be successful without trade union engagement. It is also true, however, that collective regulatory strategies have lost relevance given the unprecedented fragmentation of work patterns. Establishing new work models with shorter hours is one of the central and most prominent demands in the context of the sustainable work agenda. It requires trade unions to build broad alliances—and thus to make the political choice to make the organization of work primarily a matter of social justice, inclusion, and sustainable systemic change in order to level the field for new coalitions and to increase pressure for political reforms.

Democratizing the Economy From the Ground Up

Another front on which trade unions could create momentum for a socioecological transformation is making the fight for worker participation rights part of a bottom-up process to democratize the economy. According to Klaus Dörre, professor of work, industrial, and economic sociology at

Friedrich Schiller University in Germany, a socioecological transformation cannot be accomplished without public control of key social sectors such as energy and finance, which would liberate these sectors from the imperative of growth. Such debates might help call attention to the social and political relevance of employee empowerment and show that the lack of structures for worker participation in workplace decision making in the emerging "green sector" is ultimately unsustainable. Starting there, trade unions could reveal how their core business—the fight for workers' rights—can contribute to democratic empowerment as a key element of a sustainable reorganization of societies.[15]

Formulating concrete workplace-, industry- or sector-wide transformation plans with worker involvement would allow trade unions to build political pressure for reform and allow workers to play a critical role in decisions about the strategic direction of enterprises and the organization of work processes. Lars Henriksson, a Swedish autoworker and political activist, suggests that unions aim not to preserve unsustainable industries in the name of employment, but rather to engage workers in the development of sustainable conversion strategies. Faced with railroad privatization in 2009, for example, union representatives, environmentalists, researchers, and citizens' groups from different European countries developed RailEurope2025, a plan for a sustainable transport system. Specific goals range from a call for the expansion of bicycle infrastructure and public transport in cities to the conversion of rail systems to renewable energy sources.[16]

As this initiative shows, moving away from an oversimplified "jobs versus environment" debate enables broad social coalitions that could shift workers from being victims of change driven by unaccountable forces to being drivers of change, allowing them to take the wheel and steer. At the heart of this lies the return to solidarity and worker participation as common guiding principles. Henriksson writes: "When faced with plant closure or layoffs, unions often respond with demands for replacement jobs, severance packages or retraining. There is nothing wrong with these but they are individual solutions that more or less accept the dissolution of the workers' collective. All union strength comes from keeping the collective united.… Demanding the industry should be converted and drawing up conversion plans is a possible way of defending not only our jobs, but the world as well."[17]

Conclusion

In modern societies, work is at the center of the relationship between nature and society. It structures social relations and influences the lives of each individual. Achieving sustainable ways of living is therefore inextricably linked to the way we decide to organize work in the future. So far, there is little sign of the fundamental sociocultural change that a radical reorganization of work would require. Moreover, it remains unclear who could be—or might be willing to be—the drivers of such change.

In this situation, trade unions face a difficult balancing act. On the one hand, they have to define measures that effectively protect workers from becoming victims of the necessary, yet strongly economy-driven, change processes already under way. At the same time, they have to find ways to step out of the defensive strategy of reacting to policies that are decided elsewhere and instead become the drivers of socioecological innovation. This, however, will not proceed without friction, and requires a convincing guiding concept with the potential to mobilize and build new alliances.

This can be achieved only if trade unions redefine their role in the transformation process and renew their claim of being an emancipatory social reform movement, stressing the fact that their mandate to represent workers' interests is not confined to the workplace but extends to society at large. Whether unions will succeed in redefining their role in the process of implementing the concept of sustainability is "not only a measure of how politically relevant trade unionism will be to the challenges of life in a carbon-constrained, climate-changed world, but also how politicized it becomes as it challenges not just capitalism but also itself as part of the struggle for a 'just transition.'"[18]

Conclusion

CHAPTER 22

A Call to Engagement

Tom Prugh and Michael Renner

Sustainability is a socioecological problem. Although most of us never consider it, human society is embedded in, and completely dependent upon, the earth's natural systems. Human economic activity takes place within the matrix of these systems, both influencing and being influenced by them. In general, for most of the two or three million years of our hominid history, our share of that influencing was minimal. But at some point in the not-too-distant past, we entered what has come to be called the Anthropocene period, a time in which the sheer number of human beings and the power of human activity to shape the biosphere have exploded and, in fact, have become the main drivers of deeply troubling planet-scale changes. These now-familiar trends—a warming atmosphere and oceans, accelerating species extinctions, and so on—threaten human welfare and perhaps even the viability of human civilization.

The irony is that this is all the result of people doing what comes naturally. As John Gowdy argues in Chapter 3, humans have evolved a complex mix of traits that includes cooperation as well as competition. Human cooperation and sociality were key to our evolutionary survival in a world of fierce competitors, many with claws, teeth, speed, and other traits that we could not match. Living in small bands of hunters and gatherers, our governance institutions were commensurate with our lifestyle, i.e., relatively simple.

But sociality also became our ticket to growing populations, colonization of most of the earth's lands, and, beginning about 10,000 years ago, agriculture. When humans became farmers, we joined the small group of species (including ants and termites) that Gowdy calls *ultrasocial*. Ultrasociality is characterized by role specialization, information sharing, collective defense, and complex city-states, all in the service of production for surplus. In humans, ultrasociality has led to vast population growth, highly hierarchical societies, the domination of the planet, and an apparently perpetual mindset of More.

Tom Prugh and **Michael Renner** are codirectors of the *State of the World 2014: Governing for Sustainability* project.

Once past the turning point to ultrasociality, governance was no longer simple—and we have been struggling with it ever since. As Gowdy remarks, "Ultrasociality is an evolutionary outcome, and evolution cannot see ahead." We have only begun to be dimly aware that perhaps our evolutionary record has led us down a blind alley. Production for surplus on a planet of finite resources, the limits of which we are already crowding, is not a sound long-term survival strategy.

In this book, we use a broad definition of governance: the formal and informal mechanisms and processes that humans use to manage our social, political, and economic relationships with each other and the ecosphere. (See Chapter 2 by D. Conor Seyle and Matthew Wilburn King.) By that definition, our governance institutions are stumbling. Nowhere is this clearer than in our transnational failure to come to grips with climate change, a problem for which all nations are culpable (either in action or in aspiration, although some much more than others), which threatens all, and which requires cooperation among all to solve. But it is also evident in our collective indifference to rigorously maintaining the biological diversity that supports Earth's web of life, in the large and widening gaps between the rich and the poor within and between many countries, in the continued marginalization of indigenous peoples, and so on.

Despite our fondness for the technologies that we are so good at inventing and our hammer/nail tendency to yearn for (and apply) technical solutions to our problems, the failure of the human sustainability enterprise cannot magically be corrected in that way. Alternative, more appropriate technologies do have a role to play. But a boundless faith in techno-fixes may mislead people to believe that we can actually squeeze still more resources out of the planet and get away with it. Or that, if bad comes to worse, we can somehow just (geo-)engineer our way out of the problem. Technology *per se* is as much the problem as the solution.

Nor will simply continuing to deepen our understanding of the complexities of the earth's systems be enough. Never before in human history have we had access to so much data of all stripes as today. The Internet and the encroaching digitization of life have made accessing this information easy. But information does not equal knowledge or wisdom, even when it is essential information. As Monty Hempel points out in Chapter 4, for a variety of reasons ecoliteracy is necessary but insufficient to create action; in fact, at most universities that teach ecoliteracy it is consciously divorced from exhortations to act or from discussion of the ethical obligation to do so.

Finally, it now seems clear (particularly after the latest recession) that markets will not be riding to the rescue. Their operation without vigorous and conscientious government oversight clearly tends to be self-serving and often self-destructive. Market mechanisms are tools that need to be understood and used wisely when appropriate; they are not equipped to run the show. Among the strongest champions of unconstrained markets are multinational corporations, which have demonstrated over and over that their size and power causes them to behave according to an internal logic of their own that is very often contrary to the public's interests, and the planet's.

The problem is also not a lack of institutions and mechanisms that can handle complexity, especially of the sort that requires revamping nearly the entire economic system. Think, for example, of the organizational acumen required among commercial operations that source raw materials or other inputs from far-flung places around the globe, and that maintain a finely timed flow of products and services delivered to consumers at the other end. Or consider the operations of postal services, handling the 346.5 billion letters that were sent worldwide in 2012—nearly 1 billion pieces daily. And even in the sometimes sordid world of politics, the machinery underlying democratic elections is a marvel to behold. Millions of votes are collected in the span of mere hours, and outcomes announced in almost no time, because modern societies have come to expect virtually instant results. It speaks to the efficacy of the underlying organization that instances when things do go wrong—like the infamous "hanging chads" of the 2000 U.S. presidential elections—are the exception rather than routine.[1]

Appropriate technologies, ecoliteracy, markets in tune with the public good, organizational capacity—these are all indispensable tools in the quest for sustainability. And yet, they are not enough. The problem runs much deeper. We can only put ourselves on the path to sustainability by somehow applying what we know about good governance to the economic and political relationships that bind us to each other and to the planet we live on.

Chris Yakimov

An electronic voting booth in Almere, the Netherlands.

Improving Governance

A great deal is known about how governance fails to support sustainability and the ways it could be improved to do so more effectively. It is relatively easy, for example, to arrive at a definition of "good" governance in which most people could find much with which to agree. Conor Seyle and Matthew King, in Chapter 2, put it this way:

Whether concerned about human rights, legitimacy, or even sustainability, it appears to be the case that good governance systems need to be inclusive and participatory: they need to allow the members of the system to change the rules when needed, and have a voice in the collective decisions that are made…[S]ystems need to be accountable to processes that guarantee fair treatment and establish predictable rules that are applied equally to all members of the collective. And ultimately,…there need to be systems in place to resolve disputes and sanction those who would violate the rules and collective values of the group.

Governance, to be good, should be both efficient and legitimate, where legitimacy derives from the wide perception that the system is fair. Fairness demands equity in terms of how social and economic benefits and hardships are shared by different people, communities, and countries. But increasingly, fairness also depends on how well we respond to the worsening climate crisis so that the worst consequences are avoided for the next generations, that the costs of adjustment are shared in a reasonable manner, and that unavoidable impacts do not fall squarely on the shoulders of those who are least responsible for the calamity.

These underlying principles should not be in contention in any society that lives by defensible values. The more difficult question concerns what is needed to drive the governance process for sustainability forward. The chapters in this book examine not only the obstacles to this process, but also the multiple ideas and possibilities for needed change at different scales—from the level of individual ethics to the minutiae of international policy making:

Personal. Whether one lives in a lakeside villa or a mud hut, is a Wall Street financier or a subsistence farmer, is healthy or starving, one's initial circumstances are an accident of birth. Whatever one accomplishes begins there, and to that extent the rich no more deserve their wealth than the poor deserve their poverty. There are no self-made men or women; every human alive is helped or hindered by the legacy bequeathed him or her by the society in which he or she lives. Even prominent mainstream economists have acknowledged that most of what each of us has is due more to the wealth and assets accumulated by previous generations than to our own efforts. (See Chapter 18 by Gar Alperovitz.)

This fateful truth imposes serious obligations on those who are born into wealth. People with the great good fortune to enjoy comfortable lives have deep ethical obligations, first to be aware of how very differently their lives could have gone, and second to heed the requirements of environmental justice. The first such requirement, observes Aaron Sachs in Chapter 10, is to do no harm. While it is impossible to live a perfect or impact-free life, we each need to do what we can to minimize our own impacts, help others to achieve decent and sustainable lives, and push our own communities toward sustainability.

Local. Individual responsibility and action is indispensable, but action by individuals united into communities and movements is even more important. As Monika Zimmermann writes in Chapter 14, the current locus of activity on climate change and biodiversity preservation lies mainly within organizations of local and regional, not national, governments. Over the last 20 years or so, pioneering local governments have stepped forward on the global stage to assert their relevance to sustainability initiatives, exemplify commitments, provide and share resources, establish concrete metrics, track progress toward goals, and help spur national and international processes to do the same.

National. National governments have struggled to make collaborative progress on sustainability issues, particularly climate change, although there has been no shortage of good intentions, impassioned rhetoric, and meetings since the 1992 Rio Summit. Individual countries, with a few exceptions, have not done much better. In Chapter 11, Petra Bartosiewicz and Marissa Miley explore the congressional intransigence around climate change legislation in the United States (as well as providing an object lesson in how not to address such resistance, in the story of the U.S. environmental establishment's efforts to pass a carbon cap-and-trade bill without first building strong grassroots support for it). The European Union's carbon markets have so far proven ineffective due to a lack of government discipline in allocating permits. Sam Geall and Isabel Hilton, in Chapter 12, examine China's fractured environmental politics and note that there is emerging support among networked citizens, nongovernmental organizations (NGOs), and journalists for the country's ambitious green goals and regulations, but that structural problems, such as collusion between polluters and local officials, continue to block progress.

National governments need to do better, both in negotiations with other governments and in their own countries. The opportunities to do so are plentiful. Besides showing some spine in resisting industry efforts to undermine progress on climate, governments need to regain control of financial markets, demand corporate transparency and accountability, and sharply reduce the role of money in politics. (See Chapter 16 by Thomas Palley.)

Governments in general also can take a role in recognizing and sponsoring commons resources by means such as land trusts, cooperatives, and online peer networks for ecosystem monitoring (see Chapter 9 by David Bollier and Burns Weston), or by managing shared assets in the manner of the Alaska Permanent Fund, which allocates earnings from North Slope oil production.

International. Winston Churchill famously quipped in 1947 that "democracy is the worst form of government except all the others that have been tried." The same could be said about the United Nations with regard to international governance. The UN certainly has displayed a degree of bureaucratic inertia at times, although the larger problem is that it is often shortchanged in terms of funding and political wherewithal by the very governments that expect it to provide solutions where purely national efforts fail. And yet an international organization that provides the space to work out cooperative approaches to the sustainability challenge is more indispensable than ever.[2]

As Maria Ivanova explains in Chapter 13, governments and UN officials have come to understand that the time for addressing environmental, economic, and social dimensions separately is long past. The need to weave these policy trends together closely has been recognized in recent efforts to restructure and reinvigorate the UN's sustainability bodies, such as the UN Environment Programme.

In the same way that market mechanisms have been promoted on the national level, public-private initiatives are being pushed at the UN, sometimes in the form of a disconcerting "minilateralism" that regards self-selected groups of governments, corporations, and NGOs as key drivers. As Lou Pingeot reminds readers in Chapter 15, there is a need for much greater transparency and agreed-upon norms to ensure that minilateralism does not amount to an end run around multilateralism.

Finally, we must point out that many of the world's governing systems are still heavily male dominated and thus reflect men's values, priorities, and viewpoints much more than they do women's. Just as the emergence of more democratic forms of governance has been a slow and difficult process, so is the effort to inject greater gender balance into governance. Governments might perform better if more women held positions of leadership, although the evidence so far on this question is inevitably thin, given the continued underrepresentation of women in executive political offices and in many legislatures. (See Box 22–1.)[3]

How?

All of the above, of course, is nothing more than a wish list. Just as it is easy to list all the technologies we should be deploying rapidly to stabilize the climate, it is easy to lay out everything that governments should be doing, or doing better, to make for a sustainable world in general. Both approaches beg the question: given the stark absence of adequate movement in the right direction already, how can we make it happen?

Without question, there is no silver bullet—no single approach that will miraculously achieve what has so far eluded the determined efforts of many people. Any approach that ultimately meets with success will have to incor-

Box 22–1. Women, Governance, and Sustainability

For most of the history of civilization, it was unthinkable that women would help decide who would govern, much less themselves govern. Occasional examples of a reigning queen or empress were quirks of monarchic succession that scarcely dented men's control of government. The past century, however, has witnessed the emergence of women voters in almost all countries. The last decade has seen a gradual rise—too gradual, many would say—of women's leadership at multiple levels of government around the world. This development seems positive for governance with future generations in mind, especially if it accelerates from its currently slow pace of growth. But the evidence supporting this thesis is at best suggestive and indirect.

The numbers point to a significant emergence of women in governance and politics. Prior to 1960, women were absent from top national elective leadership, according to a timeline on women's governmental leadership produced by the International Women's Democracy Center. In that year, Siramavo Bandaranaike became the world's first woman prime minister, leading the government of Ceylon, now Sri Lanka. Within a few years, such dynamic presidents and prime ministers as Indira Gandhi in India, Golda Meir in Israel, and later Margaret Thatcher in the United Kingdom were gaining fame worldwide—and earning the reputation of being every bit as hard-nosed as the male leaders around them.

In recent years, women have achieved their nation's highest office in dozens of countries. Incumbencies in late 2013 included Angela Merkel in Germany, Dilma Rousseff in Brazil, Geun-hye Park in South Korea, Cristina Fernández de Kirchner in Argentina, Joyce Banda in Malawi, Ellen Johnson Sirleaf in Liberia, Laura Chinchilla in Costa Rica, and Dalia Grybauskaitė in Lithuania. Kosovo, not universally recognized as an indepen-

dent nation, has as its president Atifete Jahjaga.

In the United States, meanwhile, the only Democrat widely treated in the news media in late 2013 as a likely standard bearer in the 2016 election was former secretary of state Hillary Rodham Clinton, with Massachusetts senator Elizabeth Warren gaining attention as the most likely rival for her party's presidential nomination. This rivalry (at least as presented in the national media) suggests how routine it is becoming to consider that a woman could become president of the United States.

Yet in a world with 193 United Nations member states, the share of presidents who are women remains far from proportional to the share of women in the world's population. And despite gains at parliamentary, ministerial, and other levels of government, women are still vastly outnumbered in wielding governmental power. The authors of a 2007 UNICEF report concluded that at then-current rates of growth, "gender parity in national legislatures will not be achieved until 2068." Some countries still lack any women either sitting in national legislatures or carrying ministerial portfolios.

The situation appears to be comparable in corporate leadership and governance. After an initial advance in the 1970s and 80s into what was for centuries a male-only culture, women remain a small minority of chief executive officers. Just 22 CEOs among U.S. Fortune 500 companies in mid-2013 were women, according to Bryce Covert, economic policy editor for the Center for American Progress's blog ThinkProgress. Among executive officers generally, just 15 percent were women, while women comprised about 17 percent of corporate board members, according to a recent survey of the same companies by Catalyst, a research and advocacy nonprofit working to expand women's leadership.

continued on next page

Box 22–1. continued

Just one parliament—that of Rwanda—today has a majority of women members. (See Figure 22–1.) And even this example owes much to a controversial device used to jumpstart gender equity in civil governance: quota systems for candidates or sitting legislators. Critics argue that such systems weaken the equality of political opportunity, while supporters counter that they are the only way to hasten the day when government mirrors the gender balance of population. Most nations seem to agree with the supporters. According to the Quota Project, an academic and intergovernmental collaboration, more than half of UN member states have enacted some type of political gender quota system, whether voluntary by political parties or mandatory by candidacies or even reserved legislative seating. Representation of women in corporate executive suites and boardrooms improved in Norway, Spain, and Sweden after the governments of these countries set targets for such gender balance, according to Covert.

Whether women in government are more likely than men to endorse policies that promote environmental sustainability is unclear. The authors of the UNICEF report found that women policy makers are much more likely than their male counterparts to support children's well-being—a possible proxy for interest in sustainability—as well as nonviolent resolution of conflict. There is at least a smattering of evidence supporting the presumption that women on average are more collaborative and less competitive than men, and are more worried about environmental unsustainability as well. Future research may bolster a hopeful thesis about gender equality in governance: that it will make governments more likely to work with the governed to build civilizations that respect biophysical laws and still find ways peaceably to prosper and endure.

—*Robert Engelman and Janice Pratt*
Worldwatch Institute
Source: See endnote 3.

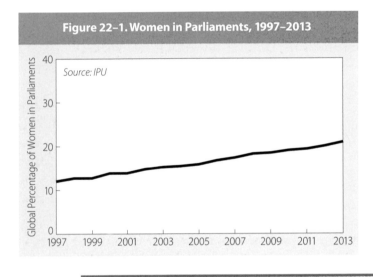

Figure 22–1. Women in Parliaments, 1997–2013

Source: IPU

porate efforts on many different levels. If there is a common theme standing behind the policy ideas and reforms explored in this book, however, it is the necessity of citizen empowerment and citizen responsibility. Call it the first law of political physics: a body at rest will remain at rest until a force is applied to it. When promising governance alternatives are known and seem worth trying out but are not yet happening, then a force needs to be applied to encourage exploratory movement in a new direction. And when governments themselves are unable to muster that force and other actors (such as corporations) are pushing in the wrong direction, an opposing vector can come only from the people.

Sustainability by *diktat* seems unlikely, given the interests—self-preservation, first and foremost—and track records of autocratic regimes in general. Sustainability therefore seems to require something like democracy, or at least a strong democratic impulse. A democracy of distributed leadership (as opposed to one that begins and ends with the ballot box) seems to be the natural home—if such a new idea as sustainability can be said to have one—for sustainability efforts. (See Box 22–2.) Where democracy is already in place, citizens and civil society organizations need to take advantage of their existing freedoms to organize, protest, deliberate, offer input to governments, and demand action. Where democracy is mainly for show or simply absent, safer tactics are required. The goal is the same: to create the irresistible force needed to elicit a positive response.[4]

Regardless of location, this is a difficult thing to do. It requires a long-term, bottom-up approach. Only a sustained mass movement has any hope of generating countervailing power to the forces that are driving the current unsustainable system. It will require courage, passion, and dedication of the sort seen in the Arab Spring uprisings and the Occupy demonstrations, but those alone are not enough; passion will burn out if it cannot be supported with dogged and determined grassroots organizing, through civil society organizations, unions, community groups, cooperatives, and concerned citizens everywhere. It is both the passion of the moment that brings people into the streets for demonstrations and the determination for the long haul that is required to make citizen empowerment a reality.

It would be naïve to assume, however, that the prospects for such a development are good or that the risks are negligible. Grassroots organizing of the sort needed may never happen or just not succeed. The physical risks in many places are significant. Such organizing will take lots of time—years, perhaps decades. During that time, many bad things are bound to happen socially and environmentally, given worsening inequality or the impacts already loaded into the climate system. And these divisive developments in turn may well lead to further repercussions that render a cooperative approach ever harder. Bottom-up organizing may be informed by values and intentions that are anything but "liberal" and "internationalist," and could instead very well end up being chauvinist, xenophobic, inward looking, or violent.

Ultimately, it seems to us, all governance begins with individuals-in-communities. Humans are no more isolated actors in politics than they are the independent molecules of mainstream economic theory. The impetus or pressure to improve governance, at every level, can come only from awakened individuals dedicated to making their communities sustainable places. From there, it may be possible to build communities of communities in a

Box 22–2. Building a Culture of Engagement

In modern nation states, democracy seems to be the most widely preferred form of government. This impulse has expressed itself again and again, most recently perhaps in the Arab Spring uprisings in the Middle East. The last quarter century has witnessed a proliferation of governments that are at least nominally democratic.

The inherent appeal of distributed and accountable power no doubt explains much of this, and is surely among democracy's strongest justifications. Is democracy also biased toward sustainability? That is, are democratic nations more likely to be sustainable than those run by other forms of government? Further, would deepening democratic engagement lead to more vigorous pursuit of sustainability? Can that deepening be accomplished outside of political theory textbooks, in the real world?

In all cases, the answer appears to be "maybe."

Strictly speaking, relatively few countries (and none in the industrialized world) are now sustainable no matter how they are governed, if that means living within per-capita carrying capacity. So to explore these questions, we have to settle for which forms of government seem most conducive to sustainability or active in pursuing it. Here, the evidence—somewhat tepidly and with many qualifications—seems to support the claim that democracies are better than autocracies or mixed forms.

There are several dimensions to this. For example, democracies are probably better equipped to cope with climate adaptation, as power inequalities tend to be less extreme and the poor are therefore less likely to suffer from related environmental harm. Democracies are generally better at disaster response (notwithstanding conspicuous counter-examples such as Hurricane Katrina in the United States), a capacity that will become more significant as a warming climate increases weather extremes. This responsiveness arises mainly from the greater need of elected leaders to answer to voters. For example, Peru suffered devastating earthquakes in 1970 and in 2001; the first killed 66,000 people, the second fewer than 150. The vastly greater 1970 death toll was due partly to higher population density, but mostly to the unresponsiveness of the ruling dictatorship compared with that of the democratically elected government 40 years later.

However, the broad, creeping challenges of sustainability, such as planetary warming and biodiversity loss, to date have not evoked the same sort of response. As political scientist Peter Burnell writes, "[w]hatever other aims democracy might serve, increase in the number of democracies does not seem an obvious solution to global warming, especially if democratization actually promotes material economic advance."

Voters everywhere are understandably concerned about their material well-being, and the very accountability that spurs democratic governments to rush aid to disaster sites also can lead them to privilege economic concerns, especially short-term ones, above all others. If voters do not clearly demand action on problems (such as climate change) that may, or may be seen to, compromise economic performance, then politicians in democratic systems have little incentive to act on those issues. For democracies to address climate change, voters—or rather, citizens, because voting is not nearly enough—must create the impetus. All the more so because, as David Orr has noted, representative democracies tend to become "ineffective, sclerotic, and easily co-opted by the powerful and wealthy" and are vulnerable to "ideologically driven factions that refuse to play by the rules of compromise, tolerance, and fair play." Perhaps even more dangerously, they can succumb to "spoiled-child psychology" that invites, in philosopher Richard Weaver's words, "a sort of contempt for realities."

If people in representative democracies

Box 22–2. continued

are contemptuous of realities, surely that has to do with their twofold isolation: from each other as political actors and from the governing processes meant to address those realities. A possible antidote to both is deliberative civic engagement (DCE), a process encompassing a variety of forms of deeper democracy that go far beyond voting to involve ordinary people in the process of collectively assessing, confronting, and solving governance problems. According to Matt Leighninger of the Deliberative Democracy Consortium, successful DCE initiatives are usually marked by:

- the bringing together of a large and diverse group of citizens,
- structured and facilitated small-group discussions combined with larger forums aimed at action,
- the opportunity for participants to consider a range of arguments, information, and policy options, and
- a final focus on concrete outcomes.

DCE initiatives have sprung up around the world, in Australia, Brazil, China, India, Nigeria, the Philippines, and South Africa, as well as in Europe and North America. Could this approach help address sustainability issues? It's an open question, but DCE has everywhere arisen as a response to urgent political and economic problems. While sustainability is a global challenge, it manifests itself in many local forms and concerns as well as in planetwide effects such as warming. To the extent that DCE becomes known as a useful approach to solving community problems, it could well take root and provide fertile ground for a culture of engagement and more permanent citizen's bodies capable of tackling problems that operate at wider scales.

Each year, the repeatedly disappointing results of the annual high-level international meetings on climate change remind us that the world's democracies are just as stuck in dealing with

sustainability as everyone else. Yet the existing research suggests that other forms of governance offer even worse prospects of coming to global grips with climate change and the other crises of sustainability. The rapid expansion of democracy around the world thus seems to offer the only kernel of hope for breaking the logjam. It is worth noting that this expansion is relatively new, having begun in earnest only in the early 1990s. Also worth noting is that most of the action on climate change seems to be taking place at the local and regional levels, where governments are closest to the people and less likely to be captured by special interests. (See Chapter 14.)

As for DCE, it has been employed mostly in temporary exercises, so it remains to be seen whether it can be established as a widespread standing practice with routine input into official decision-making processes, or perhaps even standing citizens' bodies with statutory power. There are historical examples of such bodies from hundreds or even thousands of years ago, but relatively few contemporary ones.

The potential of deliberative civic engagement is great, but it takes practice. In most cases, our deliberative muscles are not so much atrophied as never developed. Yet citizens have often proven to be committed and knowledgeable enough to take part in DCE. Research and accumulating experience are beginning to clarify which forms of DCE work best, in which circumstances and with which groups. And DCE has been found to increase citizens' civic skills, involvement, and interest in political issues, with corresponding impacts on policy. Human-authored solutions to sustainability problems seem unlikely to emerge without those—indeed, they may be the only way of deepening the responsiveness of democracies to citizens' wishes and harnessing it to the pursuit of sustainability.

—*Tom Prugh*
Source: See endnote 4.

way that affords every person on Earth a safe and fulfilling place to live, and offers future generations the same prospect. Proceeding along this course, it seems to us, is better than surrendering to the centrifugal and destructive forces now at play in the world. Perhaps Herman Daly and John Cobb, writing nearly 25 years ago in *For the Common Good*, put it best:

> On a hotter planet, with lost deltas and shrunken coastlines, under a more dangerous sun, with less arable land, more people, fewer species of living things, a legacy of poisonous wastes, and much beauty irrevocably lost, there will still be the possibility that our children's children will learn at last to live as a community among communities. Perhaps they will learn also to forgive this generation its blind commitment to ever greater consumption. Perhaps they will even appreciate its belated efforts to leave them a planet still capable of supporting life in community.[5]

Notes

Foreword

1. James Madison, "The Federalist No. 51: The Structure of the Government Must Furnish the Proper Checks and Balances Between the Different Departments," *Independent Journal*, 6 February 1788.

2. Nicholas Stern, *The Economics of Climate Change: The Stern Review* (Cambridge, U.K.: Cambridge University Press, 2006), p. xviii.

3. Nicholas Berggruen and Nathan Gardels, *Intelligent Governance for the 21st Century* (Cambridge, U.K.: Polity Press, 2012); David Runciman, *The Confidence Trap* (Princeton, NJ: Princeton University Press, 2013), pp. 318–20.

4. John Platt, "What We Must Do," *Science*, 28 November 1969, pp. 115–21.

5. Lisa-ann Gershwin, *Stung! On Jellyfish Blooms and the Future of the Ocean* (Chicago: University of Chicago Press, 2013); Intergovernmental Panel on Climate Change (IPCC), *Climate Change 2013: The Physical Science Basis*, Contribution of IPCC Working Group I (Cambridge, U.K.: 2013).

6. Adam Hochschild, *Bury the Chains* (Boston: Houghton Mifflin, 2005).

7. U. Thara Srinivasan et al., "The Debt of Nations and Distribution of Ecological Impacts from Human Activities," *Proceedings of the National Academy of Sciences*, vol. 105, no. 5 (2008), pp. 1,768–73.

8. John Stuart Mill, *Principles of Political Economy* (London: Longmans, Green, and Co., 1848/1940), pp. 746–51.

9. Kenneth E. Boulding, "The Economics of the Coming Spaceship Earth," presented at the Sixth Resources for the Future Forum on Environmental Quality in a Growing Economy, Washington, DC, 8 March 1966.

10. White House Council on Environmental Quality, "Preparing the United States for the Impacts of Climate Change," Executive Order (Washington, DC: November 2013).

11. Karl Polanyi, *The Great Transformation* (Boston: Beacon, 1967/1944), p. 73.

12. Michael Wines, "Climate Pact Is Signed by 3 States and a Partner," *New York Times*, 30 October 2013; Sadhu A. Johnston, Steven S. Nicholas, and Julia Parzen, *The Guide to Greening Cities* (Washington, DC: Island Press, 2013); Bruce Katz and Jennifer Bradley, *The Metropolitan Revolution* (Washington, DC: The Brookings Institution, 2013); Parag Khanna, "The End of the Nation-State?" *New York Times*, 12 October 2013.

13. James Howard Kunstler, *The Long Emergency: Surviving the Converging Catastrophes of the Twenty-first Century* (New York: Grove Press, 2006).

14. Madison, op. cit. note 1.

15. Alan Ryan, *On Politics*, vol. 2 (New York: Liveright Publishing, 2012), p. 1,010.

16. Runciman, op. cit. note 3, p. 316.

Chapter 1. Failing Governance, Unsustainable Planet

1. United Nations (UN) News Centre, "Typhoon Haiyan Wake-up Call to Speed Up Climate Control Efforts

– Ban" (New York: 18 November 2013); Mark Fischetti, "Was Typhoon Haiyan a Record Storm?" *Scientific American* blog, 12 November 2013; Matt McGrath, "Typhoon Prompts 'Fast' by Philippines Climate Delegate," *BBC News*, 11 November 2013.

2. Global Carbon Project, "Carbon Budget and Trends 2013," 19 November 2013, at www.globalcarbonproject. org/carbonbudget; David Biello, "400 PPM: Carbon Dioxide in the Atmosphere Reaches Prehistoric Levels," *Scientific American* blog, 9 May 2013. The yearly average for 2012 was 393.1 parts per million, a record; see World Meteorological Organization, "Greenhouse Gas Concentrations in Atmosphere Reach New Record," press release (Geneva: 6 November 2013).

3. UN Environment Programme (UNEP), *The Emissions Gap Report 2013* (Nairobi: 2013); Steiner quote from "Two-degree Global Warming Limit 'Ever-more Elusive': UN," 5 November 2013, at www.expatica.com; 3.7-degree Celsius trajectory from Bill Hare et al., "Warsaw Unpacked: A Race to the Bottom?" *Climate Action Tracker Policy Brief*, 20 November 2013, at http://climateactiontracker.org; International Energy Agency (IEA) projection from Jeremy Lovell, "Clean Energy Lags Put World on Pace for 6 Degrees Celsius of Global Warming," *Scientific American* blog, 26 April 2012.

4. The pace of national climate-related legislation in a group of 33 key industrial and developing countries has slowed substantially since 2010; see Terry Townshend et al., "How National Legislation Can Help to Solve Climate Change," *Nature Climate Change*, May 2013, pp. 430–31. Quote from Hare et al., op. cit. note 3.

5. Bill Hare et al., "Australia: Backtracking on Promising Progress," *Climate Action Tracker Policy Brief*, 13 November 2013, at http://climateactiontracker.org; "New Emissions Goal Derided as 'Bad Joke' at U.N. Climate Summit," *Japan Times*, 16 November 2013; Caroline Selle, "Poland Partners with Coal and Oil Corporate Sponsors for COP19 Climate Conference," *Desmogblog*, 18 September 2013, at www.desmogblog.com; "Polish Government Criticized for Hosting Coal Event at Same Time as UN Climate Conference," *Washington Post*, 8 November 2013.

6. Kevin Anderson and Alice Bows, "A New Paradigm for Climate Change," *Nature Climate Change*, September 2012, pp. 639–40.

7. A study by Drexel University professor Robert Brulle found that 91 "climate change counter-movement" organizations in the United States had a total income of more than $7 billion over the eight year period 2003–10, or exceeding $900 million per year, with trade associations accounting for the bulk of the total. However, the majority of the organizations are multiple-focus organizations, so that not all of this income was devoted to climate-related activities. Robert J. Brulle, "Institutionalizing Delay: Foundation Funding and the Creation of U.S. Climate Change Counter-Movement Organizations," *Climatic Change*, published online 21 December 2013.

8. UN Framework Convention on Climate Change (UNFCCC), "Glossary of Climate Change Acronyms," at http://unfccc.int/essential_background/glossary/items/3666.php. This count excludes the abbreviations of long-established international organizations relevant to the international climate process.

9. UNFCCC, Conference of the Parties, Berlin, 28 March–7 April 1995, "Directory of Participants," 6 April 1995; UNFCCC, Conference of the Parties, Nineteenth session, Warsaw, 11–22 November 2013, "Provisional List of Participants," 12 November 2013.

10. Morton Winston, "Why Hopenhagen Turned into Nopenhagen," Open Salon blog, 17 December 2009; Dan Bodansky, "[Can-cun or Can't-Cun? [That [is [not]] [might be] the Question]]," *Opinio Juris*, 9 December 2010, at http://opiniojuris.org.

11. Hans Verolme et al., *What Future for International Climate Politics? A Call for a Strategic Reset*, Publication Series Ecology, vol. 32 (Berlin: Heinrich Böll Stiftung, 2013), pp. 21–22.

12. "The Shape of a New International Climate Agreement," remarks by U.S. Special Envoy for Climate Change Todd D. Stern at Chatham House, London, 22 October 2013.

13. David Turnbull, "IPCC Says We Must Stop Digging," Oil Change International blog, 27 September 2013; Lorne Stockman, "IEA Acknowledges Fossil Fuel Reserves Climate Crunch," Oil Change International blog, 12 November 2012.

14. Carbon Tracker and the Grantham Research Institute, *Unburnable Carbon 2013: Wasted Capital and Stranded Assets* (London: 2013), p. 16; IEA, *World Energy Outlook 2012* (Paris: 2012), pp. 123–24, 230; Bloomberg New

Energy Finance and Frankfurt School–UNEP Collaborating Centre for Climate & Sustainable Energy, *Global Trends in Renewable Energy Investment 2013* (London: 2013).

15. IEA, op. cit. note 14, p. 121; Richard Heede, "Tracing Anthropogenic Carbon Dioxide and Methane Emissions to Fossil Fuel and Cement Producers, 1854–2010," *Climatic Change*, January 2014, pp. 229–41.

16. John Sousanis, "World Vehicle Population Tops 1 Billion Units," WardsAuto, 15 August 2011, at http://wardsauto.com/ar/world_vehicle_population_110815; Bill McBride, "Vehicle Sales: Fleet Turnover Ratio," CalculatedRISK: Finance and Economics blog, 24 April 2010.

17. Scrap the EU-ETS, *EU ETS Myth Busting: Why It Can't Be Reformed and Shouldn't Be Replicated*, 15 April 2013, at http://scrap-the-euets.makenoise.org/eu-ets-myth-busting; 2012 ETS share from John Upton, "Carbon Trading Is Booming in North America, No Thanks to U.S. or Canadian Governments," Grist.org, 3 January 2014.

18. The Climate Group, *Carbon Pricing*, Insight Briefing (London: May 2013), p. 3; European Energy Exchange, "EU Emission Allowances," at www.eex.com/en/Market%20Data/Trading%20Data/Emission%20Rights; Ewa Krukowska, "EON's Teyssen Urges Fix to 'Bust' EU CO2 Plan, Energy Rules," *Bloomberg*, 7 February 2012. For analysis of carbon market problems, see Carbon Market Watch website, http://carbonmarketwatch.org. Currency conversion reflects official exchange rates for the time, from European Central Bank, at www.ecb.europa.eu/stats/exchange/eurofxref/html/eurofxref-graph-usd.en.html.

19. Hare et al., op. cit. note 3; Nils Klawitter, "CO2 Emissions: Can Europe Save Its Cap-and-Trade System?" *Spiegel Online*, 3 April 2013.

20. Corporate Europe Observatory, "Revolving Door Watch," at http://corporateeurope.org/revolvingdoorwatch; ALTER EU, "Brussels: a Lobbying Paradise," 27 March 2012, at www.alter-eu.org; John O'Donnell, "Special Report – How Lobbyists Rewrite Europe's Laws," *Reuters*, 18 March 2011.

21. Evan Mackinder, "Pro-Environment Groups Outmatched, Outspent in Battle Over Climate Change Legislation," OpenSecrets blog, 23 August 2010; Mike Dorning, "Gore Says Money Influence in Politics Hacked Democracy," *Bloomberg*, 17 October 2013.

22. Corporate Europe Observatory, "The Right to Say No: EU–Canada Trade Agreement Threatens Fracking Bans," 6 May 2013, at http://corporateeurope.org.

23. UN Conference on Trade and Development, "Recent Developments in Investor-State Dispute Settlement (ISDS)," *IIA Issues Note*, May 2013; influence of threatened investor suits from Corporate Europe Observatory and Transnational Institute, *A Transatlantic Corporate Bill of Rights* (Brussels: June 2013); quote from Nathalie Bernasconi-Osterwalder and Rhea Tamara Hoffmann, *The German Nuclear Phase-Out Put to the Test in International Investment Arbitration? Background to the New Dispute Vattenfall v. Germany (II)* (Berlin and Amsterdam: Transnational Institute, Somo, and Power-Shift, October 2013), p. 3.

24. Corporate Europe Observatory, op. cit. note 22; Corporate Europe Observatory and Transnational Institute, op. cit. note 23.

25. U.S. Environmental Protection Agency, "2013 Proposed Carbon Pollution Standard for New Power Plants" (Washington, DC: 23 September 2013); Andrew Steer, "King Coal's Climate Challenge," Project Syndicate, 19 November 2013, at www.project-syndicate.org.

26. Energy Conservation Center Japan, "Final Reports on the Top Runner Target Product Standards," at www.eccj.or.jp/top_runner/index.html; UNEP, *Decoupling: Natural Resource Use and Environmental Impacts from Economic Growth* (Nairobi: 2011).

27. Jake Schmidt, "Way Too Much Public Funding Is Going into Coal Projects in Key Countries: Preliminary Findings Show," Switchboard blog (Natural Resources Defense Council), 21 November 2013; Fiona Harvey, "UK to Stop Funding Coal Projects in Developing Countries," *The Guardian* (U.K.), 20 November 2013.

28. Isabel Ortiz et al., *World Protests 2006–2013* (New York: Initiative for Policy Dialogue and Friedrich-Ebert-Stiftung, September 2013).

29. Bill McKibben, "Movements Without Leaders. What to Make of Change on an Overheating Planet," *TomDispatch*, 18 August 2013.

30. Laura Beans, "Coal Exports Face Unprecedented Opposition in the Pacific Northwest," *EcoNews*, 20 September 2013, at http://ecowatch.com; Steven Erlanger, "As Drilling Practice Takes Off in U.S., Europe Proves Hesitant," *New York Times*, 9 October 2013.

31. McKibben, op. cit. note 29.

Chapter 2. Understanding Governance

1. Elinor Ostrom, *Governing the Commons: The Evolution of Institutions for Collective Action* (Cambridge, U.K.: Cambridge University Press, 1990); UNESCO, E-Governance Capacity Building website, http://portal.unesco.org/ci/en/ev.php-URL_ID=2179&URL_DO=DO_TOPIC&URL_SECTION=201.html; Figure 1–1 from Google Scholar, online search for the terms "governance" and "government," scholar.google.com, July 2013.

2. Stephen Bell and Andrew Hindmoor, *Rethinking Governance: The Centrality of the State in Modern Society* (Cambridge, U.K.: Cambridge University Press, 2009), p. 1; James Clerk Maxwell, "On Governors," *Proceedings of the Royal Society*, no. 100 (1868); Peter Miller, "The Genius of Swarms," *National Geographic*, July 2007.

3. Gerry Stoker, "Governance as Theory: Five Propositions," *International Social Science Journal*, vol. 50, no. 155 (1998), p. 17; John G. Ruggie, "Reconstituting the Global Public Domain: Issues, Actors, and Practices," *European Journal of International Relations*, vol. 10, no. 4 (2004), p. 504; United Nations Development Programme, "UNDP and Governance: Experiences and Lessons Learned," Lessons-Learned Series No. 1 (New York: Management Development and Governance Division, 2006).

4. Lisbet Hooghe and Gary Marks, "Unraveling the Central State, But How? Types of Multi-level Governance," *American Political Science Review*, vol. 97, no. 2 (2003), pp. 233–43; Barbara Koremenos, Charles Lipson, and Duncan Snidal, "The Rational Design of International Institutions," *International Organization*, vol. 55 (2001), pp. 761–99; Sven Steinmo, *The Evolution of Modern States* (Cambridge, U.K.: Cambridge University Press, 2010); Ian S. Lustick, "Taking Evolution Seriously: Historical Institutionalism and Evolutionary Theory," *Polity*, vol. 43 (2011), pp. 179–209.

5. Allen Buchanan and Robert Keohane "The Legitimacy of Global Governance Institutions," *Ethics and International Affairs*, vol. 20, no. 4 (2006), pp. 405–37.

6. John Locke, *The Second Treatise of Civil Government* (1690).

7. Tom R. Tyler, "Psychological Models of the Justice Motive: Antecedents of Distributive and Procedural Justice," *Journal of Personality and Social Psychology*, vol. 62 (1994), pp. 850–63; Tom R. Tyler, *Why Do People Obey the Law?* (Princeton, NJ: Princeton University Press, 2006).

8. Ibid.

9. Bruce Gilley, "The Meaning and Measure of State Legitimacy: Results for 72 Countries," *European Journal of Political Research*, vol. 45 (2006), pp. 499–525; Timothy J. Power and Jennifer M. Cyr, "Mapping Political Legitimacy in Latin America," *International Social Science Journal*, vol. 60, no. 196 (2009), pp. 253–72.

10. Ostrom, op. cit. note 1; Elinor Ostrom, James Walker, and Roy Gardener, "Covenants With and Without a Sword: Self-Governance Is Possible," *American Political Science Review*, vol. 86, no. 2 (1992), pp. 404–17.

11. Ibid.

12. For a systematic review of this research, see Arun Agrawal, "Common Resources and Institutional Stability," in Elinor Ostrom, ed., *The Drama of the Commons* (Washington, DC: National Academy Press, 2002).

13. Adam Smith, *The Wealth of Nations* (1776).

14. Jared Diamond, *Guns, Germs, and Steel: The Fate of Human Societies* (New York: W. W. Norton & Company, 1999); Lane Fargher, "A Comparison of the Spatial Distribution of Agriculture and Craft Specialization in Five State-level Societies," *Journal of Anthropological Research*, vol. 65, no. 3 (2009). For perspective on how this was not necessarily a good transition, see John Gowdy, ed., *Limited Wants, Unlimited Means: A Reader on Hunter-Gatherer Economics and the Environment* (Washington, DC: Island Press, 1998).

15. Steinmo, op. cit. note 4; Lustick, op. cit. note 4.

16. See, for example, the beneficial effects of democracy to internal stability, in Barbara Walter, "Does Conflict Beget Conflict? Explaining Recurring Civil War," *Journal of Peace Research*, vol. 41, no. 3 (2004), pp. 371–78; for democracy's benefits to improved military effectiveness, see Dan Reiter and Allan Stam, "Democracy and Battlefield Military Effectiveness," *Journal of Conflict Resolution*, vol. 42 (1998), pp. 259–77.

17. United Nations Global Compact website, www.unglobalcompact.org; Thomas G. Weiss, D. Conor Seyle, and Kelsey Coolidge, *The Rise of Non-State Actors in Global Governance: Opportunities and Limitations* (Broomfield, CO: One Earth Future Foundation, 2013).

18. See, for example, Danielle Zach, D. Conor Seyle, and Jens Vestergaard Madsen, *Burden-sharing Multi-level Governance: A Study of the Contact Group on Piracy Off the Coast of Somalia* (Broomfield, CO: One Earth Future Foundation, 2013).

Chapter 3. Governance, Sustainability, and Evolution

1.   Anthony Barnosky et al., "Approaching a State Shift in Earth's Biosphere," *Nature*, 7 June 2012, pp. 52–58; Madhusree Mukerjee, "Apocalypse Soon: Has Civilization Passed the Environmental Point of No Return?" *Scientific American*, 23 May 2012, at www.scientificamerican.com; David S. Wilson and John Gowdy, "Evolution as a General Theoretical Framework for Economics and Public Policy," *Journal of Economic Organization and Behavior*, vol. 90S (2013), pp. S3–S10.

2.   David S. Wilson et al., "Evolving the Future: Toward a Science of Intentional Change," *Behavioral and Brain Sciences*, 2013, in press.

3.   Richard Thaler, "Anomalies: The January Effect," *Journal of Economic Perspectives*, vol. 1, no. 1 (1987), pp. 197–201; Richard Thaler and Cass Sunstein, *Nudge: Improving Decisions about Health, Wealth and Happiness* (New Haven: Yale University Press, 2008).

4.   Eric J. Johnson and Daniel Goldstein, "Do Defaults Save Lives?" *Science*, 21 November 2003, pp. 1,338–39.

5.   Erez Yoeli et al., "Powering Up with Indirect Reciprocity in a Large-scale Field Experiment," *Proceedings of the National Academy of Sciences*, early edition (2013), at www.pnas.org/cgi/doi/10.1073/pnas.1301210110; Jessica Nolan et al., "Normative Social Influence Is Underdetected," *Personality and Psychology Bulletin*, vol. 34, no. 7 (2008), pp. 914–23.

6.   Michael Pollitt and Irina Shaorshadze, "The Role of Behavioral Economics in Energy and Climate Policy," in Roger Fouquet, ed., *Handbook on Energy and Climate Change* (Cheltenham, U.K.: Edward Elgar, 2013).

7.   David S. Wilson, Richard Kauffman, Jr., and Miriam S. Purdy, "A Program for At-Risk High School Students Informed by Evolutionary Science," *PLoS ONE*, vol. 6, no. 11 (2011), p. e27826; Dennis Embry, "The Good Behavior Game: A Best Practice Candidate as a Universal Behavioral Vaccine," *Clinical Child & Family Psychology Review*, no. 5 (2002), pp. 273–97.

8.   Christian Cordes, "The Role of Biology and Culture in Veblenian Consumption Dynamics," *Journal of Economic Issues*, vol. XLIII (2009), pp. 115–41.

9.   Uta Frith and Chris Frith, "The Social Brain: Allowing Humans to Boldly Go Where No Other Species Has Been," *Philosophical Transactions of the Royal Society B*, vol. 365 (2010), pp. 165–75; C. Sherwood, F. Subiaul, and T. Zadiszki, "A Natural History of the Human Mind: Tracing Evolutionary Changes in Brain and Cognition," *Journal of Anatomy*, vol. 212 (2008), pp. 426–54; Bruce E. Wexler, *Brain and Culture* (Cambridge, MA: MIT Press, 2006).

10. David S. Wilson, Elinor Ostrom, and Michael Cox. "Generalizing the Core Design Principles for the Efficacy of Groups," *Journal of Economics Behavior and Organization*, vol. 90S (2013), pp. S21–S32; Christopher Boehm, *Moral Origins: The Evolution of Virtue, Altruism, and Shame* (New York: Basic Books, 2011); Martin Nowak and Roger Highfield, *Super Cooperators* (New York: Free Press, 2011); Joseph Henrich et al., "Costly Punishment Across Human Societies," *Science*, 23 June 2006, pp. 1,767–70.

11. Wilson, Ostrom, and Cox, op. cit. note 10; Wilson et al., op. cit. note 2; Elinor Ostrom and Harini Nagendra, "Insights on Linking Forests, Trees, and People from the Air, on the Ground, and in the Laboratory," *Proceedings of the National Academy of Sciences*, vol. 103 (2006), pp. 19,224–31.

12. Rebecca Adamson, Danielle Nierenberg, and Olivia Arnow, "Valuing Indigenous People," in Worldwatch Institute, *State of the World 2013: Is Sustainability Still Possible?* (Washington, DC: Island Press, 2013), pp. 113–25; Erik Assadourian, "Re-engineering Cultures to Create Sustainable Civilization," in Worldwatch Institute, idem, pp. 210–17.

13. Mark Pagel, "Evolution: Adapted to Culture," *Nature*, 16 February 2012, pp. 297–99. Box 3–1 from the following sources: Intergovernmental Panel on Climate Change (IPCC), *Climate Change 2013: The Physical Science Basis*, Contribution of IPCC Working Group I (Cambridge, U.K.: 2013); Thomas Suddendorf, Donna Rose Addis, and Michael C. Corballis, "Mental Time Travel and the Shaping of the Human Mind, *Philos Trans R Soc Lond B Biol Sci*, vol. 364, no. 1,521 (2009), pp. 1,317–24; Paul J. H. Schoemaker, "Multiple Scenario Development: Its Conceptual and Behavioral Foundation," *Strategic Management Journal*, vol. 14, no. 3 (1993), pp. 193–213; Theodsius Dobzhansky, Max K. Hecht, and William C. Steere, "On Some Fundamental Concepts of Evolutionary Biology," in Theodsius Dobzhansky, Max K. Hecht, and William C. Steere, eds., *Evolutionary Biology Volume 2* (New York: Appleton-Century-Crofts, 1968), pp. 1–34; Kevin Neville Lalan and F. J. Odling-Smee, "Niche Construction: The Forgotten Force of Evolution," *New Scientist*, 15 November 2000, pp. 42–45; Peter J. Richerson, Robert Boyd, and Brian Paciotti, "An Evolutionary Theory of Commons Management," in Elinor Ostrom et al., eds., *The Drama of the Commons* (Washington, DC: National Academy Press, 2002), pp. 403–42; Ilya Prigogine and Isabelle Stengers, *Order Out of Chaos: Man's New Dialogue with Nature* (Flamingo, 1984); Walter Truett Anderson, *Evolution Isn't What It Used to Be: The Augmented Animal and the Whole Wired World* (New York: W. H. Freeman and Company, 1996); Anne-Marie Slaughter, "Sovereignty and Power in a Networked World Order," *Stanford Journal of International Law*, vol. 40 (2004), pp. 283–327; Joseph Nye, *The Paradox of American Power: Why the World's Only Superpower Can't Go it Alone* (Oxford: Oxford University Press, 2002); Everett M. Rogers, *Diffusion of Innovations* (New York: Free Press, 1995); Candace Jones, William S. Hesterly, and Stephen P. Borgatti, "A General Theory of Network Governance: Exchange Conditions and Social Mechanisms," *The Academy of Management Review*, vol. 22, no. 4 (1997), pp. 911–45; Erik Angner, "The History of Hayek's Theory of Cultural Evolution," *Studies in History and Philosophy of Biological and Biomedical Sciences*, vol. 33 (2002), pp. 695–718; Donald T. Campbell, "Variation and Selective Retention in Socio-cultural Evolution," *General Systems*, vol. 14 (1969), pp. 69–85; Bruce G. Trigger, *Sociocultural Evolution: Calculation and Contingency* (Oxford: Blackwell Publishers, 1998).

14. Douglas P. Fry and Patrik Söderberg, "Lethal Aggression in Mobile Forager Bands and Implications for the Origins of War," *Science*, 19 July 2013, pp. 270–73.

15. Donald Campbell, "The Two Distinct Routes Beyond Kin Selection to Ultrasociality: Implications for the Humanities and Social Sciences," in Diane L. Bridgeman, ed., *The Nature of Prosocial Development: Theories and Strategies* (New York: Academic Press, 1983), pp. 11–41; John Gowdy and Lisi Krall, "Agriculture as a Major Evolutionary Transition to Ultrasociality," *Journal of Bioeconomics*, April 2014; John Gowdy and Lisi Krall, "The Ultrasocial Origins of the Anthropocene," *Ecological Economics*, vol. 95 (2013), pp. 137–47; Edward O. Wilson, *The Social Conquest of Earth* (New York: W. W. Norton & Company, 2012).

16. Gowdy and Krall, "Agriculture as a Major Evolutionary Transition to Ultrasociality," op. cit. note 15; Jean-Noël Biraben, "The Rising Numbers of Humankind," *Population & Societies*, no. 394 (2003), pp. 1–4; Peter Richerson and Robert Boyd, "The Evolution of Human Ultrasociality," in Irenaus Eibl-Eibisfeldt and Frank Salter, eds., *Ideology, Warfare, and Indoctrinability* (London: Berghahn, 1998); Peter Turchin, "Warfare and the Evolution of Social Complexity: A Multilevel-Selection Approach," *Structure and Dynamics*, vol. 4 (2010).

17. Vaclav Smil, *Harvesting the Biosphere: What We Have Taken From Nature* (Cambridge, MA: MIT Press, 2013).

18. Bert Hölldobler and Edward O. Wilson, *The Superorganism* (New York: W. W. Norton & Company, 2009), p. 4.

19. Carl Anderson and Daniel McShea, "Individual *versus* Social Complexity, with Particular Reference to Ant Colonies," *Biological Reviews*, vol. 76 (2001), pp. 211–37; Gowdy and Krall. "The Ultrasocial Origins of the Anthropocene," op. cit. note 15.

20. Hölldobler and Wilson, op. cit. note 18; Donald Campbell, "Downward Causation in Hierarchically Organized Biological Systems," in Francisco J. Ayala and Theodosius G. Dobzhansky, eds., *Studies in the Philosophy of Biology: Reduction and Related Problems* (London: MacMillan, 1974); Jared Diamond, *Collapse: How Societies Choose to Succeed or Fail* (New York: Viking Press, 2005), p. 114.

21. Richard Manning, "Bakken Business," *Harper's*, March 2013, pp. 29–37.

22. Lisi Krall and Kent Klitgaard, "Ecological Economics and Institutional Change," *Ecological Economics Reviews*, vol. 1,219 (2011), pp. 185–96.

23. Gowdy and Krall, "The Ultrasocial Origins of the Anthropocene," op. cit. note 15.

Chapter 4. Ecoliteracy: Knowledge Is Not Enough

1.   David W. Orr, *Ecological Literacy: Education and the Transition to a Postmodern World* (Albany, NY: SUNY Press, 1992).

2.   Monty Hempel, interview conducted during annual research expedition, 14 May 2013.

3.   See, for example, Joseph Tainter, *The Collapse of Complex Societies* (Cambridge, U.K.: Cambridge University Press, 1990) and Alvin Toffler, *Future Shock* (New York: Random House, 1970).

4.   Dan M. Kahan et al., "The Polarizing Impact of Science Literacy and Numeracy on Perceived Climate Change Risks," *Nature Climate Change*, vol. 2, no. 10 (2012), pp. 732–35.

5.   Albert Bandura, "Impeding Ecological Sustainability through Selective Moral Disengagement," *International Journal of Innovation and Sustainable Development*, vol. 2, no. 1 (2007), p. 32.

6.   For critiques on the question of scientific bias and objectivity, see Bruno Latour, *Science in Action: How to Follow Scientists and Engineers through Society* (Cambridge, MA: Harvard University Press, 1987) and Donna Haraway, "Situated Knowledges: The Science Question in Feminism and the Privilege of Partial Perspective," *Feminist Studies*, vol. 14, no. 3 (1988), pp. 575–99.

7.   Orr, op. cit. note 1; Fritjof Capra, *The Web of Life: A New Scientific Understanding of Living Systems* (New York: Random House, 1996); Arne Naess, "A Defence of the Deep Ecology Movement," *Environmental Ethics*, vol. 6, no. 3 (2008), pp. 265–70.

8.   Anja Kollmuss and Julian Agyeman, "Mind the Gap: Why Do People Act Environmentally and What Are the Barriers to Pro-environmental Behavior?" *Environmental Education Research*, vol. 8, no. 3 (2002), pp. 239–60; Albert Bandura, *Self-efficacy* (Hoboken, NJ: John Wiley & Sons, 1994).

9.   Alexa Spence, Wouter Poortinga, and Nick Pidgeon, "The Psychological Distance of Climate Change," *Risk Analysis*, vol. 32, no. 6 (2012), pp. 957–72.

10. Richard Louv, *Last Child in the Woods: Saving Our Children from Nature-deficit Disorder* (Chapel Hill, NC: Algonquin Books, 2006); Lenore Skenazy, *Free-Range Kids, Giving Our Children the Freedom We Had Without Going Nuts with Worry* (Wiley.com, 2009); Judith H. Heerwagen and Gordon H. Orians, "The Ecological World of Children," in Peter H. Kahn Jr. and Stephen R. Kellert, eds., *Children and Nature: Psychological, Sociocultural, and Evolutionary Investigations* (Cambridge, MA: MIT Press, 2002), pp. 29–64.

11. Mardie Townsend and Rona Weerasuriya, *Beyond Blue to Green: The Benefits of Contact with Nature for Mental Health and Well-being* (Deakin, Australia: Deakin University, 2010).

12. See, for example: Robert Gifford, "The Dragons of Inaction: Psychological Barriers that Limit Climate Change Mitigation and Adaptation," *American Psychologist*, vol. 66, no. 4 (2011), p. 290; Susanne C. Moser, "Communicating Climate Change: History, Challenges, Process and Future Direction," *Wiley Interdisciplinary Reviews: Climate Change*, vol. 1, no. 1 (2010), pp. 31–53; and Dan Kahan, "Fixing the Communications Failure," *Nature*, 21 January 2010, pp. 296–97.

13. Carl Sagan, *Demon-Haunted World: Science as a Candle in the Dark* (New York: Ballantine Books, 1997).

14. Michael Shermer, *The Believing Brain: From Ghosts and Gods to Politics and Conspiracies—How We Construct Beliefs and Reinforce Them as Truths* (New York: Times Books, 2011), pp. 5, 36; Antonio Damasio, *Descartes' Error: Emotion, Reason, and the Human Brain* (London: Penguin Books, 2005).

15. William Ophuls, *Plato's Revenge: Politics in the Age of Ecology* (Cambridge, MA: MIT Press, 2011).

16. Eugene P. Odum, "The Strategy of Ecosystem Development," *Science*, 18 April 1969, pp. 262–70.

17. Thomas E. Mann and Norman J. Ornstein, "Finding the Common Good in an Era of Dysfunctional Governance," *Daedalus*, vol. 142, no. 2 (2013), pp. 15–24.

18. Elise S. Brezis, *Globalization and the Emergence of a Transnational Oligarchy* (Helsinki: United Nations University World Institute for Development Economics Research, January 2010).

19. Lamont C. Hempel, *Environmental Governance: The Global Challenge* (Washington, DC: Island Press, 1996); John S. Dryzek, *Deliberative Global Politics: Discourse and Democracy in a Divided World* (Cambridge, U.K.: Polity, 2006).

Chapter 5. Digitization and Sustainability

1. Bolt, Beranek and Newman, Inc., *A History of the ARPANET: The First Decade* (Arlington, VA: Defense Advanced Research Projects Agency, 1981), pp. III, 24–25.

2. Amory B. Lovins, "Energy Strategy: The Road Not Taken?" *Foreign Affairs*, October 1976; Jack M. Nilles, *Managing Telework: Strategies for Managing the Virtual Workforce* (New York: John Wiley & Sons, 1998), pp. 146–50, 158–59; Bruce Bimber, *Information and American Democracy: Technology in the Evolution of Political Power* (Cambridge, U.K.: Cambridge University Press, 2003).

3. Bimber, op. cit. note 2; Richard Maxwell and Toby Miller, "What Is the Human and Environmental Cost of New Technology?" *The Guardian* (U.K.), 27 February 2013.

4. Langdon Winner, *Autonomous Technology: Technics Out-of-Control as a Theme in Political Thought* (Cambridge, MA: MIT Press, 1977).

5. Michael D. Shear and Robert Pear, "Obama Admits website Flaws on Health Law," *New York Times*, 22 October 2013.

6. Table 5–1 from the following sources: U.S. Energy Information Administration (EIA), "Table 2.1a. Energy Consumption Estimates by Sector, 1949–2011," in *Annual Energy Review 2012* (Washington, DC: 2012); U.S. Census, Population Division, "Table 1. Intercensal Estimates of the Resident Population by Sex and Age for the United States: April 1, 2000 to July 1, 2010" (Washington, DC: 2011); U.S. Census, Population Division, "Historical Population Estimates: July 1, 1900–July 1, 1999" (Washington, DC: 2000); U.S. Census, Population Division, "Table 3. Projection of the Population of Regions, Divisions and States, for 1955 and 1960, with Current Figures for 1950" (Washington, DC: 1952); U.S. Bureau of Economic Analysis, "Table 667. Gross Domestic Product in Current and Chained (2005) Dollars," in *Survey of Current Business*, April 2011.

7. EIA, *Annual Energy Outlook 2013* (Washington, DC: 2013), p. 57; R. Horace Herring, ed., *Energy Efficiency and Sustainable Consumption: The Rebound Effect* (New York: Palgrave MacMillan, 2009); L. Hilty et al., "Rebound Effects of Progress in Information Technology," *Poiesis and Praxis*, 2006, pp. 4, 19–38.

8. EIA, op. cit. note 7.

9. World Bank, "Energy and Mining," in *World Development Indicators* (Washington, DC: 2012); Pat Murphy and Faith Morgan, "Cuba: Lessons from a Forced Decline," in Worldwatch Institute, *State of the World 2013: Is Sustainability Still Possible?* (Washington, DC: Island Press, 2013).

10. Lorenz Erdmann and Lorenz M. Hilty, "Scenario Analysis: Exploring the Macroeconomic Impacts of Information and Communication Technologies on Greenhouse Gas Emissions," *Journal of Industrial Ecology*, October 2010, pp. 826–43. Sponsorships of the 11 studies reviewed in this study were determined through the author's independent research.

11. William Ophuls, "Leviathan or Oblivion?" in Herman E. Daly, ed., *Toward a Steady-State Economy* (San Francisco: W. H. Freeman, 1973); David W. Orr, "Governance in the Long Emergency," in Worldwatch Institute, op. cit. note 9, p. 288.

12. David Bollier, *Viral Spiral: How the Commoners Built a Digital Republic of Their Own* (New York: New Press, 2008), p. 3.

13. Ibid., pp. 296, 307.

14. Bimber, op. cit. note 2.

15. Clay Shirky, *Here Comes Everybody: The Power of Organizing Without Organizations* (New York: Penguin Books, 2008); Jennifer Earl and Katrina Kimport, *Digitally Enabled Social Change: Activism in the Internet Age* (Cambridge, MA: MIT Press, 2011); Steven Coleman and Peter M. Shane, eds., *Connecting Democracy: Online Consultation and the Flow of Political Communication* (Cambridge, MA: MIT Press, 2012).

16. U.S. Department of Commerce, Bureau of Economic Analysis, "Table 2.7. Investment in Private Fixed Assets, Equipment, Structures, and Intellectual Property Products by Type, 2013" (Washington, DC: 2013); U.S. Department of Commerce, Bureau of Economic Analysis, "Table 3.7S. Investment in Private Structures by Industry, 2013" (Washington, DC: 2013).

17. Pew Charitable Trusts, *Global Clean Power: A $2.3 Trillion Opportunity* (Arlington, VA: The Pew Charitable Trusts, 2013); Hazel Henderson, Rosalinda Sanquiche, and Timothy J. Nash, *Green Transition Inflection Point: Green Transition Scoreboard® 2013 Report* (St. Augustine, FL: Ethical Markets Media, 2013); World Economic Forum, *The Green Investment Report: The Ways and Means to Unlock Private Finance for Green Growth* (Geneva: 2013).

18. Sherry Turkle, *Alone Together: Why We Expect Less from Each Other and More from Technology* (New York: Basic Books, 2010); James K. Galbraith, *Created Unequal: The Crisis in American Pay* (Chicago: University of Chicago Press, 2000), p. 168.

19. Galbraith, op. cit. note 18.

20. Ryan Snyder, *The Bus Riders Union Transit Model: Why a Bus-Centered System Will Best Serve American Cities* (Los Angeles: Labor Community Strategy Center, 2009).

21. President's Council of Advisors on Science and Technology, *Sustaining Environmental Capital: Protecting Society and the Economy, Working Group Report* (Washington, DC: Executive Office of the President, 2011); Bent Flyvbjerg, Nils Bruselius, and Werner Rothengatter, *Megaprojects and Risk: An Anatomy of Ambition* (Cambridge, U.K.: Cambridge University Press, 2003.)

Chapter 6. Living in the Anthropocene: Business as Usual, or Compassionate Retreat?

1. United Nations Framework Convention on Climate Change (UNFCCC), Conference of the Parties, 15th Session, "Copenhagen Accord" (Copenhagen: 18 December 2009); Veerabhadran Ramanathan and Yangyang Xu, "The Copenhagen Accord for Limiting Global Warming: Criteria, Constraints and Available Avenues," *Proceedings of the National Academy of Sciences*, vol. 107, no. 8 (2010), pp. 8,055–62; Gordon McGranahan et al., "The Rising Tide: Assessing the Risks of Climate Change and Human Settlements in Low Elevation Coastal Zones," *Environment and Urbanization*, vol. 19, no. 1 (2007), pp. 17–37; John Barnett and W. Neil Adger, "Climate Change, Human Security and Violent Conflict," *Political Geography*, vol. 26, no. 6 (2007), pp. 639–55; Chris D. Thomas et al., "Extinction Risk From Climate Change," *Nature*, 8 January 2004, pp. 145–48.

2. The Intergovernmental Panel on Climate Change (IPCC) cited geoengineering for the first time in its 2013 draft report (approved), to be published in early 2014. See IPCC, *Climate Change 2013: The Physical Science Basis*, IPCC Fifth Assessment Report (Cambridge, U.K.: 2013).

3. David W. Keith, "Geoengineering the Climate: History and Prospect," *Annual Review of Energy and the Environment*, vol. 25 (2000), p. 245.

4. For an overview of geoengineering, see Simon Nicholson, "The Promises and Perils of Geoengineering," in Worldwatch Institute, *State of the World 2013: Is Sustainability Still Possible?* (Washington, DC: Island Press, 2013), and Alan Robock, "20 Reasons Why Geoengineering May Be a Bad Idea," *Bulletin of the Atomic Scientists*, vol. 64, no. 2 (2008), pp. 14–18.

5. Will Steffen et al., "The Anthropocene: Conceptual and Historical Perspectives," *Philosophical Transactions of the Royal Society of London A*, vol. 369 (2011), pp. 842–67; Thomas Berry, *The Great Work: Our Way Into the Future* (New York: Bell Tower, 1999); Nancy Langston, *Toxic Bodies: Hormone Disruptors and the Legacy of Des* (New Haven: Yale University Press, 2010); George Santayana, *Realms of Being* (New York: Charles Scribner's Sons, 1942).

6. William F. Ruddiman, "The Anthropocene," *Annual Review of Earth and Planetary Sciences*, vol. 41 (2013),

pp. 45–68; Dipesh Chakrabarty, "The Climate of History: Four Theses," *Critical Inquiry*, vol. 35, no. 2 (2009), pp. 197–222.

7. Erle Ellis et al., "Anthropogenic Transformation of the Biomes, 1700–2000," *Global Ecology and Biogeography*, vol. 19 (2010), pp. 589–606; Johan Rockström et al., "A Safe Operating Space for Humanity," *Nature*, 23 September 2009, pp. 472–75.

8. P. C. D. Milly et al., "Stationarity Is Dead: Whither Water Management?" *Science*, 1 February 2008, pp. 573–74; Jeremy J. Schmidt, "Integrating Water Management in the Anthropocene," *Society and Natural Resources*, vol. 26, no. 1 (2013), pp. 105–12.

9. Christopher J. Preston, "Re-thinking the Unthinkable: Environmental Ethics and the Presumptive Argument Against Geoengineering," *Environmental Values*, vol. 20 (2011), pp. 457–79; Eva Lövbrand et al., "Earth System Governmentality: Reflections on Science in the Anthropocene," *Global Environmental Change*, vol. 19 (2010), pp. 7–13.

10. Ruddiman, op. cit. note 6.

11. Max Horkheimer and Theodor W. Adorno, *Dialectic of Enlightenment: Philosophical Fragments* (Stanford, CA: Stanford University Press, 2002).

12. Will Steffen et al., "The Anthropocene: Are Humans Now Overwhelming the Great Forces of Nature?" *Ambio*, vol. 36, no. 8 (2007), pp. 614–21; Will Steffen et al., "The Anthropocene: From Global Change to Planetary Stewardship," *Ambio*, vol. 40 (2011), pp. 739–61; Kathy A. Hibbard et al., "Decadal Interactions of Humans and the Environment," in Robert Costanza et al., eds., *Integrated History and Future of People on Earth* (Cambridge, MA: MIT Press, 2006); John Rawls, *Political Liberalism*, expanded edition (New York: Columbia University Press, 2005); Wendy Wheeler, *The Whole Creature: Complexity, Biosemiotics and the Evolution of Culture* (London: Lawrence & Wishart, 2006); Val Plumwood, *Feminism and the Mastery of Nature* (New York: Routledge, 1993).

13. Andrew Dobson, "Political Theory in a Closed World: Reflections on William Ophuls, Liberalism and Abundance," *Environmental Values*, vol. 22, no. 2 (2013), pp. 241–59; Immanuel Wallerstein, *The Modern World-System Iv: Centrist Liberalism Triumphant, 1789–1914* (Berkeley, CA: University of California Press, 2011); Timothy Mitchell, *Carbon Democracy: Political Power in the Age of Oil* (London: Verso, 2011); Neil Johnson et al., "Abrupt Rise of New Machine Ecology Beyond Human Response Time," *Nature: Scientific Reports*, 11 September 2013, pp. 1–7; Chrystia Freeland, *Plutocrats: The Rise of the New Global Super-Rich and the Fall of Everyone Else* (New York: Penguin Press, 2012); Ulrich Beck, *Risk Society: Towards a New Modernity* (London: Sage Publications, 1992); Steve Lemer, *Sacrifice Zones: The Front Lines of Toxic Chemical Exposure in the United States* (Cambridge, MA: MIT Press, 2010).

14. Jane Lubchenco, "Entering the Century of the Environment: A New Social Contract for Science," *Science*, 23 January 1998, pp. 491–97; Lance Gunderson and C. S. Holling, eds., *Panarchy: Understanding Transformations in Human and Natural Systems* (Washington, DC: Island Press, 2002); James Kay, "Ecosystems as Self-Organizing Holarchic Open Systems: Narratives and the Second Law of Thermodynamics," in Sven Jørgensen and Felix Müller, eds., *Handbook of Ecosystem Theories and Management* (Boca Raton, FL: Lewis Publishers, 2000); Hugh Brody, *The Other Side of Eden: Hunters, Farmers, and the Shaping of the World* (New York: North Point Press, 2001).

15. Nancy Langston, *Forest Dreams, Forest Nightmares: The Paradox of Old Growth in the Inland West* (Seattle: University of Washington Press, 1995); Fikret Berkes, *Sacred Ecology: Traditional Ecological Knowledge and Resource Management* (Philadelphia: Taylor and Francis, 1999); Eduardo Kohn, *How Forests Think: Toward an Anthropology Beyond the Human* (Berkeley, CA: University of California Press, 2013); Aldo Leopold, *A Sand County Almanac: With Essays on Conservation From Round River* (New York: Oxford University Press, 1966); Mick Smith, *Against Ecological Sovereignty: Ethics, Biopolitics, and Saving the Natural World* (Minneapolis: University of Minnesota Press, 2011); Peter G. Brown, "Are There Any Natural Resources?," *Politics and the Life Sciences*, vol. 23, no. 1 (2004), pp. 11–20.

16. David Keith, *A Case for Climate Engineering* (Cambridge, MA: MIT Press, 2013); Steward Brand, *Whole Earth Discipline: An Ecopragmatist Manifesto* (New York: Viking, 2009).

17. Rob Bellamy et al., "'Opening Up' Geoengineering Appraisal: Multi-Criteria Mapping of Options for Tackling Climate Change," *Global Environmental Change*, vol. 23, no. 5 (2013), pp. 926–37; Andrew Dobson, *Justice and the*

Environment (New York: Oxford University Press, 1999); John Rawls, *Justice as Fairness: A Restatement* (Cambridge, MA: Belknap Press of Harvard University Press, 2001).

18. Peili Wu et al., "Anthropogenic Impact on Earth's Hydrological Cycle," *Nature Climate Change*, 2 July 2013; Eliza Harris et al., "Enhanced Role of Transition Metal Ion Catalysis During in-Cloud Oxidation of SO_2," *Science*, 10 May 2013, pp. 727–30.

19. Clive Hamilton, *Earthmasters: The Dawn of Climate Engineering* (New Haven: Yale University Press, 2013), p. 180; C. S. Holling and Gary K. Meffe, "Command and Control and the Pathology of Natural Resource Management," *Conservation Biology*, vol. 10, no. 2 (1996), pp. 328–37; Langston, op. cit. note 15; Robert F. Durant et al., eds., *Environmental Governance Reconsidered: Challenges, Choices and Opportunities* (Cambridge, MA: MIT Press, 2004).

20. Hamilton, op. cit. note 19, p. 180; Holling and Meffe, op. cit. note 19, pp. 328–37; Langston, op. cit. note 15; Durant et al., eds., op. cit. note 19; Bill Vitek and Wes Jackson, eds., *The Virtues of Ignorance: Complexity, Sustainability, and the Limits of Knowledge* (Lexington, KY: The University Press of Kentucky, 2008).

21. Donna J. Haraway, *When Species Meet* (Minneapolis: University of Minnesota Press, 2008); Eduardo Kohn, *How Forests Think: Toward an Anthropology Beyond the Human* (Berkeley, CA: University of California Press, 2013); Berkes, op. cit. note 15; Bruno Latour, *Politics of Nature: How to Bring the Sciences Into Democracy* (Cambridge, MA: Harvard University, 2004).

22. Peter G. Brown and Jeremy J. Schmidt, "An Ethic of Compassionate Retreat," in Peter G. Brown and Jeremy J. Schmidt, eds., *Water Ethics: Foundational Readings for Students and Professionals* (Washington, DC: Island Press, 2010); Vitek and Jackson, eds., op. cit. note 20; Leopold, op. cit. note 15; Steffen et al., "The Anthropocene...," op. cit. note 12, pp. 739–61; Michael Callon et al., *Acting in an Uncertain World: An Essay on Technical Democracy* (Cambridge, MA: MIT Press, 2009).

23. Tania Murray Li, *The Will to Improve: Governmentality, Development, and the Practice of Politics* (Durham, NC: Duke University Press, 2007); James Tully, *Public Philosophy in a New Key*, vols. 1–2 (Cambridge, U.K.: Cambridge University Press, 2008). For a robust view of moral consent, see Derek Parfit, *On What Matters*, vols. 1–2 (New York: Oxford University Press, 2011).

24. Christopher Stone, *Should Trees Have Standing? Towards Legal Rights for Natural Objects* (New York: Avon, 1974). For important works on law and ecology that offer ways of altering existing legal systems, see Douglas Kysar, *Regulating From Nowhere: Environmental Law and the Search for Objectivity* (New Haven: Yale University Press, 2010); Michael M'Gonigle and Paula Ramsay, "Greening Environmental Law: From Sectoral Reform to Systemic Re-Formation," *Journal of Environmental Law and Practice*, vol. 14 (2004), pp. 333–56; David Boyd, *The Environmental Rights Revolution: A Global Study of Constitutions, Human Rights, and the Environment* (Vancouver, BC: University of British Columbia Press, 2012); Cormac Cullinan, *Wild Law: A Manifesto for Earth Justice* (Totnes, U.K.: Green Books, 2011).

25. Kate Raworth, "Living in the Doughnut," *Nature Climate Change*, vol. 2 (2012), pp. 225–26; Kate Raworth, *A Safe and Just Space for Humanity: Can We Live Within the Doughnut*, Oxfam Discussion Paper (London: 2012), pp. 1–26; Paul A. Murtaugh and Michael G. Schlax, "Reproduction and the Carbon Legacies of Individuals," *Global Environmental Change*, vol. 19 (2009), pp. 14–20; Wheeler, op. cit. note 12; John Fullerton, *Redesigning Finance: Pathways to a Resilient Future: Summary of Proceedings of August 9, 2012 San Francisco Invitational Gathering* (Boston: Tellus Institute, 2012); Peter A. Victor, "Living Well: Explorations Into the End of Growth," *Minding Nature*, vol. 5, no. 2 (2012), pp. 24–31.

Chapter 7. Governing People as Members of the Earth Community

1. Suzanne Austin Alchon, *A Pest in the Land: New World Epidemics in a Global Perspective* (Albuquerque: University of New Mexico Press, 2003), p. 21.

2. Bhutan's Gross National Happiness website, www.grossnationalhappiness.com; Republic of Ecuador, "Constitution of the Republic of Ecuador," 20 October 2008; Plurinational State of Bolivia, "Ley Nº 071: Ley de Derechos de la Madre Tierra," 21 December 2010; Plurinational State of Bolivia, "Ley Nº 300: Ley Marco de la Madre Tierra y Desarrollo Integral Para Vivir Bien," 15 October 2012.

3. Cormac Cullinan, *Wild Law: A Manifesto for Earth Justice* (White River Junction, VT: Chelsea Green Publishing, 2011); Cormac Cullinan, "Earth Jurisprudence: From Colonisation to Participation," in Worldwatch Institute, *State of the World 2010* (New York: W. W. Norton & Company, 2010), pp. 143–48; Peter Burdon, ed., *Exploring Wild Law. The Philosophy of Earth Jurisprudence* (Kent Town, South Australia: Wakefield Press, 2011); Global Alliance for the Rights of Nature website, www.therightsofnature.org.

4. Council of Canadians, Fundación Pachamama, and Global Exchange, *The Rights of Nature. The Case for a Universal Declaration of the Rights of Mother Earth* (San Francisco: 2011).

5. La Via Campesina, "The Jakarta Call: Call of the VI Conference of La Via Campesina" (Jakarta: 12 June 2013); Children's Charter from Earth Junkies website, www.earthjunkies.org.

6. United Nations Harmony with Nature website, www.harmonywithnatureun.org.

7. "Constitution of the Republic of Ecuador," op. cit. note 2.

8. Natalia Greene, "The First Successful Case of the Rights of Nature Implementation in Ecuador," at http://the rightsofnature.org/first-ron-case-ecuador.

9. Community Environmental Legal Defense Fund (CELDF) website, www.celdf.org.

10. Christopher Finlayson, "Whanganui River Agreement Signed," 30 August 2012, at www.beehive.govt.nz; Whanganui Iwi and The Crown, "Tūtohu Whakatupua," 30 August 2012.

11. Ibid.

12. "Rio+20: Civil Society Protesters Upstage World Leaders," *Environment News Service*, 22 June 2012.

13. See, for example, United Nations Human Rights Council resolutions 7/23 of 28 March 2008, 10/4 of 25 March 2009, and 18/22 of 30 September 2011; United Nations Framework Convention on Climate Change, " Report of the Conference of the Parties on Its Sixteenth Session, Held in Cancun from 29 November to 10 December 2010" (Bonn: 15 March 2011), p. 4.

14. KumKum Dasgupta, "Vedanta's India Mining Scheme Thwarted by Local Objections," PovertyMatters blog (*The Guardian*), 21 August 2013; Survival, "The Dongria Kondh," at www.survivalinternational.org/tribes/dongria.

15. For examples, see CELDF, "Ordinances," at www.celdf.org/resources-ordinances.

Chapter 8. Listening to the Voices of Young and Future Generations

1. "The Great Law of Iroquois Confederacy," at www.indigenouspeople.net/iroqcon.htm.

2. World Future Council, "Who We Are," at www.worldfuturecouncil.org/about_us.html.

3. "The Constitution of the Kingdom of Norway," at www.constitution.org/cons/norway/dok-bn.html; Gro Harlem Brundtland et al., *Our Common Future* (Oxford: Oxford University Press, 1987), p. 12.

4. Kirsty Schneeberger, "Intergenerational Equity: Implementing the Principle in Mainstream Decision-making," *Environmental Law and Management*, vol. 23, no. 1 (2011), p. 25; "Award Between the United States and the United Kingdom Relating to the Rights of Jurisdiction of United States in the Bering's Sea and the Preservation of Fur Seals," 15 August 1893, at http://legal.un.org/riaa/cases/vol_XXVIII/263-276.pdf; Environmental Law Alliance Worldwide, "Philippines — Oposa et al. v. Fulgencio S. Factoran, Jr. et al. (G.R. No. 101083)," 30 July 2003, at www .elaw.org/node/1343.

5. "Pulp Mills on the River Uruguay (Argentina v. Uruguay)," *I.C.J. Reports 2009–2010* (New York: United Nations, 2010), p. 28.

6. Box 8–1 from the following sources: United Nations, *The Future We Want* (Rio de Janeiro: 2012), p. 86; United Nations Secretary General, *Intergenerational Solidarity and the Needs of Future Generations* (New York: United Nations, 2013).

7. Maya Göpel, *Ombudspersons for Future Generations as Sustainability Implementation Units* (Hamburg, Germany: World Future Council, September 2011), pp. 9–10.

8.   JNO (Office of the Parliamentary Commissioner for Future Generations) website, http://jno.hu/en.

9.   Aseem Prakash and Jennifer J. Griffin, "Corporate Responsibility, Multinational Corporations, and Nation States: An Introduction," *Business and Politics*, vol. 13, no. 3 (2012), pp. 1–10; Richard Heede, "Tracing Anthropogenic Carbon Dioxide and Methane Emissions to Fossil Fuel and Cement Producers, 1854–2010," *Climatic Change*, January 2014, pp. 229–41.

10. Allen L. White, *Who Speaks for Future Generations?* (New York: Business for Social Responsibility, December 2007), p. 7.

11. Ibid., pp. 8–9.

12. Franck-Dominique Vivien, *Le développement soutenable* (Paris: La Découverte, 2005), pp. 35–37; Robert M. Solow, "Intergenerational Equity and Exhaustible Resources," Review of Economic Studies, Symposium on the Economics of Exhaustible Resources (1974), pp. 139–52.

13. Box 8–2 from the following sources: Sovereign Wealth Fund Institute, "Kuwait Investment Authority," at www.swfinstitute.org/swfs/kuwait-investment-authority; Judith Ireland, "Future Fund Quits Tobacco Investment," *Sydney Morning Herald*, 28 February 2013; Alexander W. Cappelen and Runa Urheim, "Pension Funds, Sovereign Wealth Funds and Intergenerational Justice," Norwegian School of Economics Discussion Paper No. 19 (2012), p. 4.

14. David Roberts, "Discount Rates: A Boring Thing You Should Know About (With Otters!)," 24 September 2012, at grist.org.

15. Nicholas Stern, *Stern Review on the Economics of Climate Change*, special report to the U.K. Prime Minister and the Chancellor of the Exchequer (London: 2006); William D. Nordhaus, *The "Stern Review" on the Economics of Climate Change*, Working Paper 12741 (Cambridge, MA: National Bureau of Economic Research: December 2006).

16. United Nations Environment Programme, Tunza website, www.unep.org/tunza; United Nations Joint Framework Initiative on Children, Youth and Climate Change, *Youth in Action on Climate Change: Inspirations from Around the World* (Bonn: May 2013).

17. Global Power Shift website, http://globalpowershift.org.

18. International Youth Climate Movement (IYCM) website, http://youthclimate.org/about_youth_climate.

19. Mark Sweney, "Copenhagen Climate Change Treaty Backed by 'Hopenhagen' Campaign," *The Guardian* (U.K.), 23 June 2009.

20. Atif Ansar, Ben Caldecott, and James Tilbury, *Stranded Assets and the Fossil Fuel Divestment Campaign: What Does Divestment Mean for the Valuation of Fossil Fuel Assets?* (Oxford: Stranded Assets Programme, Smith School of Enterprise and the Environment, October 2013), pp. 10–11; U.S. National Oceanic and Atmospheric Administration, "Global Summary Information – November 2013," at www.ncdc.noaa.gov/sotc.

21. Index Mundi, "World Demographics Profile 2013," at www.indexmundi.com/world/demographics_profile.html.

22. Marx citation from Marc Lallanilla, "Inspirational Quotes for Earth Day," http://greenliving.about.com/od /greenprograms/a/earth-day-quotes.htm. To quote an activist friend of the authors: "I was at a civil society meeting on intergenerational equity the other day. I was the only person under forty there—it was disgusting."

23. Dan Milmo, "Mandela Silences 60,000 at Murrayfield Concert," *The Guardian* (U.K.), 7 July 2005.

Chapter 9. Advancing Ecological Stewardship Via the Commons and Human Rights

1.   This essay is derived from ideas in Burns H. Weston and David Bollier, *Green Governance: Ecological Survival, Human Rights, and the Law of the Commons* (Cambridge, U.K.: Cambridge University Press, 2013) and other works produced by the Commons Law Project, at www.commonslawproject.org.

2.   Ida Kubiszewski et al., "Beyond GDP: Measuring and Achieving Global Genuine Progress," *Ecological Economics*, vol. 93 (2013), pp. 57–68.

3. "The March of Protest," *The Economist*, 29 June 2013.

4. Weston and Bollier, op. cit. note l; "Universal Covenant Affirming a Human Right to Commons- and Rights-based Governance of Earth's Natural Wealth and Resources," *Journal of Human Rights and the Environment*, September 2013, pp. 215–25.

5. Sam Adelman, "Rethinking Human Rights: The Impact of Climate Change on the Dominant Discourse," in Stephen Humphreys, ed., *Human Rights and Climate Change* (Cambridge, U.K.: Cambridge University Press, Cambridge 2010), pp. 162, 167, and 173.

6. Egypt and Europe from Simon Lyster, *International Wildlife Law: An Analysis of International Treaties Concerned with the Conservation of Wildlife* (Cambridge, U.K.: Cambridge University Press, 1993), p. 11; Grotius from Kemal Baslar, *The Concept of the Common Heritage of Mankind and International Law* (Boston: Martinus Nijhoff Publishers, 1997); Prue Taylor, "The Common Heritage of Mankind: A Bold Doctrine Kept Within Strict Boundaries," in David Bollier and Silke Helfrich, *The Wealth of the Commons: A World Beyond Market and State* (Amherst, MA: Levellers Press, 2012), pp. 353–60; Antarctic Treaty System, "The Antarctic Treaty," 1 December 1959, at www.ats.aq/e/ats.htm; United Nations Office for Outer Space Affairs, "Treaty on Principles Governing the Activities of States in the Exploration and Use of Outer Space, including the Moon and Other Celestial Bodies," 27 January 1967, at www.unoosa.org/oosa/en/SpaceLaw/outerspt.html.

7. Trent Schroyer, *Beyond Western Economics: Remembering Other Economic Cultures* (London: Routledge, 2009), p. 69.

8. Garrett Hardin, "The Tragedy of the Commons," *Science*, 13 December 1968, pp. 1,243–48; International Association for the Study of the Commons, "Policy Forum, 12th Biennial Conference," press release (Gloucestershire, Cheltenham, U.K.: 14–18 July 2008).

9. Amy R. Poteete, Marco A. Janssen, and Elinor Ostrom, *Working Together: Collective Action, The Commons and Multiple Methods in Practice* (Princeton, NJ: Princeton University Press, 2010); Elinor Ostrom, "A Multiscale Approach to Coping with Climate Change and Other Collective Action Problems," *Solutions*, 24 February 2010, pp. 27–36.

10. Poteete, Janssen, and Ostrom, op. cit. note 9.

11. Michel Bauwens, "The New Triarchy: The Commons, Enterprise, the State," P2P Foundation blog, 25 August 2010, at http://blog.p2pfoundation.net/the-new-triarchy-the-commons-enterprise-the-State/2010/08/25; Peter Barnes, *Capitalism 3.0: A Guide to Reclaiming the Commons* (San Francisco: Berrett-Koehler Publishers, 2006); Michel Bauwens, "The Triune Peer Governance of the Digital Commons," in Bollier and Helfrich, op. cit. note 6, pp. 375–78.

12. These include the Open Access Data Protocols developed by Science Commons; open access journals; the BiOS license developed by CAMBIA for bioengineered products; and the suite of Creative Commons licenses for copyrightable works. See Benjamin Mako Hill, "Freedom for Users, Not for Software," in Bollier and Helfrich, op. cit. note 6, pp. 305–08.

13. David E. Martin, "Emancipating Innovation Enclosures: The Global Innovation Commons," in Bollier and Helfrich, op. cit. note 6, pp. 314–18.

14. Alaska Permanent Fund website, www.apfc.org; Peter Barnes et al., "Creating an Earth Atmospheric Trust," *Science*, 8 February 2008, p. 724; Our Children's Trust website, www.ourchildrenstrust.org; Conor Casey et al., *Valuing Common Assets for Public Finance in Vermont* (Burlington, VT: Vermont Green Tax and Common Assets Project, November 2008).

15. Gerd Wessling, "Transition Towns: Initiatives of Transformation," in Bollier and Helfrich, op. cit. note 6, pp. 239–42.

16. See Mary Christina Wood's pioneering work in explicating the public trust doctrine, especially *Nature's Trust: Environmental Law for a New Ecological Age* (Cambridge, U.K.: Cambridge University Press, 2013). Box 9–1 from the following sources: Our Children's Trust, "Legal Action," at ourchildrenstrust.org/legal; Wood, op. cit. this note; James Hansen et al., "Assessing 'Dangerous Climate Change': Required Reduction of Carbon Emissions to Protect Young People, Future Generations and Nature," *PloS ONE*, vol. 8, no. 12 (2013).

Chapter 10. Looking Backward (Not Forward) to Environmental Justice

1. "Miscellaneous City News: Edison's Electric Light. 'The Times' Building Illuminated by Electricity," *New York Times*, 5 September 1882. The author is deeply grateful to the following individuals for many different kinds of help with this chapter: Ben Cohen, Nick Howe, Jing Jin, Amy Kohout, Sarah Luria, Neil Maher, Laura Martin, Liz Mesok, Kathy Morse, Cindy Ott, Tom Prugh, Michael Renner, and Michael Smith.

2. Louis C. Hunter and Lynwood Bryant, *A History of Industrial Power in the United States, 1780–1930, Volume Three: The Transmission of Power* (Cambridge, MA: MIT Press, 1991), pp. 185–93; Amanda Little, *Power Trip: From Oil Wells to Solar Cells—Our Ride to the Renewable Future* (New York: Harper, 2009), pp. 216–21; Ernest Freeberg, *The Age of Edison: Electric Light and the Invention of Modern America* (New York: Penguin Press, 2013).

3. Edward Bellamy, *Looking Backward: 2000–1887* (New York: New American Library, 1888/2000).

4. Ibid.; Alan Trachtenberg, *The Incorporation of America: Culture and Society in the Gilded Age* (New York: Hill and Wang, 1982); Rebecca Edwards, *New Spirits: Americans in the Gilded Age, 1865–1905* (New York: Oxford University Press, 2006); Jackson Lears, *Rebirth of a Nation: The Making of Modern America, 1877–1920* (New York: HarperCollins, 2009).

5. United Nations Development Programme, *Human Development Report 2013* (New York: 2013), pp. 21–41.

6. Robert D. Bullard, *Dumping in Dixie: Race, Class, and Environmental Quality* (Boulder, CO: Westview Press, 1990); Robert D. Bullard, ed., *Confronting Environmental Racism: Voices from the Grassroots* (Boston: South End Press, 1993); Bunyan Bryant, ed., *Environmental Justice: Issues, Policies, and Solutions* (Washington, DC: Island Press, 1995).

7. Daniel Faber, ed., *The Struggle for Ecological Democracy: Environmental Justice Movements in the United States* (New York: Guilford Press, 1998); Aaron Sachs, *Eco-Justice: Linking Human Rights and the Environment*, Worldwatch Paper 127 (Washington, DC: Worldwatch Institute, 1995).

8. See, for example: James Hansen, *Storms of My Grandchildren: The Truth about the Coming Climate Catastrophe and Our Last Chance to Save Humanity* (New York: Bloomsbury, 2009); Sandra Steingraber, *Raising Elijah: Protecting Our Children in an Age of Environmental Crisis* (New York: Da Capo, 2011); and Mark Hertsgaard, *Hot: Living Through the Next Fifty Years on Earth* (Boston: Houghton Mifflin Harcourt, 2011).

9. Andrew Nikiforuk, *The Energy of Slaves: Oil and the New Servitude* (Vancouver: Greystone, 2012), pp. 1–29.

10. Some antebellum activists in the North, objecting to the southern system of sugar production, sweetened their food and drink only with local honey or maple syrup. On coal in Appalachia, see Rebecca R. Scott, *Removing Mountains: Extracting Nature and Identity in Appalachian Coalfields* (Minneapolis, MN: University of Minnesota Press, 2010), and Harry M. Caudill, *Night Comes to the Cumberlands: A Biography of a Depressed Area* (Ashland, KY: Jesse Stuart Foundation, 1963/2001). On climate change, see Hallie Eakin and Amy Lynd Luers, "Assessing the Vulnerability of Social-Environmental Systems," *Annual Review of Environment and Resources*, vol. 31 (2006), pp. 365–94; Anne Jerneck and Lennart Olson, "Adaptation and the Poor: Development, Resilience and Transition," *Climate Policy* (2008), pp. 170–82; and Paul Baer et al., *The Greenhouse Development Rights Framework: The Right to Development in a Climate Constrained World* (Berlin: Heinrich Böll Foundation, 2008).

11. Robin Mearns and Andrew Norton, eds., *Social Dimensions of Climate Change: Equity and Vulnerability in a Warming World* (Washington, DC: World Bank, 2010); Christian Parenti, *Tropic of Chaos: Climate Change and the New Geography of Violence* (New York: Nation Books, 2011); Rafael Reuveny, "Climate Change-induced Migration and Violent Conflict," *Political Geography*, vol. 26 (2007), pp. 656–73; Michael Renner, "Climate Change and Displacements," in Worldwatch Institute, *State of the World 2013: Is Sustainability Still Possible?* (Washington, DC: Island Press, 2013), pp. 343–52. The discourse we use to describe these situations can matter profoundly; if we categorize refugees as burdensome, then "adaptation" strategies could make their circumstances even worse, as industrial countries take steps to secure their borders and bolster their military readiness to repel invaders. See, for example, Giovanni Bettini, "Climate Barbarians at the Gate? A critique of apocalyptic narratives on 'climate refugees'," *Geoforum*, vol. 45 (2013), pp. 63–72, and Betsy Hartmann, "Rethinking Climate Refugees and Climate Conflict: Rhetoric, Reality and the Politics of Policy Discourse," *Journal of International Development*, vol. 22 (2010), pp. 233–46.

12. Derek Bok, *The Politics of Happiness: What Government Can Learn from the New Research on Well-Being* (Princeton, NJ: Princeton University Press, 2010); Stephanie Rosenbloom, "But Will It Make You Happy?" *New York Times*, 7 August 2010; Ann Cvetkovich, *Depression: A Public Feeling* (Durham, NC: Duke University Press, 2012).

13. Kathleen Dean Moore and Michael P. Nelson, eds., *Moral Ground: Ethical Action for a Planet in Peril* (San Antonio, TX: Trinity University Press, 2010).

14. Bill McKibben, "Global Warming's Terrifying New Math," *Rolling Stone*, 19 July 2012; on oil, see Nikiforuk, op. cit. note 9, and Daniel Yergin, *The Prize: The Epic Quest for Oil, Money, and Power* (New York: Free Press, 1991/2009). Also note Naomi Klein's recent suggestion that there is plenty of common ground between climate change activists and the Occupy movement, in Naomi Klein, "Capitalism vs. The Climate," *The Nation*, 28 November 2011.

15. John Broome, *Climate Matters: Ethics in a Warming World* (New York: W. W. Norton & Company, 2012); Henry David Thoreau, "Civil Disobedience," in Lewis Hyde, ed., *The Essays of Henry D. Thoreau* (New York: North Point Press, 2002), pp. 123–46.

16. Kelly G. Lambert, "Rising Rates of Depression in Today's Society: Consideration of the Roles of Effort-based Rewards and Enhanced Resilience in Day-to-day Functioning," *Neuroscience and Biobehavioral Reviews*, vol. 30 (2006), pp. 497–510.

17. David E. Nye, *Consuming Power: A Social History of American Energies* (Cambridge, MA: MIT Press, 1998), pp. 11–12; Martin V. Melosi, *Coping with Abundance: Energy and Environment in Industrial America* (Philadelphia: Temple University Press, 1985); Sam H. Schurr et al., *Energy in the American Economy, 1850–1975: An Economic Study of Its History and Prospects* (Westport, CT: Greenwood Press, 1960/1977).

18. U.S. Centers for Disease Control and Prevention, "Antibiotics Aren't Always the Answer," November 2013, at www.cdc.gov/features/getsmart.

19. Nye, op. cit. note 17, p. 6, and see pp. 158–215; Little, op. cit. note 2, pp. 81–109; Christopher Flavin and Nicholas Lenssen, *Power Surge: Guide to the Coming Energy Revolution* (New York: W. W. Norton & Company, 1994), pp. 195–239.

20. Broome, op. cit. note 15; Bill McKibben, *Eaarth: Making a Life on a Tough New Planet* (New York: St. Martin's Griffin, 2011), p. 151; Lewis Mumford, *The Myth of the Machine: Technics and Human Development* (New York: Harcourt, Brace, and World, 1967), p. 269.

21. Henry David Thoreau, *A Week on the Concord and Merrimack Rivers* (New York: Penguin, 1849/1998), p. 137.

22. Walter James Miller, "The Future of Futurism: An Introduction to *Looking Backward*," in Bellamy, op. cit. note 3, pp. v–xii.

Chapter 11. The Too-Polite Revolution: Understanding the Failure to Pass U.S. Climate Legislation

1. This essay was adapted from Petra Bartosiewicz and Marissa Miley, *The Too Polite Revolution: Why the Recent Campaign to Pass Comprehensive Climate Change Legislation in the United States Failed*, commissioned by the Rockefeller Family Fund in conjunction with the Columbia University Graduate School of Journalism (New York: 14 January 2013).

2. "Obama Advocates Cap-and-Trade System," *E&E Greenwire*, 9 October 2007; Michael Scherer, "Obama's Permanent Grass-Roots Campaign," *Time*, 15 January 2009.

3. Unless stated otherwise, the term "green groups" refers to those national environmental organizations that spearheaded the climate legislative effort, primarily the Environmental Defense Fund (EDF), the Natural Resources Defense Council (NRDC), the Pew Center on Global Climate Change (now the Center for Climate and Energy Solutions), the World Resources Institute, The Nature Conservancy, the National Wildlife Federation, The Sierra Club, the League of Conservation Voters, and the Alliance for Climate Protection (now the Climate Reality Project). A. Denny Ellerman and Paul L. Joskow, *The European Union's Emissions Trading System*, prepared for the Pew Center on Global Climate Change (Arlington, VA: May 2008).

4. Center for Responsive Politics, "Lobbying: Top Industries," at www.opensecrets.org/lobby/top.php?indexType=i.

5. CNNMoney, "Fortune 500 List, 2008," at http://money.cnn.com/magazines/fortune/fortune500/2008/full_list; Political Economy Research Institute, University of Massachusetts Amherst, "Toxic 100 Air Polluters, 2012 Report," at www.peri.umass.edu/toxicair20120/.

6. Michael Parr, DuPont, personal communication with authors, July 2011.

7. European Commission, "The EU Emissions Trading System (EU ETS)," fact sheet (Brussels: October 2013).

8. Richard Schmalensee and Robert N. Stavins, "The Power of Cap-and-Trade," *Boston Globe*, 27 July 2010; Daniel J. Weiss, "Today's GOP Stomp on Reagan, Father of Cap and Trade," *Think Progress*, 22 October 2010, at http://thinkprogress.org/economy/2010/10/22/173589/cap-and-reagan/.

9. Regional Greenhouse Gas Initiative, "Program Design Archive," at www.rggi.org/design/history; California Environmental Protection Agency, Air Resources Board, "Assembly Bill 32: Global Warming Solutions Act," at www.arb.ca.gov/cc/ab32/ab32.htm; Pew Center on Global Climate Change, "Economy-wide Cap-and-Trade Proposals in the 110th Congress, As of December 1, 2008," at www.c2es.org/docUploads/Chart-and-Graph-120108.pdf.

10. For one corporation's take on climate change legislation, see ConocoPhillips, "Q1 2007 Earnings Call Transcript," at http://seekingalpha.com/article/33516-conocophillips-q1-2007-earnings-call-transcript?all=true&find ="U.S.+climate+action+partnership.

11. Fallout of Btu tax from Eric Pooley, *The Climate War: True Believers, Power Brokers, and the Fight to Save the Earth* (New York: Hyperion, 2010), pp. 137–41.

12. Jigar Shah, SunEdison, personal communication with authors, July 2011.

13. Kierán Suckling, Center for Biological Diversity, personal communication with authors, April 2011.

14. Mark Dowie, *Losing Ground: American Environmentalism at the Close of the Twentieth Century* (Cambridge, MA: MIT Press, 1995), p. 33.

15. Sarah Hansen, "Cultivating the Grassroots: A Winning Approach for Environment and Climate Funders" (Washington, DC: National Committee for Responsive Philanthropy, 2012), p. 6.

16. Robert J. Brulle and Craig Jenkins, "Foundations and the Environmental Movement: Priorities, Strategies and Impact," in Daniel R. Faber and Deborah McCarthy, *Foundations for Social Change* (Lanham, MD: Rowman & Littlefield, 2005), pp. 151, 157–58.

17. California Environmental Associates, *Design to Win: Philanthropy's Role in the Fight Against Global Warming* (San Francisco: 2007), pp. 6, 19, and 44.

18. Robertson from Matthew C. Nisbet, *Climate Shift* (Washington, DC: American University School of Communications, 2011), p. 16; Robert W. Wilson Charitable Trust, IRS tax filings, 2008–2010.

19. Ron Kroese, McKnight Foundation, personal communication with authors, October 2011.

20. Betsy Taylor, personal communication with authors, November 2011. This criticism is certainly not new within the environmental community. See Dowie, op. cit. note 14, and Michael Shellenberger and Ted Nordhaus, *The Death of Environmentalism* (Oakland, CA: The Breakthrough Institute, 2004).

21. Paul Tewes, personal communication with authors, August 2011; EDF and NRDC 990 IRS filings from GuideStar USA, Inc., www.guidestar.org.

22. High-engagement giving from Katherine Fulton and Andrew Blau, *Looking Out for the Future: An Orientation for Twenty-First Century Philanthropists* (Cambridge, MA: Monitor Company Group, 2005), p. 24; Shah, op. cit. note 12.

23. Alex Kowalski, "Recession Took Bigger Bite Than Estimated," *Bloomberg News*, 29 July 2011; Louis Uchitelle, "Jobless Rate Hits 7.2%, a 16-Year High," *New York Times*, 9 January 2009.

24. Ted Glick, Chesapeake Climate Action Network, personal communication with authors, April 2011.

25. John Passacantando, personal communication with authors, April 2011.

26. John M. Broder, "House Passes Bill to Address Threat of Climate Change," *New York Times*, 26 June 2009.

27. Connor Gibson, "Koch Brothers Exposed: Fueling Climate Denial and Privatizing Democracy," Greenpeace USA blog, 12 April 2012, at http://greenpeaceblogs.org/2012/04/02/koch-brothers-exposed-fueling-climate -denial-and-privatizing-democracy.

28. Anne C. Mulkern, "Coal Industry Sees Life or Death in Senate Debate," *E&E Greenwire*, 6 July 2009; Ryan Lizza, "As the World Burns," *The New Yorker*, 11 October 2010.

29. Office of Senator Maria Cantwell, "The CLEAR Act: A Simple, Market-based, and Equitable Pathway to Energy Independence and Climate Change Mitigation," at www.cantwell.senate.gov/issues/Frequently%20Asked%20 Questions.pdf.

30. Chris Miller, personal communication with authors, May 2011.

31. Marshall Ganz, personal communication with authors, August 2011.

32. Betsy Taylor, personal communication with authors, January 2012; Gillian Caldwell, personal communications with authors, July and August 2011; Caldwell quote from Terence Samuel, "How Prospects Cooled for U.S. Global Warming Bill," *National Geographic News*, 30 July 2010, at http://news.nationalgeographic.com /news/2010/07/100731-energy-prospects-global-warming-bill.

33. Hansen, op. cit. note 15.

34. Maggie Fox, Climate Reality Project, personal communication with authors, May 2011.

35. Tewes, op. cit. note 21; David Di Martino, personal communication with authors, June 2011; Frances Beinecke, NRDC, personal communication with authors, July 2011.

36. Health Care for America Now website, http://healthcareforamericanow.org.

37. Fifty-five bills were introduced in the 2011–12 Congress that would have "blocked or hindered climate action – 40 of which would have prohibited or hindered regulation of greenhouse gas emissions, primarily by preventing EPA from regulating under the Clean Air Act," according to the Pew Center for Climate and Energy Solutions, "Legislation in the 112th Congress Related to Global Climate Change," at www.c2es.org/federal/congress/112. In the 2012 presidential race, Michele Bachmann referred to human-caused climate change as "manufactured science" and Rick Santorum called it a "hoax"; see Brad Johnson, "Michele Bachmann: Man-Made Climate Change Is 'Manufactured Science,'" *ClimateProgress*, 17 August 2011, at http://thinkprogress.org/climate/2011/08/17/297902 /michele-bachmann-man-made-climate-change-is-manufactured-science/, and climate-brad, "Santorum: 'Global Warming Hoax,'" 27 January 2012, at www.youtube.com/watch?v=3AuZ-b8bjLk. Neela Banerjee, "Obama Asks EPA to Back Off Draft Ozone Standard," *Los Angeles Times*, 2 September 2011.

38. "Text: Obama's 2013 State of the Union Address," *New York Times*, 12 February 2013; The White House, "Remarks by the President on Climate Change," press release (Washington, DC: 25 June 2013); Elizabeth Kolbert, "Paying for It," *The New Yorker*, 10 December 2012.

39. "An Open Letter to President Obama from Groups Protecting Public Health," 21 July 2011, at www.lung.org /get-involved/advocate/advocacy-documents/protecting-public-health.pdf; 350.org, "Updates," at http://gofossil free.org/updates/; 350.org, "Stop the Keystone XL Pipeline!," at http://350.org/en/stop-keystone-xl.

40. Theda Skocpol, *Diminished Democracy: From Membership to Management in American Civic Life* (Norman, OK: University of Oklahoma Press, 2003), p. 177.

41. Ibid., pp. 177, 201; C-SPAN.org, "Congress," at legacy.c-span.org/questions/weekly35.asp.

42. Naomi Klein, "Capitalism vs. Climate," *The Nation*, 28 November 2011.

43. Dowie, op. cit. note 14.

Chapter 12. China's Environmental Governance Challenge

1. Chinese Ministry of Environmental Protection, *2012 Report on the State of the Environment* (Beijing: 4 June 2013); "Global Burden of Disease Survey 2010," *The Lancet*, 13 December 2012; Edward Wong, "Air Pollution Linked to 1.2 Million Premature Deaths in China," *New York Times*, 1 April 2013.

2. PBL Netherlands Environmental Assessment Agency, "2012 Sees Slowdown in the Increase of Global CO$_2$ Emissions," press release (The Hague: 31 October 2013).

3. Jennifer Duggan, "China's Environmental Problems Are Grim, Admits Ministry Report," *The Guardian* (U.K.), 7 June 2013; Pew Research Center, "Environmental Concerns on the Rise in China," 19 September 2013, at www .pewglobal.org/2013/09/19/environmental-concerns-on-the-rise-in-china.

4. "Chinese Anger Over Pollution Becomes Cause of Social Unrest," *Bloomberg News*, 6 March 2013.

5. Kelly Ip, "Uproar Over Uranium Plant Still Smolders," *The Standard* (Hong Kong), 15 July 2013.

6. Jonathan Ansfield, "Alchemy of a Protest: The Case of Xiamen PX," in Sam Geall, ed., *China and the Environment: The Green Revolution* (London: Zed Books, 2013), pp. 136–202.

7. Steven Millward, "China Now Has 591 Million Internet Users, 460 Million Mobile Netizens," *Tech in Asia*, 17 July 2013.

8. Tang Hao, "China's 'Nimby' Protests Sign of Unequal Society," 29 May 2013, at www.chinadialogue.net/article /show/single/en/6051-China-s-nimby-protests-sign-of-unequal-society.

9. Sam Geall and Sony Pellissery, "Five Year Plans," in Sam Geall et al., eds., *Berkshire Encyclopedia of Sustainability: China, India, and East and Southeast Asia: Assessing Sustainability* (Great Barrington, MA: Berkshire Publishing, 2012).

10. Ibid.

11. See, for example, Gørild Heggelund, *Environment and Resettlement Politics in China: The Three Gorges Project* (Farham, Surrey, U.K.: Ashgate, 2004).

12. Wang Jin, "China's Green Laws Are Useless," 23 September 2010, at www.chinadialogue.net/article/show/single /en/3831--China-s-green-laws-are-useless; Alex Wang, "The Search for Sustainable Legitimacy: Environmental Law and Bureaucracy in China," *Harvard Environmental Law Review*, vol. 37, no. 365 (2013).

13. Liu Jianqiang, "China's Low-carbon Cities: From Sham to Reality," 3 November 2010, at www.chinadialogue.net /article/show/single/en/3916-From-sham-to-reality.

14. German Asia Foundation and chinadialogue, *Report on Environmental Health in the Pearl River*, 2011, at www .eu-china.net/english/Resources/chinadialogue_2011_Report-on-Environmental-Health-in-the-Pearl-River -Delta.html.

15. Sam Geall, "Data Trap," *Index on Censorship*, vol. 40, no. 4 (2011), pp. 48–58.

16. Chris Luo, "'Environmental Expert' Arrested in Anti-Rumour Campaign," *South China Morning Post*, 29 September 2013.

17. Isabel Hilton, "The Return of Chinese Civil Society," in Geall, ed., op. cit. note 6, pp. 1–15.

18. Thomas Johnson, "Environmentalism and NIMBYism in China: Promoting a Rules-based Approach to Public Participation," *Environmental Politics*, vol. 19, no. 3 (2010), pp. 430–48.

19. Geall, op. cit. note 15.

20. He Zuoxiu, "Chinese Nuclear Disaster 'Highly Probable' by 2030," 9 March 2013, at www.chinadialogue .net/article/show/single/en/5808-Chinese-nuclear-disaster-highly-probable-by-2-3; World Nuclear Association, "Nuclear Power in China," updated November 2013, at www.world-nuclear.org/info/Country-Profiles/Countries -A-F/China--Nuclear-Power/.

21. Olivia Boyd, "The Birth of Chinese Environmentalism: Key Campaigns," in Geall, ed., op. cit. note 6, pp. 40–95.

22. International Rivers, "China's Government Proposes New Dam-Building Spree," 28 February 2011, at www .internationalrivers.org/resources/china%E2%80%99s-government-proposes-new-dam-building-spree-3419.

23. Chen Zifan, "Beijing's Blue-sky Diary," 28 February 2011, at www.chinadialogue.net/article/show/single/en /4134-Beijing-s-blue-sky-diary.

24. Boyd, op. cit. note 21, pp. 40–95; Sharon LaFraniere, "Activists Crack China's Wall of Denial About Air Pollution," *New York Times*, 27 January 2012.

25. Sam Geall, "The Everyman's Science," *Solutions*, March–April 2013, pp. 18–20; "PM2.5 in Air Quality Standards, Positive Response to Net Campaign," *People's Daily*, 1 March 2012.

26. Institute of Public and Environmental Affairs, "China Water Pollution Map," at www.ipe.org.cn/En/pollution.

27. Sam Geall, "Clean As a Whistleblower," *New Statesman*, 28 June–4 July 2013, p. 16.

Chapter 13. Assessing the Outcomes of Rio+20

1. Robert L. Stivers, *The Sustainable Society: Ethics and Economic Growth* (Philadelphia, PA: Westminster Press, 1973).

2. Jim Leape, "World Wide Fund for Nature Statement to the United Nations Conference on Sustainable Development," 21 June 2012, at www.panda.org/?205343/WWF-Rio20-closing-statement; Kumi Naidoo, executive director of Greenpeace International, quoted in Brian Walsh, "What the Failure of Rio+20 Means for the Climate," *Time*, 26 June 2012; Maria Ivanova, "The Contested Legacy of Rio+20," *Global Environmental Politics*, November 2013, pp. 1–11.

3. United Nations (UN), *Declaration of the United Nations Conference on the Human Environment (Stockholm Declaration)* (Stockholm: 16 June 1972).

4. UN, *The Future We Want* (Rio de Janeiro: 11 September 2012), para. 2.

5. Table 13–1 from UN, *Millennium Development Goals Report 2012* (New York: July 2012).

6. Ibid.

7. Maria Ivanova and Natalia Escobar-Pemberthy, "Quest for Sustainable Development: The Past and Future of International Development Goals," in T. Pogge, G. Köhler, and A. D. Cimadamore, eds., *Poverty and the Millennium Development Goals (MDGs): A Critical Assessment and a Look Forward* (London: Zed Books, forthcoming 2013).

8. Tadanori Inomata, *Management Review of Environmental Governance within the United Nations System* (Geneva: UN Joint Inspection Unit, 2008).

9. Maria Ivanova, "A New Global Architecture for Sustainability Governance," in Worldwatch Institute, *State of the World 2012: Moving Toward Sustainable Prosperity* (New York: W. W. Norton & Company, 2012).

10. Ivanova, op. cit. note 2, pp. 5–6.

11. International Institute for Sustainable Development, "Summary of the 27th Session of the UNEP Governing Council/ Global Ministerial Environment Forum," *Earth Negotiations Bulletin*, 25 February 2013.

12. UN, op. cit. note 4, para. 88.

13. Stine Madland Kaasa, "The UN Commission on Sustainable Development: Which Mechanisms Explain Its Accomplishments?" *Global Environmental Politics*, August 2007, pp. 107–29; UN General Assembly, *Lessons Learned from the Commission on Sustainable Development – Report of the Secretary-General* (New York: 26 February 2013).

14. UN, op. cit. note 4.

15. UN General Assembly, *Format and Organizational Aspects of the High-Level Political Forum on Sustainable Development* (New York: 27 June 2013).

16. Maria Ivanova, "Reforming the Institutional Framework for Environment and Sustainable Development: Rio+20 Subtle but Significant Impact," *International Journal of Technology Management and Sustainable Development*, vol. 12, no. 3 (2013).

17. UN Department of Economic and Social Affairs (UN DESA), "Voluntary Commitments and Partnerships for Sustainable Development," *Sustainable Development in Action*, July 2013, p. 2.

18. Stakeholder Forum and Natural Resources Defense Council, *Fulfilling the Rio+20 Promises: Reviewing Progress Since the UN Conference on Sustainable Development* (New York: September 2013), p. 30.

19. Ibid., p. 7.

20. UN System Task Team on the Post-2015 UN Development Agenda, *Building on the MDGs to Bring Sustainable Development to the Post-2015 Development Agenda* (New York: May 2012).

21. UN, *Open Working Group of the General Assembly on Sustainable Development Goals* (New York: 15 January 2013); UN High-Level Panel of Eminent Persons on the Post-2015 Development Agenda, *A New Global Partnership: Eradicate Poverty and Transform Economies through Sustainable Development* (New York: 30 May 2013).

22. UNEP, *Global Environment Outlook 5* (Nairobi: 2012). Box 13–1 from the following sources: quote from UN High-Level Panel, op. cit. note 21, Executive Summary, item 2; on the "reality mandate," see, for example, "State of the Planet Declaration," the outcome document of the Planet Under Pressure conference, London, 26–29 March 2012, at www.planetunderpressure2012.net. The Global Footprint Network (GFN), in collaboration with countries around the world, has used the RSE approach to study the ecological demands that their societies place on the planet, and compare them with their national capacity to sustainably fulfill these demands; see GFN, "Case Stories," www.footprintnetwork.org/casestudies. In addition, the UN System of Environmental-Economic Accounting (SEEA) is providing standard concepts, definitions, classifications, accounting rules, and tables for producing internationally comparable statistics on the environment and its relationship with the economy; see http://unstats.un.org/unsd/envaccounting/seea.asp.

23. Civil Society Reflection Group on Global Development Perspectives, *Towards a Framework of Universal Sustainability Goals as Part of a Post-2015 Agenda*, *Global Policy Forum* (Berlin: Friedrich-Ebert-Stiftung, May 2013).

Chapter 14. How Local Governments Have Become a Factor in Global Sustainability

1. ICLEI–Local Governments for Sustainability (ICLEI), "ICLEI Submission for Rio+20: Contribution to the Zero Draft of the Rio+20 Outcome Document" (Bonn: 31 October 2011), p. 1.

2. United Cities and Local Governments website, www.uclg.org; ICLEI website, www.iclei.org; Metropolis website, www.metropolis.org; CITYNET website, http://citynet-ap.org; Mercociudades website, www.mercociudades.org; Eurocities website, www.eurocities.eu; C40 Cities Climate Leadership Group website, www.c40.org.

3. Network of Regional Governments for Sustainable Development website, www.nrg4sd.org; R20–Regions of Climate Action website, http://regions20.org.

4. Konrad Otto-Zimmermann, *Global Environmental Governance: The Role of Local Governments* (Boston: Sustainable Development Knowledge Partnership, 2011).

5. Virginia Sonntag O-Brien, "Local Governments Lead the Way in Combating Local Climate Change," in Jan Corfee Morlot, ed., *Climate Change: Mobilising Global Effort* (Paris: Organisation for Economic Co-operation and Development, 1997), p. 86; United Nations Framework Convention on Climate Change (UNFCCC), "Kyoto Protocol," at http://unfccc.int/kyoto_protocol/items/2830.php.

6. ICLEI, "Who Is ICLEI?" at www.iclei.org/iclei-global/who-is-iclei.html; Gro Harlem Brundtland et al., *Our Common Future* (Oxford: Oxford University Press, 1987).

7. United Nations, *Agenda 21* (New York: 1992); Box 14–1 from ICLEI, *Local Sustainability 2012: Taking Stock and Moving Forward* (Bonn: 2012).

8. Figure 14–1 from ICLEI, "Introducing the carbon*n* Cities Climate Registry (cCCR)," at www.uclg.org/sites/default/files/carbonncitiesclimate_registry.pdf.

9. Local Government Climate Roadmap website, www.iclei.org/climate-roadmap.

10. Global Cities Covenant on Climate, "The Mexico City Pact," at www.mexicocitypact.org/en/the-mexico-city-pact-2/; carbon*n* Cities Climate Registry (cCCR) website, http://citiesclimateregistry.org.

11. UNFCCC, "Report of the Conference of the Parties on Its Sixteenth Session, Held in Cancun from 29 November to 10 December 2010" (Bonn: 15 March 2011), p. 3.

12. Durban Adaptation Charter website, www.durbanadaptationcharter.org.

13. Local Government Climate Roadmap, op. cit. note 9.

14. United Nations Convention on Biological Diversity, "Subnational and Local Implementation," at www.cbd.int/en/subnational. Box 14–2 from the following sources: Mayors Conference on Local Action for Biodiversity, *Cities and Biodiversity: Bonn Call for Action* (Bonn: May 2008); City Biodiversity Summit 2010 website, www.kankyo-net.city.nagoya.jp/citysummit2010/english/index.html; "Aichi/Nagoya Declaration on Local Authorities and Biodiversity," October 2010; Convention on Biological Diversity, "COP 10 Decision X/22. Plan of Action on Subnational Governments, Cities and Other Local Authorities for Biodiversity" (Nagoya, Japan: 2010); ICLEI Cities Biodiversity Center, "The Cities for Life Summit," at http://cbc.iclei.org/cfl; *Hyderabad Declaration on Subnational Governments, Cities and Other Local Authorities for Biodiversity* (Hyderabad, India: October 2012). Box 14–3 from the following sources: United Nations Economic and Social Council, "Millennium Development Goals and Post-2015 Development Agenda," at www.un.org/en/ecosoc/about/mdg.shtml; Global Taskforce of Local & Regional Governments for Post 2015 and Habitat III, "Communique" (Istanbul: 20 March 2013); ICLEI, "UN Climate Talks Go Local: First Ever 'Cities Day' to Raise the Bar of Climate Ambition Through Local Action" (Bonn: 19 November 2013); Sustainable Development Knowledge Network, "Expert Workshop of Communitas Coalition on Sustainable Cities and Human Settlements in the SDGs," at http://sustainabledevelopment.un.org.

15. Urban Low Emission Development Strategies website, http://urbanleds.iclei.org.

16. Ibid.

17. cCCR, op. cit. note 10.

Chapter 15. Scrutinizing the Corporate Role in the Post-2015 Development Agenda

1. This chapter is adapted from Lou Pingeot, *Corporate Influence in the Post-2015 Process* (Bonn: Misereor, Global Policy Forum, and Brot für die Welt, December 2013).

2. World Economic Forum, *Everybody's Business: Strengthening International Cooperation in a More Interdependent World – Report of the Global Redesign Initiative* (Geneva: 2010), pp. 8–10, 367.

3. Oxford Martin School, *Now for the Long Term: The Report of the Oxford Martin Commission for Future Generations* (Oxford: Oxford University, 2013), p. 57.

4. Quote on the World Summit on Sustainable Development from United Nations Department of Economic and Social Affairs (UN-DESA), "Voluntary Commitments and Partnerships for Sustainable Development," Special Report of the Sustainable Development in Action Newsletter (New York: July 2013), p. 4; UN General Assembly, Resolution on "Towards Global Partnerships" (New York: 22 December 2011).

5. UN Research Institute for Social Development (UNRISD), Public-Private Partnerships for Sustainable Development conference, Copenhagen, Denmark, 15 August 2006; Friends of the Earth International, *Reclaim the UN from Corporate Capture* (Amsterdam: 2012).

6. UN General Assembly, *The Future We Want* (Rio de Janeiro: 27 July 2012); UN General Assembly, "Format and Organizational Aspects of the High-Level Political Forum on Sustainable Development" (New York: 27 June 2013).

7. UN General Assembly, *A Life of Dignity for All: Accelerating Progress Towards the Millennium Development Goals and Advancing the United Nations Development Agenda Beyond 2015*, Report of the Secretary-General (New York: 26 July 2013).

8. UN Global Compact, "Global Compact LEAD," www.unglobalcompact.org/HowToParticipate/Lead/index.html.

9. Sustainable Development Solutions Network (SDSN), *An Action Agenda for Sustainable Development: Report for the UN Secretary-General* (New York: 6 June 2013); UN Global Compact, *Corporate Sustainability and the United Nations Post-2015 Development Agenda: Perspectives from UN Global Compact Participants on Global Priorities and How to Engage Business Towards Sustainable Development Goals* (New York: 17 June 2013), p. 16.

10. Participate, "Response to the Report of the High-Level Panel on the Post-2015 Development Agenda" (Brighton, U.K.: 2012).

11. UN High-Level Panel of Eminent Persons on the Post-2015 Development Agenda, *A New Global Partnership: Eradicate Poverty and Transform Economies Through Sustainable Development* (New York: 30 May 2013).

12. Development Alternatives with Women for a New Era, "From People's Rights to Corporate Privilege: A South Feminist Critique of the HLP Report on Post 2015 Development Agenda" (Manila: undated); Participate, op. cit. note 10.

13. Georg Kell, "12 Years Later: Reflections on the Growth of the UN Global Compact," *Business & Society*, vol. 52, no. 1 (2013), pp. 31–52; Papa Louis Fall Mohamed and Mounir Zahran, "United Nations Corporate Partnerships: The Role and Functioning of the Global Compact" (Geneva: UN Joint Inspection Unit, 2010). SDSN website, http://unsdsn.org; lack of transparency on funding and other matters from author's communications with SDSN Secretariat.

14. For example, mining company Vale has actively engaged in the international climate process by lobbying the Brazilian government, both in the run-up to UN climate talks in Copenhagen in 2009 (COP 15) and as part of the Brazilian official delegation to Cancún in 2010 (COP 16); see Friends of the Earth International, *How Corporations Rule - Part 3: Vale – Leading the Corporate Lobby for Easier Offsetting and Other False 'Green' Solutions* (Amsterdam: 2012).

15. UN Global Compact, "Local Networks," at www.unglobalcompact.org/NetworksAroundTheWorld/index .html; UN Global Compact, "Top Executives Meet with UN Secretary-General to Mark Successes, Future of Sustainability Leadership Platform," press release (Davos, Switzerland: 25 January 2013).

16. SDSN numbers from SDSN, op. cit. note 13. Note that some of the thematic groups have not published their full list of members.

17. Klaus Leisinger and Peter Bakker, *The Key Challenges to 2030/2050: Mapping Out Long-Term Pathways to Sustainability and Highlighting Solutions That Should Be Scaled Up*, Background Paper for the High-Level Panel of Eminent Persons on the Post-2015 Development Agenda (New York: 16 January 2013); UN Global Compact and World Business Council for Sustainable Development, *Joint Report to the High-Level Panel of the Post-2015 UN Development Agenda* (New York: March 2013).

18. On the depoliticization of multi-stakeholder processes, see Christina Garsten and Kerstin Jacobsson, "Corporate Globalisation, Civil Society and Post-Political Regulation: Whither Democracy?" *Development Dialogue*, no. 49 (2007), p. 143; cases of corporations suing governments from Corporate Europe Observatory and Transnational Institute, *Profiting from Injustice: How Law Firms, Arbitrators and Financiers Are Fueling an Investment Arbitration Boom* (Brussels: 2012).

19. Government of Ecuador, Articulo 3: "Empresas Transnacionales y Derechos Humanos," in "Declaración en nombre de un grupo de países en la 24 a edición de sesiones del Consejo de Derechos Humanos Debate General" (Quito: September 2013); FIAN International, "Statement to the Human Rights Council on Transnational Corporations," 13 September 2013," at www.fian.org/news/article/detail/statement_to_the_human_rights _council_on_transnational_corporations/.

20. Figure 15–1 from UN System Chief Executives Board for Coordination (UN-CEB), "Agency Revenue by Revenue Type," at www.unsceb.org/content/FS-A00-01.

21. UN-CEB, "Extra Budgetary Resources Trend – Non-State Donors," at www.unsceb.org/content/extra-budg etary-resources-trend-%E2%80%93-non-state-donors.

Chapter 16. Making Finance Serve the Real Economy

1. Lawrence Mishel, Jared Bernstein, and Heidi Shierholtz, *The State of Working America 2008/2009* (Ithaca, NY: Cornell University Press); Figure 16–2 based on data from the U.S. Bureau of Economic Analysis and the U.S. Bureau of Labor Statistics.

2. Table 16–1 from Thomas I. Palley, *Financialization: The Macroeconomics of Finance Capital Domination* (New York: Macmillan/Palgrave, 2013).

3. Table 16–2 from ibid.

4. Thomas I. Palley, "Destabilizing Speculation and the Case for an International Currency Transactions Tax," *Challenge*, May/June 2001, pp. 70–89.

5. For a detailed discussion of asset-based reserve requirements, see Thomas I. Palley, "Asset Price Bubbles and the Case for Asset-based Reserve Requirements," *Challenge*, May/June 2003, pp. 53–72.

Chapter 17. Climate Governance and the Resource Curse

1. World Resources Institute, Climate Analysis Indicators Tool 2.0, at http://cait2.wri.org; Kelly Levin et al., *Playing It Forward: Path Dependency, Progressive Incrementalism, and the 'Super Wicked' Problem of Global Climate Change*, prepared for the International Studies Association Convention, Chicago, IL, 28 February–3 March 2007, updated 3 June 2010.

2. Jeffrey Sachs and Andrew Warner, *Natural Resource Abundance and Economic Growth*, Working Paper 5398 (Cambridge, MA: National Bureau for Economic Research, 1995); Terry Lynn Karl, *The Paradox of Plenty: Oil Booms and Petro-States* (Berkeley and Los Angeles, CA: University of California Press, 1997); Halvor Mahlum, Karl Moene, and Ragnar Torvik, "Institutions and the Resource Curse," *Economic Journal*, vol. 116 (2006), pp. 1–20; John L. Hammond, "The Resource Curse and Oil Revenues in Angola and Venezuela," *Science and Society*, vol. 75, no. 3 (2011), pp. 348–78.

3. Hammond, op. cit. note 2; James A. Robinson, Ragnar Torvik, and Thierry Verdier, "Political Foundations of the Resource Curse," *Journal of Development Economics*, vol. 79 (2006), pp. 447–68; Tony Hodges, "The Role of Resource Management in Building Sustainable Peace," *Accord 15*, 2004, pp. 48–53.

4. Thad Dunning, *Authoritarianism and Democracy in Rentier States* (Berkeley, CA: University of California at Berkeley, Department of Political Science, undated).

5. Figure 17–1 from World Bank, "Worldwide Governance Indicators," at http://info.worldbank.org/governance/wgi/index.aspx#home and from Reporters Without Borders, *World Press Freedom Index 2013* (Paris: 2013).

6. Amnesty International, *Nigeria: Petroleum, Pollution and Poverty in the Niger Delta* (London: 2009); Amnesty International, *Bad Information: Oil Spill Investigations in the Niger Delta* (London: 2013).

7. Erik Voeten and Michael Ross, *Oil and Unbalanced Globalization*, 28 March 2013, at http://papers.ssrn.com/sol3/papers.cfm?abstract_id=1900226;

8. Gal Luft, "The Real Reason Gas Costs $4 a Gallon," Planet Money blog (National Public Radio), 2 April 2012, at www.npr.org/blogs/money.

9. Voeten and Ross, op. cit. note 7.

10. Ibid.

11. Box 17–1 from Norwegian Ministry of Finance, "The Government Pension Fund," at www.regjeringen.no.

12. Environics Institute, "Focus Canada 2012. Climate Change: Do Canadians Still Care?" 14 December 2012, at www.environicsinstitute.org.

13. Angus Reid Global, "Less Than Half in U.S. and Britain Believe in Man-Made Climate Change," 14 May 2013, at www.angusreidglobal.com/polls.

14. Gabriel Chan et al., *Canada's Bitumen Industry Under CO_2 Constraints*, Report No. 183 (Cambridge, MA: MIT Joint Program on the Science and Policy of Global Change, January 2010).

15. Lenore Taylor, "Australia Could Be Left With No Policy on Climate Change," *The Guardian* (U.K.), 25 September 2013; John Connor and Kristina Stefanova, *Climate of the Nation 2012* (Sydney, Australia: Climate Institute, 2012); Kristina Stefanova, *Climate of the Nation 2013* (Sydney, Australia: Climate Institute, 2013).

16. Ben Packham, "Canada Praises Tony Abbott's Carbon Tax Repeal Bill, Says It Sends Important Message," *The Australian*, 13 November 2013; Amanda Hodge, "Australia, Canada Block Commonwealth Climate Fund," *The Australian*, 18 November 2013.

17. Oil Change International, "Fossil Fuel Funding to Congress: Industry Influence in the U.S.," at http://priceof

oil.org/fossil-fuel-industry-influence-in-the-u-s/; The Editors, "Rules, Revolving Doors and the Oil Industry," *New York Times*, 5 May 2010; Grant Smith, "U.S. to be Top Oil Producer by 2015 on Shale, IEA Says," *Bloomberg*, 12 November 2013.

18. United Nations Framework Convention on Climate Change website, http://unfccc.int.

19. Jan Burck, Franziska Marten, and Christoph Bals, *The Climate Performance Index 2014* (Bonn and Brussels: Germanwatch and Climate Action Network Europe, November 2013); United Nations Environment Programme, *The Emissions Gap Report 2013* (Nairobi: 2013).

20. Richard Heede, "Tracing Anthropogenic Carbon Dioxide and Methane Emissions to Fossil Fuel and Cement Producers, 1854–2010," *Climatic Change*, 22 November 2013.

21. Ibid.

22. Steven Mufson, "On Campuses, A Fossil-fuel Divestment Movement," *Washington Post*, 25 November 2013.

Chapter 18. The Political-Economic Foundations of a Sustainable System

1.   U.S. Census Bureau, "Annual Projections of the Total Resident Population as of July 1: Middle, Lowest, Highest, and Zero International Migration Series, 1999 to 2100" (Washington, DC: updated 14 February 2000).

2.   Thad Williamson, Steve Dubb, and Gar Alperovitz, *Climate Change, Community Stability, and the Next 150 Million Americans* (College Park, MD: The Democracy Collaborative at the University of Maryland, September 2010), pp. 30–41.

3.   David Dodman, "Blaming Cities for Climate Change? An Analysis of Urban Greenhouse Gas Emissions Inventories," *Environment and Urbanization*, April 2009, pp. 185–201; John Thomas with Mara D'Angelo, Stephanie Bertaina, and Rachel Friedman, *Residential Construction Trends in America's Metropolitan Regions* (Washington, DC: U.S. Environmental Protection Agency, January 2010).

4.   Seymour Martin Lipset and Noah M. Meltz, *The Paradox of American Unionism* (Ithaca, NY: Cornell University Press, 2004), p. 7.

5.   Alexander Hicks, *Social Democracy and Welfare Capitalism: A Century of Income Security Politics* (Ithaca, NY: Cornell University Press, 1999), p. 233.

6.   Economic Policy Institute, "Hourly Wage and Compensation Growth for Production/Non-supervisory Workers, 1959–2009," at www.stateofworkingamerica.org/charts/view/186; Lawrence Mishel, Jared Bernstein, and Heidi Shierholz, "Table 3.4: Hourly and Weekly Earnings of Private Production and Nonsupervisory Workers, 1947–2007 (2007 dollars)," in *The State of Working America 2008/2009* (Ithaca, NY: Economic Policy Institute, 2009). According to economists Emmanuel Saez and Thomas Piketty, the top 1 percent's share of income has increased from a low of 7.8 percent in 1970 to 17.4 percent in 2010; these figures, however, exclude capital gains income. If capital gains income is included, then the top 1 percent's share has increased from 9.03 percent to 19.77 percent and, between 2005 and 2008, it exceeded 20 percent. Thomas Piketty and Emmanuel Saez, "Income Inequality in the United States, 1913–1998," *Quarterly Journal of Economics*, vol. 118, no. 1 (2003), pp. 1–39 (tables and figures updated to 2010 in March 2012, at http://elsa.berkeley.edu/~saez/#income).

7.   According to the U.S. Census Bureau's 2011 poverty report: "The Organisation for Economic Co-operation and Development (OECD) uses a poverty threshold of 50 percent of median income. The European Union defines poverty as an income below 60 percent of national median equalized disposable income after social transfers." Using the Census Bureau's "OECD equivalence scale" for below 50 percent of family median income results in 69.1 million Americans in poverty in 2011; using the same scale for below 60 percent of family *disposable* median income results in 72.7 million Americans in poverty in 2011. Carmen DeNavas-Walt, Bernadette D. Proctor, and Jessica C. Smith, *Income, Poverty, and Health Insurance Coverage in the United States: 2011* (Washington, DC: U.S. Census Bureau, September 2012), p. 20; U.S. Census Bureau, "Current Population Survey (CPS) Table Creator," 2011, at www.census.gov/cps/data/cpstablecreator.html. The employment rate was 14.6 percent as of October 2012, per U.S. Department of Labor, Bureau of Labor Statistics, "Table A-15: Alternative Measures of Labor Underutilization," 2 November 2012, at www.bls.gov/news.release/empsit.t15.htm.

8. Richard Wilkinson and Kate Pickett, *The Spirit Level: Why More Equal Societies Almost Always Do Better* (London: Bloomsbury Press, 2009).

9. Robert M. Solow, "Technical Change and the Aggregate Production Function," *Review of Economics and Statistics,* August 1957, pp. 312–20.

10. Joel Mokyr, *The Lever of Riches: Technological Creativity and Economic Progress* (New York: Oxford University Press, 1990), p. 5.

11. Anna Bernasek, "What's the Return on Education?" *New York Times,* 11 December 2005.

12. Herbert A. Simon, "Public Administration in Today's World of Organizations and Markets," *PS: Political Science and Politics,* December 2000, p. 756.

13. Leonard T. Hobhouse, *Liberalism and Other Writings* (Cambridge, U.K.: Cambridge University Press, 1994), p. 97.

14. Luisa Kroll, "Inside the 2013 Forbes 400: Facts and Figures on America's Richest," *Forbes,* 16 September 2013.

15. Howell J. Malham, Jr., *I Have a Strategy (No, You Don't): The Illustrated Guide to Strategy* (Hoboken, NJ: John Wiley & Sons, 2013), p. 163.

16. Deborah B. Warren and Steve Dubb, *Growing a Green Economy for All: From Green Jobs to Green Ownership* (College Park, MD: The Democracy Collaborative at the University of Maryland, July 2010), pp. 43–44.

17. Pioneer Human Services, *2011 Annual Report* (Seattle: 2011); Mike Burns, CEO, Pioneer Human Services, personal communication with Steve Dubb, The Democracy Collaborative at the University of Maryland, 9 March 2006; Pioneer Human Services, *Annual Report 2010: Pioneering the Possibilities* (Seattle: 2010); John Cowen, "Pioneer Human Services: A 'Chance for Change,'" *Chronicle of Social Enterprise,* Spring 2009, p. 5.

18. Massachusetts Association of Community Development Corporations, *GOALs Initiative: Growing Opportunities, Assets, and Leaders across the Commonwealth: A Detailed Report of the Accomplishments of Community Development Corporations in Massachusetts* (Boston: July 2013), p. 3; National Alliance of Community Economic Development Associations, *Rising Above: Community Economic Development in a Changing Landscape* (Washington, DC: June 2010).

19. The Forum for Sustainable and Responsible Investment, *Report on Socially Responsible Investing Trends in the United States* (Washington, DC: 2012), p. 13; Social Investment Forum, *2007 Report on Socially Responsible Investing Trends in the United States* (Washington, DC: 2007), p. 38.

20. Irvine Community Land Trust, "About Us," at www.irvineclt.org/about.

21. John Emmeus Davis and Amy Demetrowitz. *Permanently Affordable Homeownership: Does the Community Land Trust Deliver on Its Promises?—A Performance Evaluation of the CLT Model Using Resale Data from the Burlington Community Land Trust* (Burlington, VT: Burlington Community Land Trust, May 2003); Emily Thaden, *Stable Home Ownership in a Turbulent Economy: Delinquencies and Foreclosures Remain Low in Community Land Trusts,* Working Paper (Cambridge, MA: Lincoln Institute of Land Policy, July 2011), p. 12.

22. National Center for Employee Ownership (NCEO), *A Statistical Profile of Employee Ownership* (Oakland, CA: February 2012).

23. Corey Rosen, *The Impact of Employee Ownership and ESOPs on the Costs of Unemployment to the Federal Government* (Oakland, CA: NCEO, 5 February 2013); The Democracy Collaborative at the University of Maryland, *Building Wealth: The New Asset-Based Approach to Solving Social and Economic Problems* (Washington, DC: The Aspen Institute, 2005), pp. 55–67.

24. Steven Deller et al., *Research on the Economic Impact of Cooperatives* (Madison, WI: University of Wisconsin Center for Cooperatives, March 2009).

25. David Orr, "Governance in the Long Emergency," in Worldwatch Institute, *State of the World 2013: Is Sustainability Still Possible?* (Washington, DC: Island Press, 2013), p. 290.

26. Marcelo Vieta, *The Emergence of the Empresas Recuperadas por sus Trabajadores: A Political Economic and*

Sociological Appraisal of Two Decades of Self-management in Argentina, Working Paper, no. 55|13 (Trento, Italy: European Research Institute on Cooperative and Social Enterprises, 2013). Box 18–1 from the following sources: 300 businesses from Argentinian Ministry of Labor, Employment, and Social Security, *Guia Empresas Recuperadas y Autogestionadas por Sus Trabajadores* (Buenos Aires: July 2012); 30 percent inflation from "The Price of Cooking the Books," *The Economist*, 25 February 2012; "The Take" website, www.thetake.org.

27. Steve Dubb, "Mondragón Co-op Model Gains U.S. Adherents," at http://community-wealth.org/content /mondragon-co-op-model-gains-us-adherent; Mondragón Cooperative Corporation, *Corporate Profile* 2013 (Mondragón, Spain: 2013).

28. The Democracy Collaborative at the University of Maryland, op. cit. note 23, pp. 105–10; Aziza Agia, *Innovative Significant Scale Models of Community Asset-Building: Learning from International Experience* (College Park, MD and Washington, DC: The Democracy Collaborative and the National Center for Economic and Security Alternatives, 2004); The Co-operative Group: *The Co-operative Group: Annual Report 2012* (London: 2013).

29. John Restakis, *Humanizing the Economy: Co-operatives in the Age of Capital* (Gabriola Island, CA: New Society Publishers, 2010), p. 57; Hazel Corcoran and David Wilson, *The Worker Co-operative Movements in Italy, Mondragon and France: Context, Success Factors and Lessons* (Calgary, Alberta: Canadian Worker Co-operative Federation: 31 May 2010).

30. Steve Dubb, *C-W Interview: Seikatsu Club Consumers' Co-operative Union* (College Park, MD: The Democracy Collaborative at the University of Maryland, October 2012).

31. Ibid.

32. Paul E. Peterson, *City Limits* (Chicago: University of Chicago Press, 1981).

33. James Gustave Speth, "Letters to Liberals: Liberalism, Environmentalism and Economic Growth," *Vermont Law Review*, vol. 35 (2011), p. 555.

34. Gar Alperovitz, *What Then Must We Do: Straight Talk About the Next American Revolution* (White River Junction, VT: Chelsea Green Publishing, 2013).

Chapter 19. The Rise of Triple-Bottom-Line Businesses

1. B Lab staff, personal communications with author, 2013.

2. See, for example, The Contributor.com, "GAP, Inc.'s Inaction Following Bangladeshi Garment Worker Deaths Draws Fire from TheContributor.com, The Other 98% and Indigenous Designs," press release (Minneapolis, MN: 16 October 2013); Global Reporting Initiative (GRI), *Sustainability Disclosure Database*, at database.globalreporting.org; GRI, *The GRI Reports List 1999–2013*, at https://www.globalreporting.org/resourcelibrary/GRI-Reports-List-1999-2013.zip.

3. William H. Clark, Jr. and Larry Vranka, *White Paper: The Need and Rationale for the Benefit Corporation: Why It Is the Legal Form that Best Addresses the Needs of Social Entrepreneurs, Investors, and, Ultimately, the Public* (Benefit Corp Information Center, updated 18 January 2013).

4. Ibid.

5. B Lab staff, op. cit. note 1; B Lab, "The Non-Profit Behind B Corps," at www.bcorporation.net/what-are-b -corps/the-non-profit-behind-b-corps; Benefit Corp Information Center, "State by State Legislative Status," at http://benefitcorp.net/state-by-state-legislative-status.

6. Table 19–1 from Benefit Corp Information Center, op. cit. note 5; State of Delaware, "Governor Markell Registers Delaware's First Public Benefit Corporations," press release (Wilmington, DE: 1 August 2013). Earlier, B Lab characterized the introduction of the Delaware statute as signaling "a seismic shift in corporate law," asserting that Delaware is the legal home for more than 50 percent of all public companies and about two-thirds of all Fortune 500 companies; see www.benefitcorp.net/storage/documents/Delaware_Benefit_Corporation_Legislation.pdf. Box 19–1 from "The Delaware Constitution of 1897 as amended, Title 8, Chapter 1, Subchapter XV: Public Benefit Corporations," at http://delcode.delaware.gov/title8/c001/sc15/index.shtml.

7. B Lab staff, op. cit. note 1. The author has separated out Delaware's information from data provided by B Lab. Hugo Martin, "Outdoor Retailer Patagonia Puts Environment Ahead of Sales Growth," *Los Angeles Times*, 24 May 2012; Dan D'Ambrosio, "King Arthur Flour to Begin Expansion in June," *Burlington Free Press*, 10 May 2011.

8. Kathleen Kim, "Green Merger: Method Bought by Ecover: The Companies Say Their Union Creates the World's Largest Green Cleaning Company," *Inc.*, 4 September 2012.

9. B Lab staff, op. cit. note 1.

10. B Lab, "Corporation Legal Roadmap," at www.bcorporation.net/become-a-b-corp/how-to-become-a-b-corp /legal-roadmap/corporation-legal-roadmap; B Lab staff, op. cit. note 1.

11. B Lab data as of 21 October 2013, per B Lab staff, op. cit. note 1.

12. Table 19–2 based on ibid.

13. American Sustainable Business Council website, www.asbcouncil.org; Todd Larsen, division director of Corporate Responsibility Programs, Green America, personal communications with author, 2013; Green Business Network, "Benefit Corporations," at www.greenbusinessnetwork.org/for-members-/benefit-corporations.html.

14. "The Year That Was…Just Wasn't Very Good," *HAPPI Magazine*, July 2012; Larsen, op. cit. note 13.

15. Rally Software, "Rally Software Announces Closing of Its Initial Public Offering and Full Exercise of Underwriters' Option to Purchase Additional Shares," press release (Boulder, CO: 17 April 2013).

16. "Companies Convert to Public Benefit Corporations," *Associated Press*, 5 August 2013.

17. Emily Glazer, "Danone Buys Organic Baby-Food Maker," *Wall Street Journal*, 13 May 2013.

18. Ben & Jerry's, "Ben & Jerry's Joins the Growing B Corporation Movement," press release (Burlington, VT: 22 October 2012); Unilever, "Unilever at a Glance," at www.unileverusa.com/aboutus/introductiontounilever/Uni leverataglance; B Lab, "FAQ: How Can Ben & Jerry's Be a B Corp?" at www.bcorporation.net/sites/default/files /documents/bcorps/ben_n_jerry/bj_s_faq_final_for_b_lab_site.pdf.

19. Public Disclosure Commission of Washington State, "Cash Contributions for: Grocery Manufacturers Assn Against I-522," at www.pdc.wa.gov, which included contributions reported through 17 October 2013; Non-GMO Shopping Guide, "Baby Food & Infant Formula: Plum Organics," at www.nongmoshoppingguide.com/brands /baby-food-and-infant-formula.html?bid=547.

20. For examples of nonprofit concerns, see Independent Sector, "Benefit Corporations," at www.independentsec tor.org/benefit_corporations.

21. Charlie Cray, Center for Corporate Policy, Washington, DC, personal communication with author, 6 December 2013.

22. Jay Coen Gilbert, personal communication with author, 29 October 2013.

23. Jamie Raskin, "The Rise of Benefit Corporations," *The Nation*, 8 June 2011.

Chapter 20. Working Toward Energy Democracy

1. This chapter is adapted and updated from a report prepared for the global trade union roundtable "Energy Emergency, Energy Transition" convened by the Cornell Global Labor Institute (GLI) in October 2012, in partnership with the Rosa Luxemburg Foundation and six Global Union Federations. A revised version was published in November 2012. The term "extreme energy" was first coined in Michael Klare, "The Era of Xtreme Energy. Life After the Age of Oil," *TomDispatch*, 22 September 2009, at www.tomdispatch.com/post/175127.

2. National Mining Association, "Trends in U.S. Coal Mining, 1923–2011," June 2012, at www.nma.org/pdf/c _trends_mining.pdf.

3. International Renewable Energy Agency (IRENA), *Renewable Energy Jobs & Access* (Abu Dhabi: 2012).

4. Investments from 2004 to 2012 from Bloomberg New Energy Finance and Frankfurt School–UNEP Collaborating Centre for Climate & Sustainable Energy Finance, *Global Trends in Renewable Energy Investment 2013*

(London: 2013); Liebreich quote and 2013 developments from Sally Bakewell, "Clean Energy Investment Headed for Second Annual Decline," *Bloomberg*, 13 October 2013.

5. Table 20–1 adapted from Renewable Energy Policy Network for the 21st Century (REN21), *Renewables 2013 Global Status Report* (Paris: 2013), and from Evan Musolino, "Hydropower and Geothermal Growth Slows," *Vital Signs Online* (Worldwatch Institute), 12 February 2013.

6. United Nations Environment Programme, *Keeping Track of Our Changing Environment: From Rio to Rio+20* (Nairobi: 2011); REN21, op. cit. note 5, p. 21; U.S. Energy Information Administration, *International Energy Outlook 2013* (Washington, DC: 2013).

7. See, for example, "Position of the Bolivian Climate Change Platform on Rio+20 and the Green Economy," 17 April 2012, at www.cambioclimatico.org.bo/derechosmt/052012/100512_2.pdf.

8. Table 20–2 from "Fortune Global 500," 2013, at http://money.cnn.com/magazines/fortune/global500/2013/full_list/.

9. Land Matrix Project, "Land Matrix Project Database," landportal.info/landmatrix; GRAIN, "GRAIN Releases Data Set with Over 400 Global Land Grabs," 23 February 2012, at www.grain.org; Carol Schachet, "Wind Farm Mega-Project in Oaxaca Sparks Resistance, Repression," 24 January 2013, at www.grassrootsonline.org.

10. Yemi Assefa et al., "Coal: A Key Player in Expanded U.S. Energy Exports," *Beyond the Numbers* (U.S. Bureau of Labor Statistics), February 2013; International Energy Agency (IEA), "The World Is Locking Itself Into an Unsustainable Energy Future Which Would Have Far-reaching Consequences, IEA Warns in Its Latest World Energy Outlook," press release (Paris: 9 November 2011).

11. Canadian Association of Petroleum Producers, "About Canada's Oil Sands," June 2011, at www.capp.ca; Alberta Federation of Labour, *Lost Down the Pipeline* (Edmonton, Alberta: March 2009); Matt Price, "Canadian Jobs Lost to the Tar Sands," *Huffington Post*, 5 January 2012; Tony Clarke et al., *The Bitumen Cliff: Lessons and Challenges of Bitumen Mega-Developments for Canada's Economy in an Age of Climate Change* (Ottawa, ON: Canadian Centre for Policy Alternatives, 21 February 2013), p. 8.

12. Al Weinrub, *Labor's Stake in Decentralized Energy: A Strategic Perspective*, prepared for the Cornell GLI trade union roundtable "Energy Emergency, Energy Transition," 20 September 2012, at http://energydemocracyinitiative.org/category/roundtable-papers/; Floyd McKay, "Lummi Tribe Joins the Opposition to Whatcom Coal Port," 21 September 2012, at http://crosscut.com; Amalgamated Transit Union, "Amalgamated Transit Union, ATU and TWU Oppose Approval of the Keystone XL Pipeline and Call for End of Increased Use of Tar Sands Oil," press release (Washington, DC: 19 August 2011).

13. Michael Wayland, "UAW President, Environmentalists Tout New Fuel Economy Standards as Job Creator, 'Incredible Victory'," 29 August 2012, at www.mlive.com; Nick Prigo, *A Blueprint for Greening New York City's Buildings, 1 Year: 1000 Green Superintendents* (New York: Building Service 32BJ Thomas Shortman Training Fund and Urban Green Council, September 2009).

14. Lara Skinner, *State of the U.S. Environmental Movement* (New York: Rosa Luxemburg Stiftung, forthcoming); Michael Ettlinger and Michael Linden, "The Failure of Supply-Side Economics," issue brief (Washington, DC: Center for American Progress, 1 August 2012).

15. Sharon Beder, "Critique of the Global Project to Privatize and Marketize Energy" (Seoul: Korean Labor Social Network on Energy, June 2005), pp. 177–85; David Hall, "Struggles Against Privatization of Electricity," in Kolya Abramsky, ed., *Sparking a Worldwide Energy Revolution. Social Struggles in the Transition to a Post-Petrol World* (Oakland, CA: AK Press, 2009); Greg Muttitt, *Fuel on the Fire: Oil and Politics in Occupied Iraq* (New York: New Press, 2012).

16. Ian Rutledge, "Who Owns the UK Electricity Generating Industry – And Does It Matter?" (Chesterfield, U.K.: Sheffield Energy Resources Information Services, November 2012); Freedom from Debt Coalition, *PAID Magazine*, November 2009; Daphne Wysham, "How Did Coal-Rich India End Up With Power Blackouts?" *The Nation*, 22 August 2012.

17. Public Services International Research Unit (PSIRU) website, www.psiru.org; "Heating Bills Concern 38% of UK Population, Survey Suggests," *BBC News*, 5 September 2013.

18. Carlos Crespo, Marcela Olivera, and Susan Spronk, "Struggles for Water Justice in Latin America: Public and 'Social-Public' Alternatives," in David McDonald and Greg Ruiters, eds., *Alternatives to Privatization: Public Options for Essential Services in the Global South* (Cape Town, South Africa: HSRC Press, 2012); The Platform for Public and Community Partnerships of the Americas, *Bulletin No. 1*, December 2011; "Documento Plataforma de 'Acuerdo de Cooperaction Publica/Comunitaria,' April 27–29, 2009, Uruguay," www.aguayvida.org; Food & Water Watch and the Cornell University ILR School GLI, *Public-Public Partnerships* (Washington, DC and New York: January 2012).

19. Community Power Network, "Rural Electric Cooperatives and Renewables: The Future of Distributed Generation?" 31 October 2013, at http://communitypowernetwork.com.

20. Matthias B. Krause, "Thousands of German Cities and Villages Looking to Buy Back Their Power Grids," 11 October 2013, at greentechmedia.com; David Hall, Emanuele Lobina, and Philip Terhorst, *Re-municipalisation in Europe* (London: PSIRU, 2012).

21. Dieter Reiter, "Welcome Address," 10th Munich Economic Summit, 19–20 May 2011, p. 3, at www.cesifo-group.de/DocDL/Forum-3-2011.pdf.

22. Pablo Gonzalez and Camila Russo, "Chevron $1.24 Billion Deal Leads YPF Post-Repsol Shale Hunt," *Bloomberg*, 17 July 2013.

23. Sikonathi Manshantsha, "Escom Boosts CEO's salary 9.6%," 15 June 2012, at moneyweb.co.za; Congress of South African Trade Unions, "NUMSA Condemns Eskom Extravagant and Opulent Parties!" press release (Johannesburg: 10 July 2012).

24. Zhou Yan and Chen Limin, "Sinopec to Continue Overseas Investment," *China Daily*, 27 March 2012; "A Lesson in Capitalism," *The Economist*, 5 April 2001; Jeffrey Jones, "Sinopec to Pay $4.65 Billion in Oil Sands Deal," *Reuters*, 12 April 2010.

25. National Union of Metalworkers of South Africa, Statement from International Conference on Building a Renewable Energy Sector in South Africa, Johannesburg, 4–8 February 2012; Trade Unions for Energy Democracy, "Canadian Union of Public Employees Says Public Ownership of Energy Is Key to Winning the War Against Climate Change," 17 April 2013, at http://energydemocracyinitiative.org.

26. John Farrell, *Democratizing the Electricity System: Vision for a 21st Century Grid* (Washington, DC: Institute for Local Self Reliance, June 2011); Weinrub, op. cit. note 12.

27. Ibid.

28. Michael Moynihan, *Electricity 2.0 Unlocking the Power of the Open Energy Network (OEN)* (Washington, DC: NDN and the New Policy Institute, 4 February 2010).

29. IEA, *World Energy Outlook 2007: China and India Insights* (Paris: 2007), p. 573; Alliance for Progressive Labour, "Fight for Our Future: No Price on Nature: Asian Movements' Statement on the Green Economy," 23 June 2012, at www.apl.org.

30. REN 21, op. cit. note 5; David Hall, *Electrifying Africa Through the Public Sector* (Greenwich, U.K.: PSIRU, 2007).

31. IRENA, op. cit. note 3.

32. Jan-Christoph Kuntze and Tom Moerenhout, *Local Content Requirements and the Renewable Energy Industry: A Good Match?* (Geneva: International Centre for Trade and Sustainable Development (ICTSD), 2013).

33. ICTSD, "WTO Appellate Body Rules Against Canada in Renewable Energy Case," *Bridges Weekly Trade News Digest*, 8 May 2013.

Chapter 21. Take the Wheel and Steer! Trade Unions and the Just Transition

1. Anabella Rosemberg, "Building a Just Transition. The Linkages Between Climate Change and Employment," in International Labour Organization (ILO), "Climate Change and Labour. The Need for a 'Just Transition,'" *International Journal of Labour Research*, vol. 2, no. 2 (2010), pp. 125–56. Box 21–1 from idem.

2. Nora Räthzel and David Uzzell, "Mending the Breach Between Labour and Nature: A Case for Environmental Labour Studies," in Nora Räthzel and David Uzzell, eds., *Trade Unions in the Green Economy. Working for the Environment* (London: Routledge, 2013), p. 10.

3. Rosemberg, op. cit. note 1.

4. Monowar Islam and Fazle Rabbi Sadeque Ahmed, "Climate Change Impact on Employment and Labour Market," presentation at the ILO Tripartite National Conference on "Green Jobs: The Way Forward," Dhaka, Bangladesh, 18 September 2011.

5. The estimated employment growth rate for industrialized countries was extrapolated to total employment worldwide. ILO International Institute for Labour Studies, *World of Work Report 2009: The Global Jobs Crisis and Beyond* (Geneva: 2009).

6. ILO International Institute for Labour Studies, *World of Work Report 2012: Better Jobs for a Better Economy* (Geneva: 2012).

7. Nina Netzer, *A Global Green New Deal. Response to Crisis or Paradigm Shift Towards Sustainability?* (Berlin: Friedrich-Ebert-Stiftung International Policy Analysis, May 2011). Table 21–1 from Nina Netzer and Judith Althaus, "Green Economy. Turning Over A New Leaf Towards Sustainable Development?" *FES Perspective*, June 2012.

8. Table 21–2 from Netzer and Althaus, op. cit. note 7.

9. See, for example, the ILO agreement on "Sustainable Development, Decent Work and Green Jobs" developed during the 102nd ILO Conference, Geneva, Switzerland, 15–20 June 2013.

10. International Trade Union Confederation (ITUC) and Trade Union Advisory Committee to the OECD (TUAC), "Trade Union Statement to COP 13" (Bali, Indonesia: 3–14 December 2007), p. 6.

11. On the counterfactual validity of the normative conditions of the capitalist labor market, see Axel Honneth, "Arbeit und Anerkennung. Versuch einer theoretischen Neubestimmung," in Axel Honneth, *Das Ich im Wir. Studien zur Anerkennungstheorie* (Berlin: Suhrkamp, 2010), pp. 78–102.

12. Begoña María-Tomé Gil, "Moving Towards Eco-unionism. Reflecting the Spanish Experience," in Räthzel and Uzzell, eds., op. cit. note 2, p. 68.

13. Beate Littig, "Von Rio 1992 zu 'Rio+20': Arbeit im Kontext der aktuellen Nachhaltigkeitsdiskussion," *WSI Mitteilungen*, August 2012; Beate Littig and Markus Spitzer, *Arbeit neu. Erweiterte Arbeitskonzepte im Vergleich. Literaturstudie zum Stand der Debatte um erweiterte Arbeitskonzepte*, Working Paper 229 (Düsseldorf: Hans Böckler Stiftung, 2011).

14. Hans Böckler Stiftung, ed., *Pathways to a Sustainable Future. Results from the Work & Environment Interdisciplinary Project* (Düsseldorf: 2001).

15. Klaus Dörre, "Kapitalismus im Wachstumsdilemma: Die Verdrängung der ökologischen Krisendimension und ihre Folgen," *WSI Mitteilungen*, February 2013, p. 151.

16. The most prominent example at workplace level is the attempt by workers to convert Lucas Aerospace in the 1970s. Lars Henriksson, "Cars, Crisis, Climate Change and Class Struggle," in Räthzel and Uzzell, eds., op. cit. note 2, pp. 78–86.

17. Ibid.

18. John Barry, "Trade Unions and the Transition Away from 'Actually Existing Unsustainability.' From Economic Crisis to a New Political Economy Beyond Growth," in Räthzel and Uzzell, eds., op. cit. note 2, p. 238.

Chapter 22. A Call to Engagement

1. Universal Postal Union, "Development of postal services in 2012. A few key preliminary figures…," at www.upu.int/en/resources/postal-statistics/2012-results.html.

2. Churchill quote from "Parliament Bill, HC Deb 11 November 1947 vol 444 cc203-321," at http://hansard.millbanksystems.com/commons/1947/nov/11/parliament-bill#column_206.

3. Box 22–1 from the following sources: International Women's Democracy Center, "Women in Politics: A Time-line," at http://iwdc.org/resources/timeline.htm; Dan Balz, "Democrats Paving Way for Clinton 2016," *Washington Post*, 17 November 2013, p. A2; Noam Scheiber, "Hillary's Nightmare? A Democratic Party That Realizes Its Soul Lies with Elizabeth Warren," *The New Republic*, 10 November 2013; Patricia Moccia, ed., *The State of the World's Children 2007: Women and Children—The Double Dividend of Gender Equity* (New York: United Nations, 2007); Bryce Covert, "Number of Women CEOs at Major Companies Jumps by 4 Percent," *ThinkProgress*, 8 July 2013, at http://thinkprogress.org; Catalyst, "Catalyst 2013 Census of Fortune 500: Still No Progress After Years of No Progress," 10 December 2013, at www.catalyst.org; The Quota Project website, www.quota project.org; Robert Engelman, *State of World Population 2009: Facing a Changing World—Women, Population and Climate* (New York: United Nations, 2009). Figure 22–1 from Inter-Parliamentary Union, "Women in National Parliaments," at www.ipu.org/wmn-e/classif-arc.htm.

4. Box 22–2 from the following sources: proliferation of democratic governments from Monty Marshall and Benjamin Cole, *Global Report 2011: Conflict, Governance and State Fragility* (Vienna, VA: Center for Systemic Peace, 2011), p. 10; definition of sustainability from Global Footprint Network, *Ecological Footprint Atlas 2010* (Oakland, CA: 2010), pp. 19, 20; coping with climate adaptation from Peter Burnell, "Democracy, Democratization, and Climate Change: Complex Relationships," *Democratization*, October 2012, p. 828; disasters and democracy from Alastair Smith and Alejandro Quiroz Flores, "Disaster Politics: Why Earthquakes Rock Democracies Less," *Foreign Affairs*, 15 July 2010; quote from Burnell, op. cit. this note, p. 833; Orr and Weaver quotes both from David Orr, "Governance in the Long Emergency," in Worldwatch Institute, *State of the World 2013: Is Sustainability Still Possible?* (Washington, DC: Island Press, 2013), p. 287; Matt Leighninger, "Mapping Deliberative Civic Engagement," in Tina Nabatchi et al., *Democracy in Motion: Evaluating the Practice and Impact of Deliberative Civic Engagement* (Oxford: Oxford University Press, 2012); pp. 20, 28, 29, and 31; expansion of democratic regimes from Marshall and Cole, op. cit. this note; for historical examples of DCE, see, for example, Thomas Prugh, Robert Costanza, and Herman Daly, *The Local Politics of Global Sustainability* (Washington, DC: Island Press, 2000), Chapter 6; Lauren Collingwood and Justin Reedy, "Listening and Responding to Criticisms of Deliberative Civic Engagement," in Nabatchi et al., op. cit. this note, pp. 256–57.

5. Herman E. Daly and John B. Cobb, Jr., *For the Common Good: Redirecting the Economy toward Community, the Environment, and a Sustainable Future* (Boston: Beacon Press, 1989), p. 400.

Index